ROCKY MOUNTAIN BIRDS

Rocky Mountain Birds

Birds and Birding in the Central and Northern Rockies

Paul A. Johnsgard

School of Biological Sciences
University of Nebraska–Lincoln

Zea E-Books
Lincoln, Nebraska
2011

ISBN 978-1-60962-016-5 paperback
ISBN 978-1-60962-017-2 e-book

Set in Zapf Elliptical types.
Design and composition by Paul Royster.

Zea E-Books are published by the
University of Nebraska–Lincoln Libraries.

Electronic (pdf) edition available online at http://digitalcommons.unl.edu/zeabook/
Print edition can be ordered from http://www.lulu.com/spotlight/unllib

Contents

Preface

&

Acknowledgments

The Rocky Mountain region has fascinated me ever since I traveled to Glacier and Yellowstone National Parks as a teenager, and saw for the first time such wonderful birds as ospreys, American dippers, and Lewis's woodpeckers. Grand Teton National Park later became my special favorite, and was the subject for an earlier book (Johnsgard, 1982) on the region's natural history Shortly after I published that book, Roberts Rinehart, director of Colorado Associated University Press, asked me if I would be interested in writing a book on the birds of Rocky Mountain National Park. I instead proposed a book covering the entire northern Rockies region, encompassing all the U.S. parks north to the Canadian border. I later moved the boundaries of coverage north to the 52nd parallel, thus also encompassing the Canadian montane parks north through Jasper National Park. The western boundary of Idaho became the region's westernmost limits, the eastern boundaries of Montana, Wyoming and Colorado its eastern limits, and the 40th parallel in Colorado (encompassing Rocky Mountain National Park) its southern limits. That book (Johnsgard, 1986) individually discussed 354 species, including all of those that had then been reported from any of the nine included national parks.

I decided that a new book should have a broader geographic coverage, namely the entire northern Rocky Mountain region, would be a more attractive subject, and its text directed both toward surveying the birds of the region and informing regional birders. The geographic coverage has been extended south to the New Mexico border, and includes all of Colorado's southern Rockies, thereby adding three national parks and two national monuments. Descriptions of several shortgrass sites in eastern Montana, Wyoming and Colorado have also been added.

This book is in part based on my earlier *Birds of the Rocky Mountains*, but over a third of the original text has been eliminated. The rest has been updated, expanded and modified to be less technical and more useful to birders, especially in Wyoming, where many of our greatest national treasures and wildlife attractions are located. These attractions include two national parks, two national monuments, two national recreation areas, five national forests, and one national grassland. The state's national forests collectively comprise about 8.3 million acres, and its national parks, national monuments, national recreation areas, national wildlife refuges and national grassland add another three million acres. Additionally, the Bureau of Land Management manages 18 million acres of public lands. Thus, Wyoming has almost 50 million acres of public land in federal ownership, or nearly half of its total land area, as well as more than 100,000 acres of state-owned public lands. As a result, opportunities for birding or other outdoor activities in Wyoming are among the greatest of any state in the nation. For comparison, less than three percent of adjacent Nebraska's total land area is state- or federally-owned and offers free public access.

All of the regularly occurring Wyoming bird species are individually discussed in this book. Relative abundance and distributional information is provided for about 40 additional species that have been reported from one or more of the region's national parks, but that do not regularly occur in Wyoming, for a total inclusion of 328 species. Although this book should be especially useful to bird enthusiasts in Wyoming, persons in Colorado, Montana and Idaho should also find much helpful information in it. I hope that our book will also be of value to park visitors in the Canadian Rockies too, since some information on the birds of five Canadian national parks is also provided. Viewing locations and species abundance during the summer period are emphasized in this volume, since in this mountainous region tourists and birders are mostly active during these months. Relevant contact addresses are also included. Although contact telephone numbers and website addresses often change or disappear over time, they too have been included in the knowledge that they will eventually become obsolete.

National Breeding Bird Survey trend data mentioned here are for the four-decade period 1966–2007 (Sauer, Hines and Fallon, 2008); significant trends are those having a probability of at least 90%. If the regional Rocky Mountain trend data also seemed significant, long-term trend estimates for this region (U.S.F.W.S. Region 6) are also reported for a few species. Total population estimates for North America (Rich et al., 2004) have been adjusted to exclude any populations occurring south of the Mexico border.

Additional information on Rocky Mountain species, such as their identification, periods of migration and breeding, and general behavior, can be found in my earlier book. There are also book-length surveys of the birds of Idaho (Burleigh, 1972), Colorado (Andrews and Righter, 1992) and Wyoming (Faulkner, 2010). A comprehensive state bird book for Montana is currently in preparation.

In writing my book, I received much help. Besides the persons mentioned in my previous book, I would like to thank Darrin Pratt, director of the University of Colorado Press, who urged me to consider revising *Birds of the Rocky Mountains*, which was published in 1985 by Colorado Associated University Press. Although I no longer had that book's original manuscript, Paul Royster of the University of Nebraska provided me with a scanned version of the original text, which has been incorporated into the University of Nebraska's Digital Commons library (Johnsgard, 2009). This resource greatly eased my work on text revisions and additions. With regard to locating and describing birding sites, books by Jane and Robert Dorn (Wyoming), Hugh Kingery (Colorado), Terry McEneaney (Montana, Yellowstone National Park), Bert Raynes and Darwin Wile (Grand Teton–Jackson Hole), Oliver Scott (Wyoming), and Dan Svingen and Kas Dumroese (Idaho) were invaluable. People who helped me personally included Tom Mangelsen, Terry McEneaney, Benj. Sinclair, Bert Raynes, and Bruce and Donna Walgren.

I must especially thank Linda Brown, who as usual helped me with all the stages of data-gathering. manuscript preparation, and editing. I also owe a special debt of gratitude to Jackie Canterbury, my last graduate student at the University of Nebraska, who provided the idea for collaborating on a book on birds and bird-finding in Wyoming. After a year of planning, it became apparent that we should write a separate publication on the birds and birding in Wyoming's Big Horn Mountains, and that I would take on a much broader geographic approach, encompassing the central and northern Rocky Mountains. I also especially thank Paul Royster and the help of the University of Nebraska's Digital Commons program, for making this publication possible.

Paul A. Johnsgard
August 7, 2011.

Maps

Tables

Figures

Chapter 1

Habitats, Ecology and Bird Geography in the Rocky Mountains

Vegetational Zones and Bird Distributions in the Rocky Mountains

All bird species, by virtue of their variably specialized niche adaptations, are most abundant in, or may even be completely restricted to, particular habitats. It is thus particularly important that birders pay attention to the habitat in which they are observing birds, for this very often provides important clues as to the species that are most likely to be encountered. Except for such specialized habitats as rocky outcrops, mud flats, and other special substrates, the majority of biological habitats are most easily described in terms of their dominant plants. Furthermore, there is a rather remarkable consistency in the vertical stratification of vegetation throughout the central and northern Rocky Mountains. Thus, there is a sequence of plant community types that typically occurs sequentially from the plains and foothills upward toward the mountain tops. The exact altitudes at which a particular community type occurs vary greatly, depending on latitude and on such local influences as directional exposure to sunlight and winds, and soil characteristics. However, most vegetational community types on Rocky Mountain slopes occur as broad belts or "zones," ranging in altitudinal width from a few hundred feet to about two thousand feet (Table 1). These biological communities were classified as "Life Zones" by such early ecologists as Cary (1917). However, both the terminology and climatic criteria for defining life zones greatly oversimplified the environmental factors that affect plant and animal distributions over broad geographic regions.

Because of such limitations, the life-zone concept has been increasingly abandoned in favor of more precise ecologically defined terms, based on the dominant plant species or vegetational life-form present in each major biological community type. These vegetational units are usually called "plant associations" (if defined by the specific dominant plants, such as the pinyon–juniper association) or "plant

1

Table 1. Vertical Distribution of Major Plant Communities in the Central and Northern Rocky Mountains

Physiography	Vegetation Zones	Traditional Life-Zones*	Approximate Average Altitude in Feet		
			Colorado	Wyoming, S. Idaho	Montana, S. Alberta
Alpine	Tundra	Arctic-Alpine	11,000-14,000+	10,300-12,000+	6600-8000+
Subalpine	Engelmann spruce, Subalpine fir	Hudsonian	10,000-11,000	9500-10,300	6000-6600
Montane	*Climax Phase* Douglas-fir Western redcedar, Western hemlock Ponderosa pine *Successional Phase* Lodgepole pine Quaking aspen	Canadian	8000-10,000	7500-9500	4500-6000
Foothills and Mesas	Pinyon, Juniper Oak, Mountain mahogany Sagebrush scrub Saltbush, Greasewood	Transition	6000-8000	5500-7500	4000-4500
Plains and Valleys	Shortgrass Plains Riparian Deciduous Forest	Upper Sonoran	under 6000	under 5000	under 4000

* Wyoming data based mostly on Cary (1917); see also Porter (1962).

formations" (if defined by the general life-form of the dominant plants, such as tundra, grassland, shrubland or coniferous forest). Within a single plant formation there may be many different plant associations, and both temporary (successional) and stable or long-lasting (climax) community types.

Although many persons may not be interested in making such fine botanical distinctions as distinguishing specific plant associations while birding, in many cases this is not necessary for minimal habitat identification. Many birds respond to forest habitats relative to their general dominant life-form characteristics (e.g., mature hardwood forest, mixed coniferous-deciduous forest, coniferous forest), and thus detailed botanical identification may not be necessary. Nevertheless, bird habitat associations are identified in this text as accurately as possible, and thus some familiarity with the usual vegetational zonation patterns in the Rocky Mountains can be very useful in finding particular birds.

Climate, Landforms, and Vegetation

The entire Rocky Mountain region is characterized by continental climate, with great seasonal and daily changes in temperature, and fairly short and cool summers. In general, the higher precipitation levels occur in the northwestern portion of the area. In northern Idaho and northwestern Montana, moist winter air from the Pacific Northwest spills inland to produce the lush western red-cedar–western hemlock forests on the west-facing slopes. Most of the regional precipitation is orographic in nature. That is, it is related to montane topography, with the heaviest precipitation levels typically occurring on the western slopes, and the eastern slopes and valleys often showing reduced precipitation or "rain-shadow" effects. The Big Horn Mountains of Wyoming represent an exception to this general trend; their eastern slope receive some precipitation from Great Plains sources, while their western slopes are in a major rain-shadow resulting from high mountain ranges farther west. The driest parts of the area are in low-altitude basins such as the Snake River Basin of southern Idaho, the Bighorn Basin of Wyoming, and especially the Red Desert region of southeastern Wyoming, where annual precipitation is sometimes less than ten inches.

The predominant landforms of the region are, of course, associated with mountainous topography. These consist of three relatively discrete montane regions. The northern Rockies extend from Canada south into western Montana and Idaho as far as the Snake River Basin. The central Rockies are centered on the Yellowstone Plateau of northwestern

Wyoming and adjacent portions of southern Montana and extreme eastern Idaho, and comprise the Greater Yellowstone ecoregion. The southern Rockies extend from southern Wyoming and Colorado southward into New Mexico. These are all very high mountain ranges that collectively form the Continental Divide, with the maximum elevations being 12,972 feet in British Columbia, 12,294 feet in Alberta, 12,665 feet in Idaho, 12,850 feet in Montana, 13,785 feet in Wyoming, and 14,431 feet in Colorado.

The Rocky Mountains not only form the continental watershed throughout, separating the Great Plains to the east from the Great Basin to the west, but also provide the headwaters for such major river systems as the Snake, Colorado, and Missouri. Only in central Wyoming does the Continental Divide drop below 7,000 feet. There, in the Red Desert region, it separates and encloses the arid and alkaline Great Divide Basin before rising and leaving the state in the Sierra Madre range.

Besides the primary Rocky Mountain chain, there are a number of smaller subsidiary ranges, including the Bighorn Mountains of Wyoming, and several smaller groups of mountains in eastern Wyoming and Montana (Map 1). The geologic forces that shaped the area of the northern and central Rocky Mountains are complex, but except for the volcanic influences in the Yellowstone region the mountains are largely the result of folding and thrust-faulting of sedimentary layers, starting in late Cretaceous times some seventy million years ago. Lateral pressures on these layers caused folding, buckling, and faulting to occur, with large areas being lifted upwards and subsequently eroded away.

After the Cretaceous layers had been eroded away, progressively earlier layers of Mesozoic and Paleozoic deposits were exposed, until finally the Precambrian core levels were locally exposed. At the start of Cenozoic times (roughly 65 million years ago) the mountains were perhaps as rugged as the present-day Rockies, but were generally much lower and far more subtropical in climate. Folding and thrust-faulting actions terminated by about the end of the Eocene some sixty million years ago, but during various later parts of the Cenozoic several periods of volcanic activity resulted in the deposition of great lava plateaus in the Columbia Plateau, as well as tremendous depositions of wind-blown ash, filling mountain valleys and basins. In mid-Cenozoic times the entire Rocky Mountain area began to rise gradually, until the range reached its present-day elevation. This lifting was countered by continued erosion, especially during Pleistocene times, when actions of ice, water, and winds removed several thousands of feet from the exposed strata, and left evidence of such glacial handiwork as U-shaped valleys,

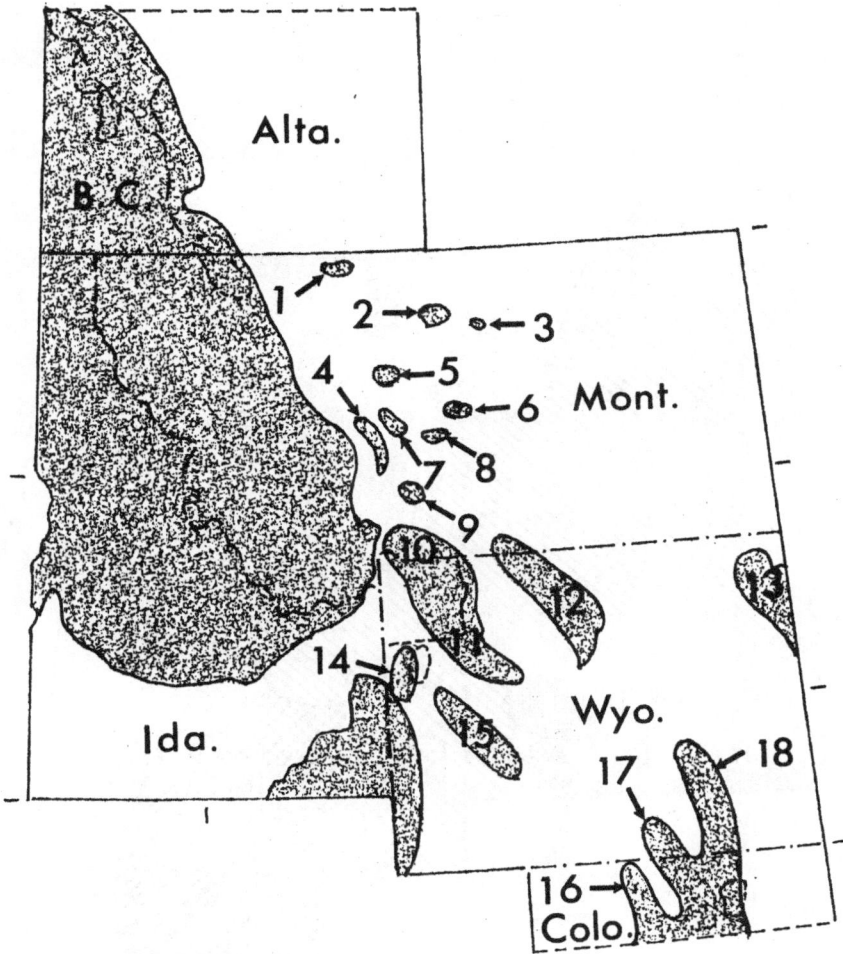

Map 1. Map of northern Rocky Mountain region, showing state or provincial boundaries and locations of mountain ranges. The ranges include 1: Sweetgrass Mts. 2: Bearpaw Mts. 3: Little Rocky Mts. 4: Big Belt Mts. 5: Highwood Mts. 6: Judith Mts. 7: Little Belt Mts. 8: Big Snow Mts. 9: Crazy Mts. 10: Snowy Mts. 11: Absaroka Mts. 12: Bighorn Mts. 13: Black Hills 14: Teton Mts. 15: Wind River Mts. 16: Park Mts. 17: Medicine Bow, Snowy and Sierra Madre Mts. 18: Laramie Mts. This map does not include the southern parts of Colorado that are also covered in the text. Map by author.

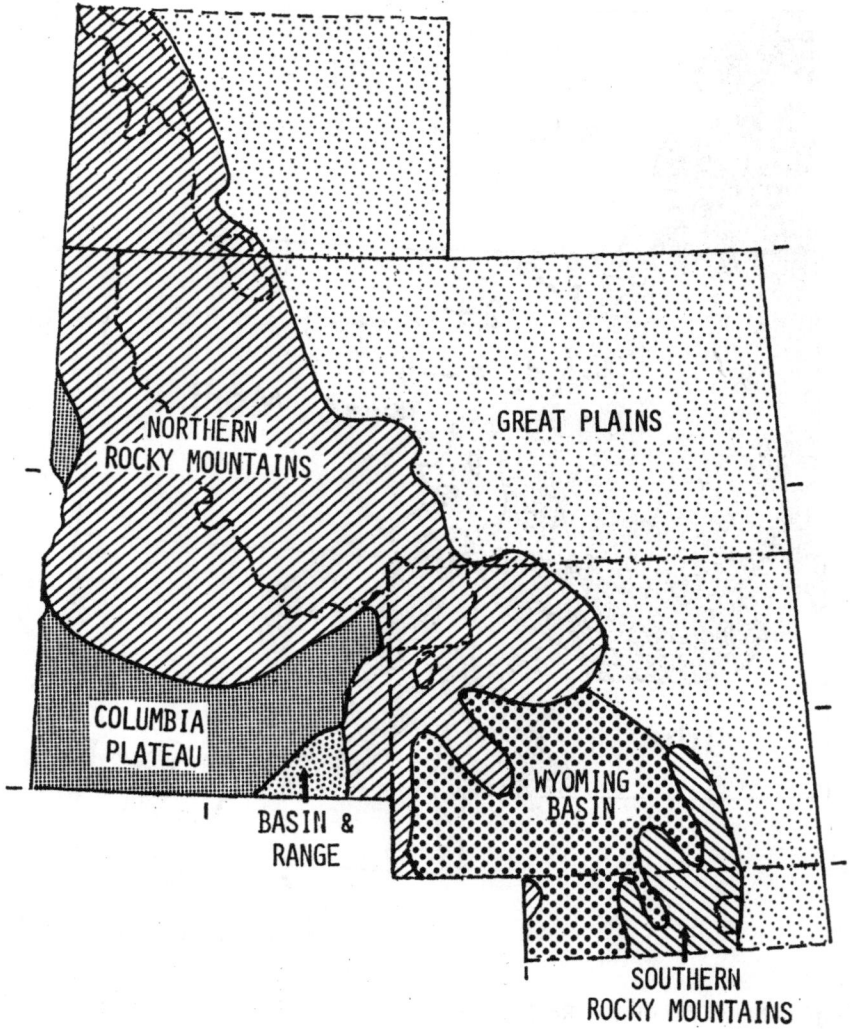

Map 2. Map of the northern Rocky Mountain region, showing distributions of major landscape features. Map by author.

cirques, and moraine-formed lakes (Map 2).

The cold climate of Pleistocene times brought with it a host of northern plant and animal species, which moved variably southward along the Rocky Mountain chain. Subsequently, many of the less mobile species became isolated on high peaks, as the climate ameliorated and a period of drying and warming began. This, of course, was much more true for plant life than for the more mobile animals. Yet, to some degree it can also be observed in such locally variable species as the rosy-finches and dark-eyed juncos, with their distinctive isolated populations extending from Canada southward. In any case, tundra areas were fragmented and isolated during post-Pleistocene times, and became increasingly restricted to the highest mountains. Likewise, progressively more arid-adapted montane communities and more varied patterns of geographic distributions developed, reflecting differences in altitude, precipitation, slope, soils, and other environmental factors (Map 3).

On a typical mountain slope these ecological zones often overlap and variably merge, especially on gradual slopes. For example, in southwestern and south-central Montana, the usual vertical sequence of tree species from lower slopes to alpine timberline is as follows: ponderosa pine, limber pine, Douglas-fir, lodgepole pine, spruces (Engelmann & white) and subalpine fir (Pfister *et al.*, 1977). Birders should be aware of the major vegetational zones in the Rocky Mountains, both as a means of predicting the occurrence of particular birds and also as a way of more fully appreciating the complex ecological interactions discernable in the region.

The major Rocky Mountains plant communities, and some of their associated botanic and avian characteristics, are as follows:

Alpine Tundra: Tundra occurs above timberline where trees are absent or confined to exceptionally protected locations, and dominated by perennial herbs (grasses and sedges) and often some low-growing shrubs. Summers are very short, wind effects are pronounced, and very few breeding birds occur here, including the American pipit, rosy-finches, and white-tailed ptarmigan.

Subalpine Zone (timberline zone): This area of generally low and often twisted trees ("krummholz") is typically dominated by subalpine fir (*Abies lasiocarpa*) and Engelmann spruce (*Picea engelmannii*). In some areas of Wyoming and Colorado the whitebark pine (*Pinus albicaulis*) is primarily a timberline species, while in Yellowstone Park whitebark pine also grows at lower altitudes, among stands of lodgepole pines. Somewhat farther south, limber pine (*Pinus flexilis*) is a

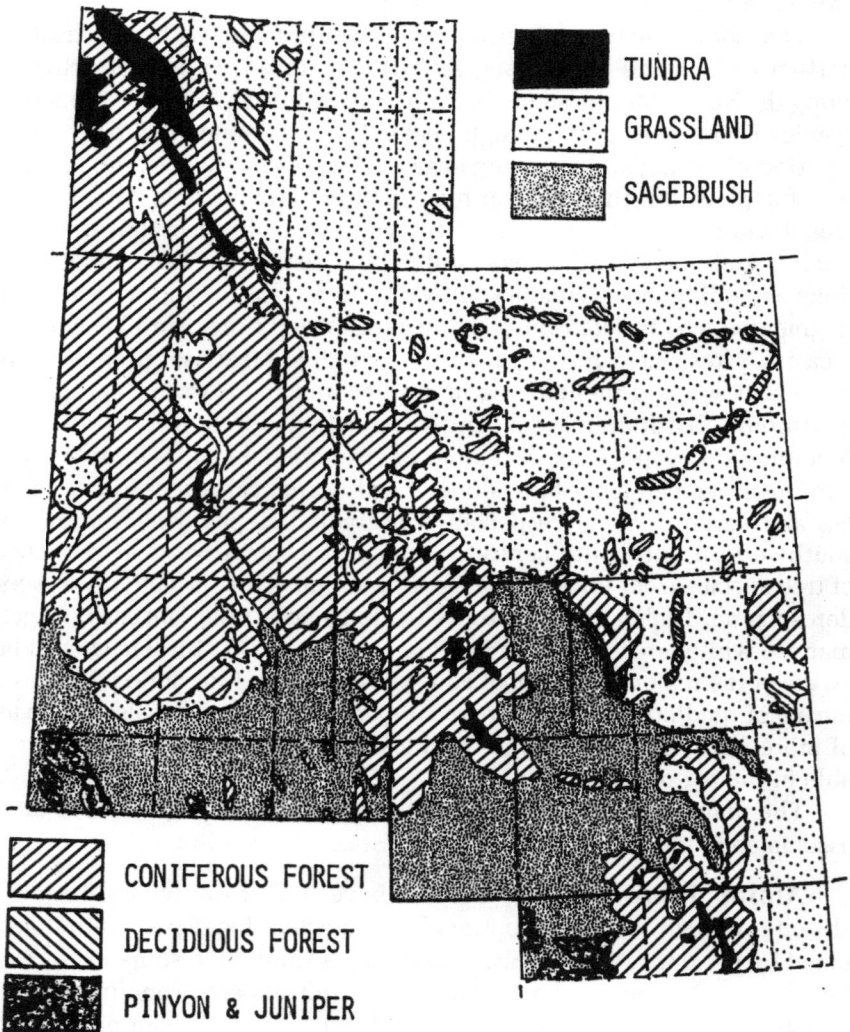

Map 3. Map of the northern Rocky Mountain region, showing distributions of natural vegetation community types. This map does not include those parts of southern Colorado that are also covered in the text. Map by author.

characteristic timberline species that also occurs on drier foothill sites in the shortgrass plains east of the Rockies. The whitebark and limber pines are both small-stature, five-needled pines that are similar in appearance. Both of these pines also produce large and nutritious seeds that are relished by seed-eating birds such as Clark's nutcrackers, and mammals ranging in size from mice to grizzly bears. The mature cones of whitebark pines often fall to the ground intact, where they may be quickly found and their seeds consumed by rodents, grouse, and many other birds.

Whitebark seedlings are shade-intolerant, so they may be unable to survive dense and mature spruce-fir forests. However, their heavy seeds are avidly collected and may be carried up to 15 miles by Clark's nutcrackers, which prefer to cache their seeds in bare or burned areas. Up to about 150 whitebark seeds can be carried by a Clark's nutcracker, and it can remember many of its cache sites for as long as nine months. This caching behavior by nutcrackers facilitates fairly rapid regrowth of trees after burns, as a result of germination by those seeds that the nutcrackers forget to recover. Some of these extremely slow-growing trees may reach many centuries of age (the apparent record is 1,270 years). In the Rocky Mountain region whitebark pines are seriously impacted by blister rust (*Cronartium*), bark beetles, and the parasite dwarf mistletoe (*Arceuthobium*). The blister rust came to America from Europe, via Canada, and has killed about ten percent of Yellowstone's whitebark pines, and up to 100 percent in northwest Montana.

The cones of limber pine are even larger than whitebark seeds, and may remain attached to the tree for prolonged periods, where crossbills may seek them out and pry them open for their large seeds. Eventually, however, the cones open and drop their seeds where they are often quickly found and consumed.

Lodgepole Pine Forest: Vast areas of the middle portions of the montane forest are covered by lodgepole pine (*Pinus contorta*) in the central and northern Rockies; for example, most of Yellowstone Park is dominated by such forests, which typically are regenerated only following fires. Lodgepole stands are typically very crowded, shady, and of nearly uniform ages, often as a result of having germinated after a forest fire, when their resin-sealed cones are burned and their seeds released. Because of their dense crowding and thin bark, lodgepole pines are easily killed by forest fires. However, they regenerate rapidly because of their fire-adapted seed-dispersal and a very rapid growth rate. Lodgepole pine seedlings are shade-intolerant, but do well on low-nutrient soils where few other trees can survive. In the Greater Yellowstone region these are often volcanic-based (rhyolitic) soils that have few nutrients and can support little other vegetation, resulting in low plant species diversity and a very limited food supply for most birds and mammals.

Douglas-fir Forest: This lower- to middle-altitude zone is dominated by Douglas-fir (*Pseudotsuga menziesii*), sometimes forming dense, single-species stands, but also often sharing dominance in the northern Rockies with Engelmann spruce (*Picea engelmannii*) or white spruce (*Picea glauca*). From Wyoming southward, blue spruce (*Picea pungens*) replaces white spruce, and reaches lower altitudes than does Engelmann spruce. Mature Douglas-firs have thick bark and are quite fire-resistant, sometimes forming magnificent stands several centuries old. These trees are also slow-growing and the seedlings are shade-tolerant, and thus are able to survive in the densely shaded conditions of lodgepole pine forests, ultimately replacing them.

Western Red-cedar–Western Hemlock Forest: On the moist western slopes of Glacier Park, and elsewhere in the northern Rockies, this distinctive community is locally found at lower to middle altitudes, and is dominated by these massive and beautiful forest giants. The western red-cedar (*Thuja plicata*) has a fluted trunk with thick bark and fern-like foliage. The western hemlock (*Tsuga heterophylla*) is a similarly beautiful and an important timber tree, so few of these forests remain except in protected areas such as parks. Seeing birds in these dense and tall temperate rainforests is very difficult, but they support some very attractive birds, and simply to wander through their cathedral-like groves can be a quasi-religious experience.

Ponderosa Pine Forest: This sometimes open and savanna-like forest is extremely widespread throughout the Rockies, from Canada to Mexico. It often forms the lower edge of the montane coniferous forest, and frequently extends out into the high plains in scattered groves on mesas or other favored sites. The dominant, and sometimes only, tree is ponderosa pine (*Pinus ponderosa*), which typically grows in fairly open rather than dense groves, with considerable grassy or shrubby cover between the trees. When mature the ponderosa pine is fairly fire-resistant, and can survive most of the ground fires that are common on the plains. In eastern Wyoming the limber pine is an important additional component of this semi-arid forest type, whereas in western Wyoming and Colorado various junipers replace limber pines to some degree. Diem and Zeveloff (1980) listed more than 110 bird species associated with ponderosa pine forests in western North America, although many have only marginal relationships with ponderosa forests. Eleven of the birds (mostly woodpeckers) are self-excavating cavity-nesting species, while 17 species use natural cavities or pre-excavated ones, often those excavated by woodpeckers.

Riparian Deciduous Forest: The upper reaches of the Yellowstone, Missouri, North Platte, and other major rivers of the Great Plains bring west into the region an important biota that is especially rich in eastern bird life, as noted earlier. Cottonwoods (*Populus angustifolia, P. acuminata* and their hybrids) are important riparian-zone trees, and their large size provides many foraging niches and nesting sites for songbirds. Other common trees are alders (*Alnus*), and willows (*Salix*), which sometimes form dense shoreline thickets, depending on the amount and seasonality of water availability. Riparian forests often extend for long distances along rivers passing through non-forested regions, which may allow for range expansion and dispersal across these barriers by forest-adapted bird species.

Aspen Woodland: Quaking aspen (*Populus tremuloides*) groves occur widely in the central and northern Rockies, either as a transitional ("seral") community following fire or logging, or as a long-term "climax" community in low hillsides too dry to support coniferous forests. Aspens are sensitive to fires and depend mostly on suckering for regeneration and local expansion; large areas in the Rockies are essentially clones of a single genetic type. Aspen woodlands often are rich in bird life, particularly woodpeckers, as well as various cavity-nesting birds that often exploit old woodpecker excavations for their own nesting sites. Sapsuckers are particularly attracted to aspens, and in turn their drillings attract other sap-eating species of birds, mammals, and insects. Woodpeckers also eat many insects that live under the bark of various coniferous forest trees, such as bark and wood-boring beetles. Their bark-flaking activities may also expose the insects and their larvae or pupae to other predators or parasitcs. However, woodpecker drilling activities may also expose the trees to infections by pathogens.

Pinyon–Juniper Woodland: On foothills and other areas below the coniferous forest a low forest composed of various species of junipers (*Juniperus monosperma, J. scopulorum, J. occidentalis*, etc.) and arid-adapted pines (*Pinus monophylla, P. edulis*, etc.) locally occurs. This arid-adapted woodland is poorly represented in this region, but locally extends north to the Snake River of Idaho. These pines produce nut-like seeds that are important foods for many rodents and some birds, such as the Clark's nutcracker and various jays.

Oak–Mountain Mahogany Scrub: Like the last community type, this is also an arid-adapted community better developed in the southern Rockies than in northern areas. It is largely limited to Colorado and extreme southern Idaho, and consists of several species of scrubby oaks (*Quercus gambelii* primarily), mountain mahogany (*Cercocarpus parvif/orus, C. ledifolius*, etc.) and woody shrubs such as serviceberry (*Amelanchier utahensis*). These trees and shrubs typically grow in clumps, separated by grassy areas, forming a chaparral-like community. Like the pinyon–juniper community, this one attracts a diverse bird life, especially those with southwestern geographic affinities.

Sagebrush Scrub: Over vast areas of the Intermountain West the land is dominated by sagebrush, especially big sagebrush (*Artemisia tridentata*). In some areas sagebrush shares dominance with various grasses, which are more palatable than are the leaves of sage. Only a few species of birds and mammals have adapted to living on a sage diet, most obviously the greater sage-grouse and pronghorn (*Antilocapra americana*).

Saltbush–Greasewood Scrub: In the Great Divide Basin area of Wyoming, and locally elsewhere, the highly alkaline soils allow only for the growth of this arid-land community type. The vegetation is scattered, shrubby, and bunch-like, the dark green color of the greasewood (*Sarcobatus vermiculatus*) contrasting with the more grayish shadscale (*Atriplex canescens*) and saltbush (*A. confertifolia*). Like the sagebrush scrub, edible foods are few, drinkable water is rare, and bird and mammal diversity is very low in this community type.

Grasslands: Within and to the west of the Rocky Mountains the grasses tend to be perennial bunchgrasses rather than the continuous sod-forming grasses typical of the Great Plains, with various wheatgrasses (*Agropyron*), needle-and-thread (*Stipa*) and fescues (*Festuca*) often dominant. On higher plateaus, such as the Yellowstone Plateau, and on mid-level mountains, perennial grasslands often dominate on fertile soils. For example they cover about 20 percent of the Bighorn National Forest, and are a major component of Yellowstone Park's northern range, the heart of the park's elk and bison summer habitat. The vegetation on Yellowstone's northern range mostly consists of perennial grasses, mainly Idaho fescue (*Festuca idahoensis*) and wheatgrasses , interspersed with open stands of big sagebrush. In the Big Horn Mountains, these grasslands are similarly dominated by wheatgrasses, pom-pelly brome (*Bromus pompellinaus*), Idaho fescus and inland bluegrass (*Poa interior*). Seed-eating species of birds and mammals are common, such as Brewer's sparrows and Uinta ground squirrels.

The vast shortgrass-dominated plains lying to the east of the Rocky Mountains are less bunch-forming, and instead are dominated by low, perennial grasses that produce continuous cover wherever there is adequate precipitation. These arid prairies have numerous species of such as grama (*Bouteloua* spp.) and buffalo grass (*Buchloe dactyloides*), as well as other taller grass species in protected or ungrazed areas. Heavy grazing or extended droughts favor invasion by cacti and other non-edible plants. Mountain plovers and McCown's longspurs are typical breeding birds of shortgrass prairies, and black-tailed prairie dogs (*Cynomys ludoviciana*) were historic keystone mammals over much of the Great Plains (Johnsgard, 2005).

Typical Birds of Rocky Mountain Habitats

Forest-adapted birds probably are affected by other environmental aspects of this altitudinal and microclimatic gradient, such as specific food plants, and potential nesting sites as much as or more than they are by the botanic components of the vegetation. The highest zone and the most climatically extreme conditions for life occur on mountaintops, where the growing seasons are very short, and breeding conditions are usually marginal. As a result, species diversity is very low in this zone. In the alpine tundra zone from Colorado north to Jasper National Park, the most typical breeding birds are the white-tailed ptarmigan and the rosy-finches. In the southern Rockies of Colorado, the breeding rosy-finch is the brown-capped, whereas in the central Rockies of Wyoming it is the black, and still farther north in the northern Rockies of Montana and Canada it is the gray-headed. The white-crowned sparrow is a timberline nester, frequenting low willows and sedges, and nesting in low shrubs or on the ground (Braun, 1980).

In the subalpine spruce-fire zone just below timberline the Brewer's sparrow is a typical breeding species. It also breeds disjunctively in sagebrush scrub at much lower-altitudes, perhaps as a distinct species. The pine siskin, red and white-winged crossbill, are frequent breeders. At somewhat lower altitudes, where the trees grow taller, breeders include the northern pygmy-owl, black-backed and American three-toed woodpeckers, common raven, Steller's jay and Clark's nutcracker. Golden-crowned and ruby-crowned kinglets also nest in these taller trees, as do the Townsend's warbler and pine grosbeak.

Among the non-passerine birds that are often associated with mature mesic or middle-level montane coniferous forests, dominated by Douglas-fir, are such attractive non-passerines as the northern goshawk, great gray owl, boreal owl, pileated woodpecker, and Williamson's sapsucker. Passerine species include the gray and Steller's jays, Clark's nutcracker, Wilson's, yellow-rumped, and MacGillivray's warblers, golden-crowned and ruby-crowned kinglets, mountain chickadee, red-breasted nuthatch, Townsend's solitaire, varied, hermit, and olive-backed thrushes, and dark-eyed junco. Smith (1980), summarizing data from 12 spruce-fir studies during the breeding season, determined that the mountain chickadee and yellow-rumped warbler was observed in all 12, the hermit thrush, ruby-crowned kinglet and pine siskin in 11, the Clark's nutcracker in ten and the American robin in nine. In eight studies the red-breasted nuthatch, brown creeper, Townsend's solitaire, pine grosbeak and chipping sparrow were detected, in seven the northern flicker, red crossbill and dark-eyed junco, and in six the American three-toed woodpecker, dusky flycatcher, and Cassin's finch were present. The total number of species observed in all 12 studies ranged from 12–30; the three studies from the northern Rockies of Colorado, Wyoming and Montana reported 12–19 species.

Sanderson, Bull and Edgerton (1980) reported that 26 bird species of the interior Northwest breed only in mature or old-growth mixed conifer forests. The nonpasserines are the wood duck, Barrow's goldeneye, bufflehead, hooded merganser, northern goshawk, golden eagle, bald eagle, osprey, merlin, barn owl, barred owl, Vaux's swift, pileated woodpecker, Williamson's sapsucker, and white-headed woodpecker. Passerines include the Hammond's flycatcher, Clark's nutcracker, white-breasted nuthatch, pygmy nuthatch, brown creeper, hermit thrush, ruby-crowed kinglet, Townsend's warbler, evening grosbeak, pine grosbeak and red crossbill. By comparison, a total of ten species were reported to breed in the grass-forb stage, 30 in the shrub-seedling stage, 38 in the pole-sapling stage, 59 in the young forest stage, and 83 in the mature forest stage.

In the moist and highly shaded Pacific-slope red-cedar–hemlock forests of northwestern Montana and Idaho, birds that might be found typically include the spruce grouse, Williamson's sapsucker, chestnut-backed and boreal chickadees, Townsend's warbler, pine grosbeak and white-winged crossbill. Engeline (1980) reported that old-growth stands in the Sierras support the golden-crowned kinglet, chestnut-backed chickadee, pileated woodpecker and great gray owl. Verner (1980) similarly found that Sierra Nevada birds breeding in large-tree mixed-conifer forests with at least 70 percent canopy cover included the spotted and great gray owls, pileated woodpecker, Hammond's flycatcher, chestnut-backed chickadee, red-breasted nuthatch, winter wren, hermit thrush, golden-crowned kinglet and purple finch. Most of these species can also be found in the western-slope old-growth forests of Glacier National Park.

The birds of Rocky Mountain lodgepole pine forests are in general much like those of the Douglas-fir and other mid-level coniferous forest communities. Several woodpecker species are typical of lodgepole forests, as well as some closely associated cavity-nesting songbirds. The hermit thrush, Cassin's finch and red crossbill are also typical of this zone, and these same species often extend higher into the subalpine zone. In the Jackson Hole area, the common breeding birds of lodgepole pine forests are the northern flicker, downy woodpecker, gray jay, mountain chickadee, hermit thrush, yellow-rumped warbler, dark-eyed junco and chipping sparrow (Anderson, 1980).

In Colorado, the most common breeding species in lodgepole forests were yellow-rumped warbler, ruby-crowned kinglet, dark-eyed junco, hermit thrush, black-capped and mountain chickadees, pine siskin, gray jay and Townsend's solitaire. All these species tend to have broad niches, and are common in other coniferous communities. The number of breeding species ranged from 8–14 in four studies in lodgepole pine stands and from 10–20 in five studies of mixed lodgepole and other conifers. Species density was also substantially higher in the mixed forest stands. In a previously burned Colorado study site, both species diversity and breeding densities were somewhat higher in the burned site, and the once-present hermit thrush, ruby-crowned kinglet and the two chickadees were replaced by the American robin, mountain bluebird, broad-tailed hummingbird, northern flicker and *Empidonax* flycatchers (Hein, 1980).

In the much drier, lower and more park-like ponderosa pine forests very widespread, mature ponderosa pines are quite fire-resistant, and their open, savanna-like community structure results in a forest type that can easily survive and thrive after periodic ground fires. These forests support many typical western birds. Diem and Zeveloff (1980) reported that at least 113 bird species reside in ponderosa pine forests. Insectivorous species that selectively nest in coniferous trees are the olive-sided flycatcher, ruby- and golden-crowned kinglets, western tanager and olive, yellow-rumped, black-throated and gray-rumped warblers. Omnivorous species nesting selectively in coniferous trees include the Clark's nutcracker and the gray and pinyon jays. The band-tailed pigeon is the only granivorous species that nests selectively in conifers. Tree-nesting species using either coniferous or deciduous trees are numerous (34 species) in these forests. Another 28 species are cavity-nesters, 15 species nest in bushes or small trees, 15 nest on the ground and 11 nest on cliffs, in caves, or on rocks or among talus. Many of these species also occur in other regional community types.

The aspen forests typically have highly diversified breeding avifauna, including such species as the ruffed grouse, flammulated owl, northern pygmy-owl, several woodpeckers, and warbling vireo. Because of the excavating actions of woodpeckers on soft-wooded trees such as aspens and birches, many cavity-nesting birds such as the tree and violet-green swallows, house wren, black-capped and mountain chickadees, and mountain bluebirds, are attracted to this forest type. Canopy-nesters such as the warbling vireo, yellow-rumped warbler, and western wood-pewee are common. Shrub-nesters include the yellow warbler, various *Empidonax* flycatchers (dusky, Hammond's and cordilleran), black-headed grosbeak and MacGillivray's warbler. Common ground-nesters include the dark-eyed junco, hermit thrush, Townsend's solitaire and white-crowned sparrow (Flack, 1976). Scott (1957) found that 60 percent of the breeding bird species in aspen forests of the Jackson Hole region were cavity nesters, as compared with 54 percent for spruce-fir forests, and 19–31 percent for lodgepole or lodgepole and spruce or spruce–fir types. The aspen sites also had substantially higher breeding bird densities (523 pairs/100 acres) than did any of the other Rocky Mountains forest types analyzed by Scott, Whelan and Svoboda (1980).

Riparian woodlands developed along montane drainages often have a mix of coniferous and deciduous species, and this plant diversity also generates considerable bird diversity. Typical birds are a mix of eastern and western species and include broadly distribution species such as the willow flycatcher and warbling vireo, and such western birds as the spotted towhee, lazuli bunting, Bullock's oriole and black-headed grosbeak. Eastern counterpart species of the deciduous forest, such as eastern towhee, indigo bunting, Baltimore oriole and rose-breasted grosbeak, or their interspecies hybrids, are often a part of the mix. Other eastern species that extend variably westward along river systems include the yellow-billed cuckoo, eastern screech-owl, red-headed and red-bellied woodpeckers, least flycatcher, eastern phoebe, eastern bluebird, brown thrasher, red-eyed vireo, and orchard oriole. In the Jackson Hole area, the common birds of riparian aspen groves are the northern flicker, downy and hairy woodpeckers, western wood-pewee, tree swallow, black-capped chickadee, mountain bluebird, warbling vireo, yellow warbler, black-headed grosbeak, white-crowned sparrow and Lincoln's sparrow (Anderson 1980). Western riparian communities are not only fairly high in species diversity, but population densities may reach 600–1,300 pairs per 100 acres (Johnson *et al.*, 1980).

The dry woodlands of pinyon pine, juniper, oak, and mountain mahogany, which are best developed to the south and west of the region under consideration here, carry into the region a distinctive group of birds such as the common poor-will, northern saw-whet owl, pinyon jay, gray flycatcher, juniper titmouse, blue-gray gnatcatcher, western bluebird, Bewick's wren, black-throated gray warbler and Grace's warbler. Balda and Masters (1980) identified five "obligatory" species of this community as consisting of the western screech-owl, gray flycatcher, western scrub-jay, juniper (previously "plain") titmouse and gray vireo. Pinyon-juniper plots of 1015 acres may support 20 or so breeding species, a species diversity comparable to ponderosa pine or mixed conifer forests, but having far lower breeding densities, namely 40–80 pairs, versus 200–350 pairs, per 100 acres (Johnson *et al.*, 1980).

The still drier and lower sagebrush scrub and alkaline-associated saltbush and greasewood communities likewise have a few highly distinctive breeding species. Plots of 10–15 acres of saltbush-greasewood habitat often support only three or four breeders and 15–20 pairs per 100 acres, whereas a similar plot of sagebrush may support 10–12 species and 50–100 pairs (Johnson *et al.*, 1980). The sagebrush community supports several distinctive species such as greater sage-grouse, Gunnison sage-grouse, sage thrasher, green-tailed towhee, Brewer's sparrow, and sage sparrow. In recent years much of this community has been converted to agriculture through irrigation, and strongly sage-dependent species as the sage-grouse have suffered accordingly.

The birds of the shortgrass prairies east of the foothills typically have wide breeding distributions on the Great Plains (Johnsgard, 1979, 2001). They include such attractive species as the sharp-tailed grouse, long-billed curlew and upland sandpiper. Several grassland sparrows are also present, such as the grasshopper, lark and vesper sparrows, lark bunting, McCown's longspur and chestnut-collared longspur. Some of these same grassland species are associated with the bunchgrass prairies of the Pacific Northwest, but the longspurs and lark bunting are lacking, and others are very rare.

Even in national parks not all the habitats are pristine; historical and recent forest fires have placed much of the area of Yellowstone Park in various stages of vegetational transition ("succession") dominated by lodgepole pine, for example. Ranching activities in Grand Teton National Park and national forests in the Greater Yellowstone region have also influenced grassland and shrub succession in non-forested areas, as a result of overgrazing. Browsing of elk has greatly affected aspen distribution and survival, and damming of streams by beavers has resulted in the formation of unique beaver-pond communities, with an interesting and diverse associated plant, bird, and mammal life. Indeed, such species as the trumpeter swan and sandhill crane are largely dependent upon beaver activity in the Greater Yellowstone region for the production and maintenance of suitable breeding habitat.

Human-caused effects on the environment are apparent everywhere throughout the Rocky Mountains, as grazing, lumbering, agriculture, mining, energy development, road-building, and other familiar symbols of modern civilization have left their marks on the landscape. In general, the influence of man is to reduce environmental diversity, either fortuitously or purposefully, by eliminating unwanted species in favor of more economically profitable uses for the land. As a result, some bird species have become extremely rare throughout the region, even in national parks. These include several mammalian and avian predators, and many birds closely associated with native grasslands or shrublands, such as the sharp-tailed grouse and greater sage-grouse.

Some species, such as the peregrine falcon, trumpeter swan and sandhill crane, have responded to protection and management, and have moved back into areas from which they have been absent for many decades. Between 1980 and 1991 more than 500 peregrines were released in the Greater Yellowstone ecosystem, and at least 75 nesting pairs were established. Still others have benefited indirectly from human activities, and have become extremely abundant in and around human activity centers. These include such introduced species as the house sparrow, European starling, and the still rapidly expanding Eurasian collared-dove. Various native species including the American robin, common grackle, great-tailed grackle and brown-headed cowbird have also benefited from human-caused habitat changes by adapting to urban parks and gardens.

On the basis of their widespread occurrence in the central and northern Rocky Mountain national parks, a collective list of 55 typical Rocky Mountain birds is shown below, which provides a nuclear group of species that might be observed while visiting almost any of these parks.

Water-dependent Species
Mallard
Ring-necked Duck
Barrow's Goldeneye
Common Merganser
Osprey
Spotted Sandpiper
Wilson's Snipe
Belted Kingfisher
American Dipper
American Pipit
Common Yellowthroat

Forest-dependent Species
Spruce Grouse
Dusky (Blue) Grouse
Ruffed Grouse
Great Horned Owl
Northern Pygmy-Owl
Red-naped Sapsucker
American Three-toed Woodpecker
Olive-sided Flycatcher
Gray Jay
Steller's Jay
Clark's Nutcracker
Common Raven
Black-capped Chickadee
Mountain Chickadee
Red-breasted Nuthatch
Brown Creeper
Ruby-crowned Kinglet
Mountain Bluebird
Townsend's Solitaire
Hermit Thrush
Yellow-rumped Warbler
Wilson's Warbler
Western Tanager
Lincoln's Sparrow
Fox Sparrow
White-crowned Sparrow
Dark-eyed Junco
Pine Grosbeak

Widespread and Other Species
Killdeer
Common Nighthawk
Willow Flycatcher
Tree Swallow
Violet-green Swallow
Northern Rough-winged Swallow
Cliff Swallow
Barn Swallow
American Robin
Yellow Warbler
MacGillivray's Warbler
Chipping Sparrow
Vesper Sparrow
Savannah Sparrow
Brewer's Blackbird
Rosy-finches (3 spp.)

Many of the national parks have a number of distinctive bird species present, and the Appendix provides a listing of the relative summer abundance of 278 Rocky Mountain birds in nine U.S. and Canadian National Parks. Some of these species are regarded as threatened, endangered or sensitive by one of more of the states (Table 2).

Recent Changes in Rocky Mountain Ecology and Avifauna

Since the 1986 publication of my *Birds of the Rocky Mountains*, many ecological changes have occurred in the region. These include many significant forest fires, providing clear and frightening evidence of significant global warming and associated regional drying influences. Outbreaks of some serious coniferous forest insect pests in the region, such as bark beetles (*Dendroctonus, Ipis* and *Polygraphus*, family Scolytidae), are partly related to the weakening of forest trees owing to water stress. Nearly two dozen species of *Dendroctonus* exist in western North America, of which 19 occur north of Mexico. About 6.4 million acres have been seriously affected by bark beetles in recent decades throughout the West, including 2.5 million acres in the Forest Service's Rocky Mountain region, and especially in Wyoming, Colorado and western South Dakota. One species of bark beetle (the western pine beetle, *D. brevicomis*), selectively attacks the ponderosa pine while a close relative (mountain pine beetle, *D. ponderosae*) additionally attacks other pines, including lodgepole and whitebark. Another Rocky Mountain bark beetle (Engelmann spruce beetle, *D. engelmanni*) selectively attacks spruces. The Douglas-fir beetle (*D. pseudotsugae*) attacks Douglas-fir, as well as western larches (*Larix* spp.). During the long-term drought of the early 2000's the Rocky Mountain conifers were greatly weakened, and became highly vulnerable to attack by bark beetles. Although bark beetles have long been present in the West, the unusually mild winters of recent years have also reduced the winter mortality of the beetles, setting the stage for widespread coniferous forest disasters.

After emergence from their pupae, adult bark beetles disperse by constructing an exit hole from their pupal cells by boring through the host tree's bark and flying to the next host, which may be some distance away, but is usually close by. How they identify appropriate host trees is still uncertain, but visual clues as well as chemical stimuli may be used, such as by detecting host-plant monoterpenes (which deter most plant-eating insects but can be detoxified by bark beetles) or the beetle's own "aggregation pheromones." On entering the bark of a new host, many attacking beetles must be present in order to overcome the tree's chemical defenses. This mass colonization behavior is achieved by airborne dispersal of pheromones produced by the newly arriving beetles, as well as recognizing defensive volatile substances produced by the tree under attack. Once a tree is invaded, egg-deposition galleries are excavated under the bark, which also serve for ventilation.

Infected trees fight back by exuding copious amounts of pitch and resin, which are thought to have toxic properties, and unless they are weakened by drought and unable to produce enough defensive sap, they may be able to ward off small infestations. Massive attack by the beetles may kill mature trees in a year or two through biomass reduction, by reducing the rate of the trees' photosynthesis (Stark, 1982), or perhaps as a result of a fungus ("blue stain") that the beetle introduces into the tree via small specialized pouch (Stephenson, 2011). Large numbers of dead trees add to the danger of uncontrolled wildfires, especially during drought years.

Woodpeckers are one of the most effective means of controlling bark beetle infestations, especially when populations are still sparse. With regard to the black-backed and American three-toed woodpeckers, it appears that the black-backed specializes on wood-boring beetles (mainly Cerambicidae) following fires, while the American three-toed species is more closely associated with spruce forests and consuming various bark beetles (Scolytidae) that attack these trees (Leonard, 2001, Dixon and Saab, 2000). The most important species attacking the spruce beetle are the hairy, downy and American three-toed woodpeckers, while the hairy, downy and pileated woodpecker are known to be important predators of the western pine beetle.

Studies have found that woodpeckers may consume 20–30 percent of bark beetle larvae or pupae, and up to 98 percent in epidemic populations. Several bird species also attack adult beetles during their dispersal stages. Depending on the species, woodpeckers puncture, drill or flake the bark, which reduces the chances of survival for beetles that remain in the tree or are dislodged off along with the bark. Bark-flaking also exposes the insects to other bark-gleaning predators, such as brown creepers and black-and-white warblers, as well as various insect predators and parasitoids (Dahlsten, 1982).

During severe outbreaks, which are associated with severe drought periods that stress the trees, the beetles are so destructive that large forest areas become increasingly vulnerable to devastating wildfires. The mountain pine beetle has impacted more than 1.5 million acres of coniferous forests in northern Colorado and southern Wyoming, and has especially affected lodgepole pines and whitebark pines. In Wyoming

Table 2. Rocky Mountain Birds of State, Regional and National Conservation Concern

	USFWS Status*	State Status**				USFS Status***
		CO	ID	MT	WY	
Family Anatidae: Ducks, Geese & Swans						
Trumpeter Swan, *Cygnus buccinator*		p	p	p	p(PI)	S:R1
Harlequin Duck, *Histrionicus histrionicus*		p	p	p	p	S:R1-4
Family Phasianidae: Partridges & Grouse						
Greater Sage-grouse, *Centrocercus urophasianus*	C	p	p	p	p	
Columbian Sharp-tailed Grouse, *T. phasianellus columbianus*		p	p	p?	p	S:R4
Family Gaviidae: Loons						
Common Loon, *Gavia immer*		p	p	p	p(PI)	S:R1-4
Family Ardeidae: Bitterns & Herons						
American Bittern, *Botaurus lentiginosus*		p	p	p	p(PII)	S:R3
Family Threskiornithidae: Ibises & Spoonbills						
White-faced Ibis, *Plegadis chihi*		p	p	p	p(PI)	S:R3
Family Accipitridae: Kites, Hawks & Eagles						
Osprey, *Pandion haliaetus*		p	p	p	p	S:R3
Bald Eagle, *Haliaeetus leucocephalus*	T	p	p	p	p	T
Northern Goshawk, *Accipiter gentilis*		p	p	p	p	S:R3
Ferruginous Hawk, *Buteo regalis*		p	p	p	p(PIII)	S:R1,3,4
Family Falconidae: Falcons						
Peregrine Falcon, *Falco peregrinus*	T	p	p	p	p	T
Family Gruidae: Cranes						
Greater Sandhill Crane, *Grus canadensis tabida*		p	p	p	p	
Whooping Crane, *Grus americana*	E	p		p	p	E
Family Charadriidae: Plovers						
Snowy Plover, *Charadrius alexandrinus*		p		p	p	S:R3
Interior Piping Plover, *Charadrius melodus circumcinctus*	E	p		p	p	E
Mountain Plover, *Charadrius montanus*		p		p	p	S:R3,4

	USFWS Status*	State Status** CO	ID	MT	WY	USFS Status***
Family Scolopacidae: Sandpipers & Phalaropes						
Upland Sandpiper, *Bartramia longicauda*		p	p	p	p	S:R3
Long-billed Curlew, *Numenius americanus*		p	p	p	p(PIII)	S:R3
Family Laridae: Gulls & Terns						
Interior Least Tern, *Sterna antillarum athalassos*	E				p	E
Black Tern, *Chlidonias niger*		p	p	p	p(PII)	S:R3
Family Cuculidae: Cuckoos & Anis						
Yellow-billed Cuckoo, *Coccyzus americanus*	C	p	p	p	p(PII)	S:R3
Family Strigidae: Typical Owls						
Flammulated Owl, *Otus flammeolus*		p	p	p	p	S:R3
Mexican Spotted Owl *Strix occidentalis lucida*	T	**p**				**S:R3,4**
Boreal Owl, *Aegolius funereus*		p	p	p	p	S:R1,3,4
Western Burrowing Owl, *Speotyto cunicularia hypogea*		p	p	p	p(PII)	
Family Picidae: Woodpeckers						
American Three-toed Woodpecker, *Picoides dorsalis*		p	p	p	p	S:R4
Black-backed Woodpecker, *Picoides arcticus*		p	p	p	p(PII)	
Family Tyrannidae: Tyrant Flycatchers						
Southwestern Willow Flycatcher, *Empidonax traillii extimus*		p				S:R3,4
Family Emberizidae: Towhees & Sparrows						
Baird's Sparrow, *Ammodramus bairdii*		p		p	p	S:R3

* U.S. Fish & Wildlife Service Symbols: E = Endangered, T = Threatened, C = Candidate for Federal Listing

** Current or recent presence of species in state is indicated by a "p"; Wyoming conservation categories PI, PII, and PIII indicate descending conservation priority levels.

*** U.S. Forest Service Symbols: S = Sensitive, R1 = Region 1 (MT, northern ID, ND & northwestern SD), R2 = Region 2 (CO & WY); R3 = Region 3 (AZ & NM), R4 = Region 4. (NV, UT, southern Idaho & western Wyoming). T & E symbols as with U.S. Fish & Wildlife Service. Forest Service listings are based on a summary by Finch (1992).

the infestation started about 2000, with the mountain pine beetle, the spruce beetle and the Douglas-fir beetle (Stephenson, 2011).

It has been estimated that by 2012 nearly all the mature lodgepole pines in that region will have been killed, and by that time well over half of the whitebark pines in Yellowstone Park will have died. Rare nesting birds typical of dense lodgepole forests, such as northern goshawk and boreal owl, have probably already been seriously affected, and species such as the Clark's nutcracker ar heavily dependent on the seeds of whitebark pines for their survival. Forest-thinning has been one means of trying to control beetle outbreaks, but that approach is not very effective, as the adult beetles are somewhat mobile. Global warming has also reduced winter mortality in beetle populations, and its many ecological effects are likely to become an increasingly serious problem in the future. The blister-rust fungus (*Cronartium*) has also long had had a devastating impact on western pine forests (mainly affecting western white, ponderosa and lodgepole), especially in Montana, Idaho and Wyoming.

Aspen woodlands have also declined in the Rocky Mountains during recent decades, for reasons not yet fully understood. Some aspen diseases are facilitated as a result of elk browsing, whose activities damage the twigs and open a path for pathogen entry. Heavy browsing by elk probably also weakens the trees, which is one reason why the introduction of wolves into the Greater Yellowstone Ecosystem has had a beneficial effect on aspens. This local aspen recovery in turn has probably affected the many species of birds that are associated from their presence. These include at least seven regional species of woodpeckers (which favor drilling in aspens over trees with harder woods), and the secondary use of woodpecker nesting cavities by more than a dozen species of cavity-nesting Rocky Mountain birds. There are at least 72 cavity-nesting bird species in western North America (Scott, Whelan and Svoboda, 1980), and many of them are largely or entirely dependent on pre-existing cavities for their nesting.

Forest fragmentation has been a serious problem in the western states, as a result of clear-cutting forestry methods in intact forests, and the effects of forest fires and post-fire salvage logging. The bird species most likely to benefit from recent unsalvaged burns are the black-backed and American three-toed woodpeckers. Others, such as the Lewis' woodpecker and American kestrel may tolerate or even benefit from partial or light salvage logging, provided that large snags and trees that they tend to select for nesting are left uncut. Some species (Williamson's sapsucker, Lewis' woodpecker, northern flicker) appear to respond most positively to larger nest-tree diameters during site-selection. Many others (hairy, black-backed and American three-toed woodpeckers, brown creeper, western and mountain bluebirds and red-breasted nuthatch) evidently respond more strongly to higher snag densities at nest sites (Kotliar *et al*, 2002).

Some species respond very rapidly to post-fire forest conditions; the black-backed and American three-toed woodpeckers often colonize burns within one or two years post-burn, but become rare after about five years. The olive-sided flycatcher also appears immediately after fires, and remains as long as snags remain available. Some seed-eaters (Clark's nutcracker, pine siskin, Cassin's finch and red-crossbill) may be attracted for a short time to the sudden availability of exposed seeds or ashy mineral nutrients. Some insect-eaters are also typically more abundant in burned than unburned sites, such as western wood-pewees, house wrens and tree swallows. Lewis's woodpeckers may remain abundant in post-burn areas up to 25 years. Mountain and western bluebirds typically decline in mid-successional stages of forest regrowth after about 50 years. The same appears to be true of cordilleran and dusky flycatchers, and American robins also begin to decline in mid- to late-successional stages. Mountain chickadees, ruby-crowned kinglets and both Swainson's and varied thrushes reach their peak abundance in late stages of forest succession. Species that are typically most abundant in mature, unburned forests include the Steller's and gray jays, plumbeous, Cassin's and warbling vireos, ruby- and golden-crowned kinglets, brown creeper, red-breasted nuthatch, mountain chickadee, Townsend's warbler and the Swainson's and varied thrushes (Kotliar *et al*, 2002).

Rocky Mountain birds that are notably sensitive to fires (and logging) because of their apparent need for large tracts of mature coniferous forest include the Vaux's swift, boreal owl, white-headed and pileated woodpeckers, mountain and chestnut-backed chickadees, red-breasted and white-breasted nuthatches, brown creeper, winter wren, golden-crowned and ruby-crowed kinglets, Swainson's, hermit and varied thrushes, Townsend's warbler and pine grosbeak (Wuerthner, 2006).

Taylor and Barmore (1980) studied the effects on birds in lodgepole pine and spruce-fir-pine forest burns in Yellowstone and Grand Teton National Parks, using post-burn sites ranging from one year to more than three centuries since burning. Their data suggest that black-backed and American three-toed woodpeckers are most abundant during the first three years following fires, while hairy woodpeckers also arrive early and are present until about 40 years post-fire. Mountain bluebirds and tree swallows are the most common breeders during years 5–29, but disappear after about 50 years. White-crowned sparrows are also gone by about 30 years post-burn, and both the Clark's nutcracker and Cassin's finch are gone by about 40 years. Species that were more abundant in unburned or partly burned forests than in severely burned ones include the western tanager, both kinglets, red-breasted nuthatch, mountain chickadee and yellow-rumped warbler. With the completion of plant succession to mature spruce-fir forests, the brown creeper, mountain chickadee and red-breasted nuthatch become the common timber-searching species. Foliage-insect-eating species of such mature forests include the ruby-crowned and golden-crowned kinglets, yellow-rumped warbler and western tanager. In contrast, the American robin, dark-eyed junco, chipping sparrow, and gray jay were present over nearly the entire three-century successional series.

Logging in coniferous forests of the Rocky Mountains affects bird species quite differently. Tobalske, Shearer and Hutto (1991) studied logging effects on birds breeding in larch–Douglas-fir forests of Montana. Of 32 species analyzed, ten exhibited statistically significant population differences associated with logging. Tree swallows were very frequently found in clear-cuts, but were almost absent elsewhere. However, the golden-crowned kinglet, Swainson's and varied thrushes, and Townsend's warbler were significantly less abundant in clearcuts or partly cut forests, and the ruby-crowned kinglet and fox sparrow were least abundant in clearcuts. Other species that were less abundant in clearcuts, but not at a statistically significant level, were the brown creeper, winter wren, and MacGillivray's warbler. Species that were more abundant in logged than in intact forests were the olive-sided flycatcher, Townsend' solitaire and American robin.

Where to Search for Specific Rocky Mountain Birds

A table listing nine of the national parks covered in this book, their seasonal status, and relative summer (or overall) abundance of their birds is presented in Appendix 1. Many of the more attractive, elusive or charismatic birds of the Rocky Mountain region can also be found in the region's national wildlife refuges, national forests or other public-access sites. Sixty of these species and some suggested locations for finding them within the four northern Rocky Mountain states are shown below, exclusive of the separately summarized montane national parks. Unless otherwise indicated, these species are likely to be present at least during summer, if not throughout the year. Abbreviations are: N. F. = National Forest; N. M. = National Monument; N. P. = National Park; N. R. A. = National Recreation Area; N. W. R. = National Wildlife Refuge; S. P. = State Park; W. P. A. = Waterfowl Production Area. For all sites the relevant state is indicated parenthetically.

Trumpeter Swan: Bridger-Teton N. F., (WY), Camas N. W. R. (ID), Harriman S. P. (ID), National Elk Refuge (WY), Red Rock Lakes N. W. R. (MT),

Harlequin Duck: Bridger-Teton N. F., (WY), Lolo N. F. (MT), Priest Lake S. P. (ID).

Spruce Grouse: Beaverhead-Deer Lodge N. F. (MT), Flathead N. F. (MT) Routt N. F. (CO), Kootenai N. W. R. (ID), Sawtooth N. R. A. (ID).

Dusky Grouse: Beartooth Highway (WY/MT), Black Canyon N. P. (CO), National Bison Range (MT), Flathead N. F. (MT), McCrosky S. P. (ID), Red Rock Lakes N. W. R. (MT), Wasatch-Cache N. F. (WY).

Greater Sage-Grouse: Bowdoin N. W. R. (MT), Bridger-Teton N. F. (WY), Craters of the Moon N. M. (ID), Grays Lake N. W. R. (ID), Routt N. F. (CO).

Gunnison Sage-Grouse: Gunnison N. F. (CO).

Sharp-tailed Grouse: Benton Lake, N. W. R. (MT), Bowdoin N. W. R. (MT), C. M. Russell N. W. R. (MT), Freezeout Lake W. M. A. (MT).

White-tailed Ptarmigan: Arapaho-Roosevelt N. F. (CO), Wasatch-Cache N. F. (WY).

Western & Clark's Grebes: Bear Lake N. W. R. (ID), Bowdoin N. W. R. (MT), Camas N. W. R. (ID), Freezeout Lake W. M. A. (MT), Medicine Lake N. W. R. (MT), Minidoka N. W. R. (ID), Ninepipe N. W. R. (MT), Red Rock Lakes N. W. R., (MT).

White-faced Ibis: Bear Lake N. W. R. (ID), Benton Lake, N. W. R. (MT), Browns Park N. W. R. (CO), Camas N. W. R. (ID), Freezeout Lake W. M. A. (MT), Medicine Lake N. W. R. (MT), Minidoka N. W. R. (ID), Oxford Slough W. P. A. (ID), Seedskadee N. W. R. (WY).

Sandhill Crane: Alamosa N. W. R. (CO), Bear Lake N. W. R. (ID), Camas N. W. R. (ID), Harriman Wildlife Refuge (ID), Lee Metcalf N. W. R. (MT), Grays Lake N. W. R. (ID), Red Rock Lakes N. W. R. (MT).

Osprey: Canyon Ferry W. M. A. (MT), Canyon Ferry W. M. A. (MT) C. M. Russell N. W. R. (MT), Deer Flats N. W. R. (ID), Heyburn S.P. (ID), Lee Metcalf N. W. R. (MT), Medicine Lake N. W. R. (MT), Red Rock Lakes N. W. R. (MT).

Bald Eagle: Browns Park N. W. R. (CO), C. M. Russell N. W. R. (MT), Deer Flat N. W. R. (ID), Kootenai N. W. R. (ID), Lee Metcalf N. W. R. (MT), Mesa Verde N. P. (CO), Minidoka N. W. R. (ID), Monte Vista N. W. R. (CO), National Elk Refuge (WY), Red Rock Lakes N. W. R. (MT), Swan River N. W. R. (MT).

Golden Eagle: Arapaho N. W. R. (CO), Beartooth Highway (WY/MT), Beaver Creek Park (MT), Black Canyon N. P. (CO), Browns Park N. W. R. (CO), C. M. Russell N. W. R. (MT), Grays Lake N. W. R. (ID), Minidoka N. W. R. (ID), Monte Vista N. W. R. (CO), National Bison Range (MT), National Elk Refuge (WY), Red Rock Lakes N. W. R. (MT), Seedskadee N. W. R. (WY), Terry Badlands (MT).

Prairie Falcon: Beartooth Highway (WY/MT), Beaver Creek Park (MT), C. M. Russell N. W. R. (MT), Bighorn Canyon N. R. A. (WY/MT), Grays Lake N. W. R. (ID), Makoshika S. P. (MT), Medicine Rocks S. P. (MT), Red Rock Lakes N. W. R. (MT), Seedskadee N. W. R. (WY), Snake River Birds of Prey Area (ID), Terry Badlands (MT).

Merlin: Browns Park N. W. R. (CO), C. M. Russell N. W. R. (MT), Snake River Birds of Prey Natural Area (ID).

Peregrine: Bighorn Canyon N. R. A. (WY/MT), Colorado N. M. (CO), Mesa Verde N. P.(CO), Red Rock Lakes N. W. R. (MT), Seedskadee N. W. R. (WY), Snake River Birds of Prey Natural Area (ID).

American Avocet: Alamosa N. W. R. (CO). Bear Lake N. W. R. (ID), Benton Lake N. W. R., Bowdoin N. W. R. (MT), C. M. Russell N. W. R. (MT), Freezeout Lake W. M. A. (MT), Grays Lake N. W. R. (ID), Hutton Lake N. W. R. (WY), Medicine Lake N. W. R. (MT), Minidoka N. W. R. (ID), Ninepipe N. W. R. (MT), Red Rock Lakes N. W. R. (MT), Seedskadee N. W. R. (WY).

Black-necked Stilt: Alamosa N. W. R. (CO), Bear Lake N. W. R. (ID), Benton Lake N. W. R. (MT), Freezeout Lake W. M. A. (MT).

Long-billed Curlew: Benton Lake N. W. R. (MT), Bowdoin N. W. R. (MT), Bruneau Dunes S. P. (ID), Camas N. W. R. (ID), Freezeout Lake W. M. A. (MT), Grays Lake N. W. R. (ID), Harriman Wildlife Refuge (ID), Minidoka N. W. R. (ID), Red Rock Lakes N. W. R. (MT).

Upland Sandpiper: Benton Lake N. W. R. (MT), C. M. Russell N. W. R. (MT), Medicine Lake N. W. R. (MT).

Marbled Godwit: Benton Lake N. W. R. (MT), Bowdoin N. W. R. (MT), Freezeout Lake W. M. A. (MT), Red Rock Lakes N. W. R. (MT).

Band-tailed Pigeon: Alamosa N. W. R. (CO) Monte Vista N. W. R. (CO).

Flammulated Owl: Bitterroot N. F. (ID), Boise N. F. (ID), Kootenai N. W. R. (ID), Lee Metcalf N. W. R. (MT).

Burrowing Owl: Alamosa N. W. R. (CO) Arapaho N. W. R. (CO), Benton Lake N. W. R. (MT), C. M. Russell N. W. R. (MT), Deer Flat N. W. R. (ID), Rocky Mountain Arsenal N. W. R. (CO).

Great Gray Owl: Beaverhead-Deerlodge N. F. (WY), Boise N. F. (ID), Bridger-Teton N. F., (WY), Deer Flats N. W.R. (ID), Harriman S. P. (ID), Lolo N. F. (MT), Red Rock Lakes N. W. R. (MT), Sawtooth N. R. A. (ID).

Boreal Owl: Beaverhead-Deerlodge N. F. (WY), Bridger-Teton N. F., (WY) Flathead N. F. (MT), Lolo N. F. (MT).

Northern Pygmy-Owl: Boise N. F. (ID), Heyburn S. P. (ID), Lee Metcalf N. W. R. (MT), Sawtooth N. R. A. (ID).

Northern Saw-whet Owl: Beaverhead-Deerlodge N. F. (WY), Boise N. F. (ID), C. M. Russell N. W. R. (MT), Lee Metcalf N. W. R. (MT), Mesa Verde N. P. (CO), Red Rock Lakes N. W. R. (MT), Wasatch-Cache N. F. (WY).

Black Swift: Black Canyon N. P. (CO), Kootenai N. W. R. (ID), Lee Metcalf N. W. R. (MT), National Bison Range (MT), Treasure Falls (CO), Rifle Falls S. P. (CO).

White-throated Swift: Black Canyon N.P. (CO), Black Hills N. F. (SD/WY), Browns Park N. W. R. (CO), C. M. Russell N. W. R. (MT), Colorado N. M. (CO), Great Sand Dunes N. P. (CO), Mesa Verde N. P. (CO).

Black-chinned Hummingbird: Browns Park N. W. R. (CO), Grays Lake N. W. R. (ID), Kootenai N. W. R. (ID), Swan River N. W. R. (MT).

Rufous Hummingbird: Arapaho N. W. R. (CO), Beaverhead-Deerlodge N. F. (WY), Browns Park N. W. R. (CO), Kootenai N. W. R. (ID), Great Sand Dunes N. P. (CO), Mesa Verde N. P. (CO), Red Rock Lakes N. W. R. (MT), Swan River N. W. R. (MT).

Broad-tailed Hummingbird: Black Canyon N.P. (CO), Great Sand Dunes N. P. (CO), Kootenai N. W. R. (ID), Mesa Verde N. P. (CO), Red Rock Lakes N. W. R. (MT), Seedskadee N. W. R. (WY).

Calliope Hummingbird: Beaverhead-Deerlodge N. F. (WY), Grays Lake N. W. R. (ID), Kootenai N. W. R. (ID), Lee Metcalf N. W. R. (MT).

Lewis's Woodpecker: Boise N. F. (ID), Craters of the Moon N. M. (ID), Great Sand Dunes N.P. (CO), Kootenai N. W. R. (ID), Lee Metcalf N. W. R. (MT), Mesa Verde N. P. (CO), National Bison Range (MT), Seedskadee N. W. R. (WY).

Pileated Woodpecker: Bitterroot N. F. (ID), Boise N. F. (ID), Farragut S. P. (ID), Kootenai N. W. R. (ID), Lee Metcalf N. W. R. (MT), Lolo N. F. (MT), Mc-Crosky S. P. (ID), Round Lake S. P. (ID), Winchester S. P. (ID).

Williamson's Sapsucker: Lolo N. F. (MT), Routt N. F. (CO), Red Rock Lakes N. W. R. (MT).

American Three-toed Woodpecker: Black Hills N. F. (SD/WY), Bridger-Teton N. F., (WY), Flathead N. F. (MT), Lolo N. F. (MT), Sawtooth N. R. A. (ID), Wasatch-Cache N. F. (WY).

Black-backed Woodpecker: Black Hills N. F. (SD/WY), Boise N. F. (ID), Bridger-Teton N. F., (WY) Flathead N. F. (MT), Lolo N. F. (MT), Kootenai N. W. R. (ID), Wasatch-Cache N. F. (WY).

Clark's Nutcracker: Beartooth Highway (WY/MT), Kootenai N. W. R. (ID), Mc-Crosky S. P. (ID), National Elk Refuge (WY), National Bison Range (MT), Pine Butte Swamp Preserve (MT), Red Rock Lakes N. W. R. (MT).

Chestnut-backed Chickadee: Heyburn S.P. (ID), Kootenai N. W. R. (ID), Swan River N W. R. (MT).

Pygmy Nuthatch: Black Hills N. F. (SD/WY), National Bison Range (MT), Red Rock Lakes N. W. R. (MT).

Bushtit: City of Rocks National Reserve (ID), Colorado N. M. (CO), Dinosaur N. M. (CO), Great Sand Dunes N.P. (CO).

Rock Wren: Deer Flats N. W. R. (ID), Grays Lake N. W. R. (ID), Minidoka N. W. R. (ID), National Bison Range (MT).

Canyon Wren: Colorado N. M. (CO), Dinosaur N. M. (CO), Seedskadee N. W. R. (WY),

American Dipper: Bridger-Teton N. F. (WY), Browns Park N. W. R. (CO), Kootenai N. W. R. (ID), National Bison Range (MT), Red Rock Lakes N. W. R. (MT), Rifle Falls S. P. (CO).

Varied Thrush: Bitterroot N. F. (ID), Browns Park N. W. R. (CO), Flathead N. F. (MT), Kootenai N. W. R. (ID), Swan River N. W. R. (MT).

Western Bluebird: Arapaho N. W. R. (CO), Farragut S. P. (ID), Florissant Fossil Beds N. M. (CO).

Mountain Bluebird: Arapaho N. W. R. (CO), Bowdoin N. W. R. (MT), C. M. Russell N. W. R. (MT), Kootenai N. W. R. (ID), Monte Vista N. W. R. (CO), National Bison Range (MT), National Elk Refuge (WY), Red Rock Lakes N. W. R. (MT), Seedskadee N. W. R. (WY).

Townsend's Solitaire: Black Hills N. F. (SD/WY), C. M. Russell N. W. R. (MT), Grays Lake N. W. R. (ID), Heyburn S. P. (ID), Kootenai N. W. R. (ID), Mc-Crosky S. P. (ID), Rifle Falls S. P. (CO).

American Pipit: Arapaho-Roosevelt N. F. (CO), Beartooth Highway (WY/MT).

Townsend's Warbler: Farragut S. P. (ID), Heyburn S. P. (ID), McCrosky S. P. (ID), Rifle Falls S. P. (CO), Swan River N. W. R. (MT).

Black-throated Gray Warbler: Colorado N. M. (CO), Dinosaur N. M. (CO), Mesa Verde N. P.(CO), Oxford Slough W. P. A. (ID), Flaming Gorge N. R. A. (WY).

MacGillivray's Warbler: Black Hills N. F. (SD/WY), Colorado N. M. (CO), Farragut S. P. (ID), Grays Lake N. W. R. (ID), Great Sand Dunes N. M. (CO), Kootenai N. W. R. (ID), McCrosky S. P. (ID), Mesa Verde N. P. (CO).

Virginia's Warbler: Black Canyon N. P. (CO), Castlewood Canyon S. P. (CO), Golden Gate Canyon S. P. (CO), Mesa Verde N. P. (CO).

Western Tanager: Camas N. W. R. (ID), C. M. Russell N. W. R. (MT), Minidoka N. W. R. (ID), National Bison Range, (MT), Red Rock Lakes N. W. R. (MT).

Gray-crowned Rosy-Finch: National Elk Refuge (WY), Red Rock Lakes N. W. R. (MT),

Black Rosy-Finch: Beartooth Highway (MY/WY), Wasatch-Cache N. F. (WY).

Brown-capped Rosy-Finch: Medicine Bow Mountains (WY), Guanella Pass (CO), Mount Evans (CO).

Cassin's Finch: Grays Lake N. W. R. (ID), Red Rock Lakes N. W. R. (MT), Seedskadee N. W. R. (WY), Wasatch-Cache N. F. (WY).

Synopsis of Major Birding Locations in the Rocky Mountains Region

U.S. National, State and Local Sites

Colorado

ALAMOSA NATIONAL WILDLIFE REFUGE

This 11,000-acre refuge near Alamosa consists mostly of Rio Grande riverbottom wetlands in a high mountain valley. This refuge and nearby Monte Vista National Wildlife Refuge collectively attract at least 183 bird species, including 70 breeders. Alamosa is recognized as one of the state's Important Bird Areas. An on-line seasonally organized refuge checklist that also includes the 14,000-acre Monte Vista refuge is available: http://www.npwrc.usgs.gov/resource/birds/ chekbird/r6/alamonte.htm. For more information contact the refuge manager, 9383 E. Rancho Lane, Alamosa, CO 81101 (719/589-4021).

ARAPAHO-ROOSEVELT NATIONAL FOREST

This 1.8 million-acre national forest is situated on the Continental Divide in north-central Colorado includes part or all of the **Cache la Poudre Wilderness** (9,400 acres) **Comanche Peak Wilderness** (67,000 acres), **Indian Peaks Wilderness** (73,000 acres), **Mt. Evans Wilderness** (70,000 acres), **Neota Wilderness** (10,000 acres), **Never Summer Wilderness** (7,800 acres) and **Rawah Wilderness** (74,000 acres). The 36,000-acre **Arapaho National Recreation Area** is on the western slope of the Divide. The forest has a range in altitude that extends up to 14,260 feet, and it is possible to drive up to the Guanella Pass (south of Georgetown on County Road 381) at 11,669 feet. This pass reaches the summer home of the white-tailed ptarmigan, American pipit and brown-capped rosy-finch. Loveland Pass, on U.S. Hwy 6, reaches 11,990 feet, and has these same alpine species. The Mount Evans Highway (Colorado Hwy. 5) is the highest paved road in the U.S., and terminates below the summit of Mt. Evans (14,264 feet), where the ptarmigan, pipit and rosy-finch might be seen in summer. There are ranger stations at Boulder, Ft. Collins, Granby and Idaho Springs, The Arapaho-Roosevelt Forest's bird list includes over 200 species and is available from the Sulfur Ranger District Office, Arapaho-Roosevelt National Forest, 62429 Highway 40, P.O. Box 10, Granby, CO 80446 (303/887-3331).

ARAPAHO NATIONAL WILDLIFE REFUGE

This 15,000-acre refuge in north-central Colorado is located along the Continental Divide, at about 8,300 feet elevation, in an arid rain-shadow area only about

60 miles away from Rocky Mountain National Park. The refuge is located in the North Park region of Colorado; mountain "parks" are montane meadows or relatively treeless areas in an otherwise generally forested region. The refuge attracts at least 198 bird species, including at least 70 breeders. An on-line seasonally organized checklist for the refuge is available: http://www.npwrc.usgs. gov/resource/birds/chekbird/r6/arapaho.htm. For further information, contact the refuge manager, 953 JC Road #32, Walden, CO 80480 (970/723-8202).

BENT'S OLD FORT NATIONAL HISTORIC SITE

This small (148-acre) historic site along the old Santa Fe Trail is located east of La Junta in southeastern Colorado. The site has a very high level of species diversity for its small area, and a marsh that is especially notable for its elusive rails (black, Virginia and sora). A non-seasonal on-line site checklist is available that includes 137 species (29 breeders): http://www.npwrc.usgs.gov/resource/ birds/chekbird/r6/bentfort.htm. For information contact Bent's Old Fort National Historic Site, 35110 Highway 194 East, La Junta, CO 81050 (719/373-5010).

BLACK CANYON OF THE GUNNISON NATIONAL PARK

This 20,700-acre national park is located in western Colorado, straddling a 2,700-foot canyon of the Gunnison River. The park's vegetation mostly consists of scrub oaks and serviceberry, as well as pinyon–juniper woodlands. Its notable birds include dusky grouse, golden eagle, broad-tailed hummingbird, white-throated swift and violet-green swallow. Other typical western or southwestern species include the broad-tailed hummingbird, Steller's jay, western scrub-jay, dusky flycatcher, Virginia's and MacGillivray's warblers and blue-gray gnat-catcher. For further information, contact the Black Canyon of the Gunnison National Park, 102 Elk Creek, Gunnison, CO 81230 (970/641-2337). Not far from Black Canyon Park, about 25 miles south of Montrose off US Highway 550, is Box Cañon Park. Black swifts can be seen there around their nesting sites, which are situated behind a 500-foot waterfall. For information contact Box Cañon Park, City of Ouray, Ouray, CO (970/325-7080).

BROWN'S PARK NATIONAL WILDLIFE REFUGE

This refuge of more than 13,000 acres is located in extreme northwestern Colorado along the Green River and adjoining Dinosaur National Monument. The refuge consists mostly of mountain meadows and rocky slopes, bluffs and marshy habitats as well as the Green River. The refuge attracts at least 181 bird species, including 78 breeders. It is recognized as one of the state's Important Bird Areas. A non-seasonal on-line bird checklist for the refuge is available: http://www.npwrc.usgs.gov/resource/birds/chekbird/r6/brown.htm. For more information, contact the refuge manager, 12318 Highway 318, Maybell, CO 81640 (970/365-3613).

COLORADO NATIONAL MONUMENT

This national monument of 20,400 acres is located in western Colorado along the Grand Valley of the Colorado River. The monument mostly consists of pin-yon-juniper, as well as sagebrush and some coniferous forest, along steep rim-rock and canyon topography. Its raptors include breeding peregrines, kes-trels and other cliff-loving raptors, as well as the white-throated swift, canyon wren and common raven. Scrub areas attract the gray vireo, bushtit, west-ern scrub and pinyon jays, juniper titmouse, black-throated gray warbler and black-throated sparrow. It is recognized as one of the state's Important Bird Areas. A seasonally organized on-line checklist includes 127 species: http:// www.npwrc.usgs.gov/resource/birds/chekbird/r6/fruita.htm. For further in-formation, contact Colorado National Monument, Fruita, CO 81521-9530 (970/858-3617).

COMANCHE NATIONAL GRASSLAND

This 435,000-acre region of shortgrass and mixed-grass prairie in southeast-ern Colorado also includes some steep canyons and wetlands. The bird list includes at least 235 species, of which the lesser prairie-chicken is notable. A public-access lek is located near Campo, and Cassin's sparrows are common in this area. The nearby Cimarron National Grassland in southwestern Kansas has a similar and better-studied avifauna, including more than 330 species (Boyle and Wauer, 1994; Cable, Seltman and Cook, 1996). For information contact the Forest Service, P.O. Box 127, Springfield, CO 81073 (303/523-6591).

DINOSAUR NATIONAL MONUMENT

This national monument is located on the Colorado-Utah border along the Green and Yampa rivers, and comprises 320 square miles. The vegetation and topography are not typical of the Rocky Mountains, but instead are an exten-sion of the basin and range topography of Utah and Nevada. As a result, the bird life is distinctly arid-adapted, and includes many species otherwise occur-ring only to the west and south. Park specialties include the juniper titmouse, bushtit, canyon towhee, black-throated gray warbler, canyon wren, house finch and lesser goldfinch. There is a printed bird list available for the monument, and the on-line checklist for the adjacent Brown's Park National Wildlife Ref-uge (see above) is probably also applicable. For further information, contact the monument superintendent, 4545 East Highway 40, Dinosaur, CO 81610 (970/374-3000).

FLORISSANT FOSSIL BEDS NATIONAL MONUMENT

This 5,900 acre-national monument west of Colorado Springs in central Colo-rado is mainly of interest for its Oligocene plant fossils. However, 15 miles of trails through its ponderosa pine forest offer good birding possibilities for species such as mountain and western bluebirds, Williamson's sapsucker and red crossbill. Shrubby areas have sage-adapted species such as the green-tailed towhee. A non-seasonal on-line site checklist of about 110 species is available: http://www.npwrc.usgs.gov/resource/birds/chekbird/r6/floriss.htm. For more information contact Florissant Fossil Beds National Monument, P.O. Box 185, Florissant, CO 80816 (719/748-3253).

GREAT SAND DUNES NATIONAL PARK

This national park of 38,600 acres in southwestern Colorado is noted for its enormous sand dunes along the base of the Sangre de Cristo Mountains. Surrounded by shrublands and pinyon–juniper woodlands, the park has a number of southwestern birds, including the bushtit and juniper titmouse. Other attractive species include MacGillivray's warbler, western tanager, Lewis' woodpecker, broad-tailed and rufous hummingbirds, white-throated swift, pinyon jay and western scrub-jay. An adjacent wetland area attracts many unexpected species such as sandhill cranes. A non-seasonal on-line site checklist has about 155 species: http://www.npwrc.usgs.gov/resource/birds/chekbird/r6/gdunes.htm. For more information, contact Great Sand Dunes National Park, 11999 Highway 150, Mosca, CO 81146-9798 (719/378-6399).

GRAND MESA-UMCOMPAHGRE AND GUNNISON NATIONAL FORESTS

These national forests collectively cover 3.16 million acres of Colorado's West Slope. Elevations in these forests range from 5,800 to 14,309 feet (Umcompahgre Peak), and include parts or all of ten wilderness areas. Grand Mesa-Umcompahgre National Forest has over 250 lakes and reservoirs in its 1.2-million acres, and includes **Big Blue** (98,000 acres) and **Mt. Sneffels** (16,000 acres) wilderness areas. Nearby Gunnison National Forest has 27 peaks above 12,000 feet of elevation, and encompasses over 1.6 million acres. Its wilderness areas include **West Elk** (176,000 acres) and parts of **Collegiate Pass** (160,000 acres), **Maroon Bells-Snowmass** (174,000 acres) and **Raggeds** (59,000 acres), The range of the rare and declining Gunnison sage-grouse, the most recently discovered North American bird species, is centered in the Gunnison Valley. This species is so rare that viewing opportunities are very limited. One lek near Doyleville is accessible to the public. For information, contact the Sisk-a-dee Environmental Organization, 323 N. Wisconsin St., Gunnison, CO 81230. Bird lists for these national forests are not yet available, but county lists from the Colorado Field Ornithologists group might suffice (see below). There are ranger stations at Collbran, Grand Junction, Montrose, Norwood and Paonia. For more information, contact the forest supervisor, 2250 Highway 50, Delta, CO 81416 (970-874-6717).

MESA VERDE NATIONAL PARK

This national park of 52,000 acres is located in extreme southwestern Colorado in typical canyon-and-mesa country, with a mixture of sagebrush, pinyon–juniper and coniferous forests. The park's diverse bird life includes raptors such as the red-tailed hawk, golden eagle, peregrine, northern saw-whet owl and the nationally endangered Mexican spotted owl. The white-throated swift is abundant, as are the broad-tailed and black-chinned hummingbirds. Pinyon-adapted species include the bushtit, pinyon jay, western scrub-jay, Virginia's warbler and black-throated gray warbler. The park is recognized as one of Colorado's Important Bird Areas. For more information, contact Mesa Verde National Park, P.O. Box 8, Mesa Verde National Park, CO 81330 (970/529-4465).

MONTE VISTA NATIONAL WILDLIFE REFUGE

This 14,000-acre refuge is located south of the town of Monte Vista, in an arid mountain valley. The refuge is notable for the large flocks of greater sandhill cranes that stage here each spring and fall. It and nearby Alamosa Refuge attract at least 183 bird species, including 70 breeders. A seasonally organized bird checklist that also includes the Alamosa refuge is available on-line: http://www.npwrc.usgs.gov/resource/birds/chekbird/r6/alamonte.htm. For more information contact the Alamosa refuge manager, 9383 El. Rancho Lane, Alamosa, CO 81101 (719/589-4021).

PAWNEE NATIONAL GRASSLAND

This 193,000-acre area in northeastern Colorado is famous for its shortgrass birds, including the mountain plover, both grassland longspurs, lark bunting, Cassin's and Brewer's sparrows, and several prairie raptors. For information contact the Pawnee National Grassland office at 2150 Centre Ave., Bldg. E, Fort Collins, CO 80526-819 (970/295-6600 <www.fed.us/r2/arnf>.

RIFLE FALLS STATE PARK

This 92-acre park near Rifle in western Colorado's White River National Forest is one of the few places where black swifts may be seen nesting. The American dipper, ruby-crowned kinglet and Townsend's solitaire may also be seen here, as well as some birds of the pinyon–juniper plant community. For information on Rifle Falls State Park, contact Colorado State Parks (970/625-1607). For information on White River National Forest, contact the Forest Service office, 9th & Grand, Glenwood Springs, CO (907/945-2521).

ROCKY MOUNTAIN ARSENAL NATIONAL WILDLIFE REFUGE

This 17,000-acre refuge is located just east of Denver, and is a rehabilitated military facility that is now mostly comprised of grassland habitats. The refuge's summer birds include the burrowing owl and ferruginous hawk, as well as some mostly eastern-oriented species, and is recognized as one of Colorado's Important Bird Areas. A seasonally organized on-line bird checklist includes 227 species: http://www.npwrc.usgs.gov/resource/birds/chekbird/r6/arsenal.htm. For more information contact the Rocky Mountain Arsenal National Wildlife Refuge, Building 121, Commerce City, CO 80020 (303/289-0930).

ROCKY MOUNTAIN NATIONAL PARK

This national park encompasses about 417 square miles of Colorado's magnificent Front Range, and it ranges in altitude range from about 7,800 feet at Estes Park to 14,255 feet at Longs Peak. Tourists can drive over the Continental Divide, at 12,183 feet, on Trail Ridge Road, the highest road in any national park. About one-third of the park exceeds 11,000 feet in elevation, and there are over 50 square miles of alpine tundra vegetation. This park not only includes many areas of alpine tundra that are easily accessible, but there are also large tracts of montane coniferous forest, mainly of ponderosa pine, plus Engelmann spruce and subalpine fir at subalpine levels. At least 262 bird species have been reported in the Park, which is recognized as one of Colorado's Important Bird Areas. Park specialties (some fairly rare) include the band-tailed pigeon, common poorwill, American three-toed woodpecker, black swift, pygmy nuthatch, western bluebird and brown-capped rosy-finch. Notable breeders include the ring-necked duck and white-tailed ptarmigan. Hugh Kingery's 2007 book *Birding Colorado* has a five-page account of Rocky Mountain National Park, as well as descriptions of all the Colorado sites described below. A checklist of park birds may be obtained at the headquarters, and a seasonally organized on-line checklist of 260 species (including the Arapaho National Recreation Area, Granby and Estes Park) is also available: http://www.npwrc.usgs.gov/resource/birds/chekbird/r6/rockymt.htm. For more information on Rocky Mountain National Park, contact the Park Superintendent, Estes Park, CO 80517 (303/586-2371).

ROOSEVELT NATIONAL FOREST

This national forest of 790,000 acres is on the eastern slope of the Rocky Mountains. There are ranger stations in Boulder, Estes Park, Fort Collins and Greeley. The forest is now jointly operated with Arapaho National Forest; see the Arapaho–Roosevelt Forest description above. For information contact the ranger station at 240 W. Prospect, Fort Collins, CO 80526 (303/482-5155).

ROUTT NATIONAL FOREST

This national forest of 1.1 million acres is located along the central Rockies' Continental Divide, and has a range in elevation from 6,750 to 13,533 feet. There are ranger stations at Steamboat Springs, Walden and Yampa. It includes **Mount Zirkel Wilderness** (140,000 acres, with 64 lakes and 14 peaks above

12,000 feet), as well as parts of **Never Summer** (7,800 acres) and **Flat Tops** (235,000 acres) wildernesses. Notable nesting or resident birds are the bufflehead, dusky grouse and greater sage-grouse, northern goshawk, Williamson's and red-naped sapsuckers, broad-tailed hummingbird, and at least four warblers. Boreal owls have been reported at Rabbit Ears Pass, as well as Hammond's and dusky flycatchers. The forest bird list includes over 200 species and is available from the Forest Service office, 29587 W. U.S. Highway 40, Suite 20, Steamboat Springs, CO 80487 (303/879-1722).

SAN ISABEL NATIONAL FOREST

This national forest of 1.2 million acres is located in south-central Colorado. It includes the **Mt. Massive** (28,000 acres) and **Sangre de Cristo** (227,000 acres) wildernesses. There are ranger stations at Cañon City, Leadville and Salida. A bird list is not yet available. For more information contact the supervisor, 2840 Kachina Dr., Pueblo, CO 81008 (719/553-1400).

SAN JUAN NATIONAL FOREST

This national forest of 1.8 million acres is located in southwestern Colorado. Its elevations range from 5,000 to more than 14,000 feet, and the forest includes parts of the **Lizard Head** (106,000 acres), **South San Juan** (127,000 acres), and **Weminuche** (460,000 acres, with several peaks above 14,000 feet) wilderness areas. A bird list is not yet available. There are ranger stations at Bayfield, Dolores, Durango, and Pagosa Springs. For more information contact the forest supervisor, 15 Burnett Court, Durango, CO 81301 (970/247-4874).

WHITE RIVER NATIONAL FOREST

This enormous national forest of 2.3 million acres is located in Colorado's central Rockies. The forest contains all of **Hunter–Frying Pan Wilderness** (74,000 acres), and part of six other wilderness areas. These are **Collegiate Peaks** (160,000 acres), **Eagles Nest** (134,000 acres), **Flat Tops** (235,000 acres), **Holy Cross** (116,000 acres), **Maroon Bells–Snowmass** (174,000 acres) and **Raggeds** (59,000 acres). The forest encompasses many peaks above 14,000 feet, including the 14,431-foot Mount Elbert, the highest peak in the Rocky Mountains. Independence Pass on Colorado Highway 63 east of Aspen, reaches a dizzying 12,095 feet, and is a good place for finding alpine birds such as brown-capped rosy-finches and white-tailed ptarmigans. There are ranger stations at Aspen, Carbondale, Eagle, Meeker, Minturn, Rifle and Silverthorne. No bird list is yet available. For more information contact the forest supervisor, 900 Grand Ave., Glenwood Springs, CO 81601 (970/945-2521).

Further information on Colorado's ten national forests (totaling over 14 million acres) and two national grasslands (totaling 612,000 acres) may be obtained from the Forest Service's Rocky Mountain Region Office, 1177 W. 8th Ave., P.O. Box 25227, Lakewood, CO 80225 (303/236-9433).

The Bureau of Land Management (BLM) administers 8.3 million acres of public lands in Colorado, including 22 special management recreation areas. Locations with deep canyons likely to be of special interest to raptor-enthusiasts include the **Black Ridge Canyon**, a 74,000-acre region between Black Ridge and the Colorado River (contact the Grand Junction office for information), **Irish Canyon**, a 13,000-acre canyon area at the eastern edge of the Uinta Mountains (contact the Craig office for information) and **Phantom Canyon**, a 28,000-acre area between Victor and Venice (contact the Cañon City office for information). The State BLM office address is 2850 Youngfield St., Lakewood, CO 80215 (303.239-3800). There are district offices in Cañon City (719/269-8500), Craig (970/824-8261), Montrose (970/249-7791) and Grand Junction (970/244-3000). A seasonally organized bird checklist that includes non-federal land in Mesa and western Garfield counties and totals 322 species is available on-line at: http://www.npwrc.usgs.gov/resource/birds/chekbird/r6/grandjct.htm or from the Grand Junction Resource Area at the Grand Junction District Office (2815 H. Road, Grand Junction, CO 81506).

The land area represented by Colorado's national forests, national grasslands and BLM lands total over 36,000 square miles. In addition, Colorado has three national parks of about 250,000 acres, six national monuments of about 150,000 acres, and five national wildlife refuges; these lands collectively total about 750 square miles. More than a third of the state's overall area thus consists of public-access federal lands.

Colorado also has over 40 state parks that collectively total about 150,000 acres. In easternmost Colorado, **Bonny Lake State Park** (970/354-7306) near Idalia, is a good place for finding such eastern birds as red-bellied woodpecker, eastern bluebird, brown thrasher, northern cardinal and Baltimore oriole. **Barr Lake State Park** (303/659-6005), near Brighton and Denver, has a local bird list of more than 350 species that includes many eastern birds, but the site attracts the western Bullock's rather than Baltimore oriole. Also near Denver, **Chatfield State Park** (303/791-7275) and adjacent Waterton Canyon have a remarkably large bird list of more than 360 species. **Roxborough State Park** (303/979-395), **Castlewood Canyon State Park** (303/688-52420), and **Golden Gate Canyon State Park** (303/582-3707) are all near Denver, and provide habitats for cliff- and canyon-adapted species at lower elevations. Young (2000) and Kingery (2007) described these sites and their birds in detail.

Information on these and other Colorado's state parks and public-access lands may be obtained from the Colorado Division of Wildlife, 6060 Broadway, Denver, CO 80216 (303/297-1802), or the Colorado Tourism Board, 5500 S. Syracuse, Suite 267, Englewood, CO 8011 (800/433-2656). Kingery (2007) provided contact information for 15 local Audubon chapters and other bird-related groups in Colorado.

The Rocky Mountain Bird Observatory is a major research organization in the state, Their website is at: http://www.rmbo.org. The Colorado Field Ornithologists group (P.O. Box 181, Lyons CO 80540) has produced individual bird checklists for all of Colorado's counties: http://www.coloradocountybirding. com/checklists/. Their website is www.cfo-link.org. Birding site information for many Colorado locations can be found at: www.coloradobirdingtrail.com. The state has recognized 54 Important Bird Areas, of which two have been classified as being of global significance.

Idaho

BEAR LAKE NATIONAL WILDLIFE REFUGE

This 17,600-acre refuge in extreme southeastern Idaho adjoins **Bear Lake State Park** and consists of marshes and grasslands at about 5,900 feet of elevation. The refuge attracts at least 161 bird species, of which at least 60 are breeders, including the black-necked stilt, American avocet, black tern, various herons, and up to 5,000 white-faced ibises and 13,000 Franklin's gulls. The refuge is recognized as a globally significant Important Bird Area, and listed as one of the state's "Blue Ribbon" birding sites. A seasonally organized bird checklist for the refuge is available on-line: http://www.npwrc.usgs.gov/resource/birds/chekbird/r1/bearlake.htm. For more information, contact the refuge manager, 370 Webster St., Montpelier, ID 83254 (208/847-1757).

BOISE NATIONAL FOREST

This national forest covers 2.6 million acres north and east of Boise. Mostly consists of ponderosa and lodgepole pines, with some Douglas-fir forests. There are ranger offices in Boise, Emmett, Idaho City, Mountain Home, and Lomar. A seasonally organized bird checklist of 310 species, with about 180 breeders (and including 13 gallinaceous birds, 12 owls, 10 woodpeckers and 13 finches) is available on-line: http://www.npwrc.usgs.gov/resource/birds/chekbird/r1/boise.htm. The forest headquarters are at 1249 Vinnell Way, Suite 200, Boise, ID 83709 (208/373-4100).

BRUNEAU DUNES STATE PARK

This 4,800-acre state park just north of Bruneau (and 18 miles south of Mountain Home) is noted for its sand dunes up to 470 feet high. The vegetation is mostly sagebrush and grassland, and the land birds include the long-billed curlew, sage thrasher, sage sparrow, and other grassland sparrows such as lark and (rarely) black-throated. Other summer birds include yellow-breasted chat, lazuli bunting, black-headed grosbeak and western tanager. There is also a large, marshy lake that attracts migrating shorebirds and waterfowl. For more information, contact the park headquarters, Bruneau, ID 83604 (208/366-7919). Only a few miles to the north is the 13,200-acre **C. J. Strike Wildlife Management Area**, which surrounds a reservoir on the Bruneau River. The lake and its delta attract a host of water birds, including the American white pelican, double-crested cormorant, western and Clark's grebes, black-crowned night-heron, American avocet, black-necked stilt as well as several gulls and terns. The site is managed by the Idaho Dept. of Fish and Game office in Grandview (208/845-2324)

CAMAS NATIONAL WILDLIFE REFUGE

This is a beautiful refuge of more than 10,000 acres, is located about four miles northwest of Hamer, in southeastern Idaho. The refuge consists of a diverse array of habitats ranging from wetlands to prairie, irrigated meadows, and sagebrush. It attracts at least 177 bird species, of which at least 80 are breeders. There are many colonial nesters, including three egret species, black-crowned night-heron and white-faced ibis, as well as the trumpeter swan and peregrine. Identified as one of the state's "Blue Ribbon" birding sites. A seasonally organized on-line bird checklist for the refuge is available: http://www.npwrc.usgs.gov/resource/birds/chekbird/r1/camas.htm. For more information contact the refuge manager, Hamer, ID 83425 (208/662-5423).

CARIBOU–TARGHEE NATIONAL FOREST

This 3,000,000-acre forest (previously the Caribou, National Forest, of 987,000 acres, and the Targhee National Forest, of 1,642,000 acres), is partly in Utah and Wyoming, and ranges from grasslands to high peaks. There are ranger stations at Ashton. Driggs, Dubois, Idaho Falls, Island Park, Malad City, Montpelier, Pocatello and Soda Springs. No bird list is yet available, but the Grand Teton–Jackson Hole list (see the Bridger-Teton National Forest description in Wyoming section) should apply. For information contact the forest office at 250 4th. Ave, Pocatello, ID 83201 (208/236-6700), or 1405 Hollipark Dr., Idaho Falls ID 83401 (208/524-7500).

CHALLIS NATIONAL FOREST

See Salmon–Challis National Forest.

CITY OF ROCKS NATIONAL RESERVE

This 14,400-acre preserve in south-central Idaho (south of Burley) consists of sheer granite cliffs and pinnacles, with scattered pinyon pines (which are rare in Idaho) and junipers. The refuge attracts pinyon and western scrub jays, Virginia's warbler, green-tailed towhee, bushtit, gray flycatcher, juniper titmouse, blue-gray gnatcatcher, and other arid-adapted birds. The area's raptors include the golden eagle, prairie falcon, western screech-owl and northern pygmy-owl. It is recognized as one of the state's Important Bird Areas. The reserve is located about four miles west of Almo. A seasonally organized on-line site checklist containing 156 species is available: http://www.npwrc.usgs.gov/resource/birds/ chekbird/r1/rocks.htm. The reserve is administered by the National Park Service (208/733-8398). Information might also be available from the BLM office, Rte. 3, Box 1, Burley ID 83118 (208/678-5514).

CLEARWATER NATIONAL FOREST

This 1.7 million-acre forest includes the hisoric Lolo Pass of Lewis and Clark fame, as well as part of the **Selway-Bitterroot Wilderness** (totaling 1.3 million acres, 20 percent of which is in Montana). No bird list is yet available. For more information contact the Forest Service office, 12730 Hwy. 12, Orofino, ID 83544 (208/476-4541).

COEUR D'ALENE NATIONAL FOREST

The Coeur d'Alene National Forest covers 726,000 acres in northern Idaho. Its headquarters are at 3815 Schreiber Way, Coeur d'Alene, ID 83815 (208/765-723), and it is now part of the collective Idaho Panhandle National Forest. A non-seasonal on-line bird checklist includes 242 species: http://www.npwrc.usgs.gov/ resource/birds/chekbird/r1/coeur.htm. The state-owned **Coeur d'Alene River Wildlife Management Area** (located 1.4 miles northeast of Harrison) provides public access to typical breeding habitats of species such as the osprey, black tern, and pied-billed, horned and red-necked grebes. The **Mineral Ridge Recreation Trail** (332 acres) between two bays of Lake Coeur d'Alene attracts up to 70 bald eagles in winter. It is managed by the BLM office in Coeur d'Alene (208/765-7356).

CRATERS OF THE MOON NATIONAL MONUMENT

This 53,000-acre site in south-central Idaho 18 miles southwest of Arco, is mostly a basaltic lava landscape, with often-scanty sagebrush and grass cover. A 640,000-acre **Craters of the Moon National Preserve** is adjacent to the monument, and includes a 43,000-acre **Craters of the Moon Wilderness**. The monument is recognized as one of the state's Important Bird Areas and supports species that include the golden eagle, dusky grouse, greater sage-grouse, common poor-will, Lewis' woodpecker, common raven, rock wren and green-tailed towhee. A seasonally organized on-line site checklist containing 182 species is available, and includes over 70 probable breeders: http://www.npwrc. usgs.gov/resource/birds/chekbird/r1/cratmoon.htm. The site is administered by the National Park Service (208/527-3257)

CURLEW NATIONAL GRASSLAND

This 48,000-acre area of sage scrub and grassland in southern Idaho (20 miles west of Malad City) has a bird list of at least 129 species, many of which are of Great Basin zoogeographic orientation. It includes such arid-adapted and southwestern species as the black-throated and Brewer's sparrows, gray flycatcher, western scrub-jay, pinyon jay and juniper titmouse. In spite of its name, the site supports few if any long-billed curlews. For information contact the Forest Service office, Malad, ID 83252 (208/766-4743).

DEER FLAT NATIONAL WILDLIFE REFUGE

This 11,500-acre refuge in southwestern Idaho consists of marshes, sagebrush flats and riparian woodlands. The refuge's impressive bird list includes 215 species, with at least 86 breeders, such as western and Clark's grebes, Caspian tern, and seven species of owls. There are also three *Empidonax* flycatchers, and the Townsend's, MacGillivray's and Wilson's warblers. The refuge also hosts large numbers of fall and winter raptors, such as the peregrine and both eagle species. It is located about five miles west of Nampa, and is recognized as one of the state's Important Bird Areas. A seasonally organized on-line bird checklist for the refuge is available: http://www.npwrc.usgs.gov/resource/birds/chekbird/r1/deer.htm. For more information, contact the refuge manager. P.O. Box 448, Nampa, ID 83651 (208/467-9278).

FARRAGUT STATE PARK

This 4.000-acre state park is located on Lake Pend Oreille, and is about 25 miles north of Pend Oreille. The refuge consists of coniferous forest growing on steep rocky slopes, as well as shoreline vegetation. The lake attracts a wide variety of ducks and grebes. Forest birds include the pileated woodpecker, ruffed grouse, Swainson's thrush, Cassin's vireo, western and mountain bluebirds, Townsend's and MacGillivray's warblers, western tanager, black-headed grosbeak, and cordilleran flycatcher. A bird list of over 160 species, including 69 breeding species, is available. For more information, contact the park headquarters at 13400 East Ranger Rd., Athol, ID 83801 (208/683-2425).

GRAYS LAKE NATIONAL WILDLIFE REFUGE

This refuge of some 15,000 acres in southeastern Idaho is located about 35 miles north of Soda Springs. It supports the densest known breeding population of greater sandhill cranes, and up to 40,000 Franklin's gulls. Other nesting marsh birds include the eared grebe, Virginia rail, sora, American avocet, Wilson's phalarope, Forster's tern and marsh wren. Long-billed curlews and bobolinks nest in surrounding meadows. The refuge's bird list totals at least 199 species, including 69 breeders, and the refuge is recognized as one of the state's Important Bird Areas. A seasonally organized on-line bird checklist for the refuge is available: http://www.npwrc.usgs.gov/resource/birds/chekbird/r1/gray-lake.htm, For more information, contact the refuge manager, Box 837, Soda Springs, ID 83276 (208/574-2755).

HARRIMAN WILDLIFE REFUGE & STATE PARK

Harriman Wildlife Refuge encompasses the 4.700-acre Harriman State Park, and also includes about 8,000 acres around the park, both of which are located at the western edge of the Greater Yellowstone ecosystem. The refuge is located 18 miles north of Ashton, in Fremont County. Henrys Ford River and other wetlands attract nesting by species such as the trumpeter swan, red-necked grebe, sandhill crane, long-billed curlew, and great gray owl. The refuge is a major wintering site for up to 3,000 trumpeter swans. It is recognized as one of the state's Important Bird Areas, and has been identified as one of its "Blue Ribbon" birding sites. Not far to the north is **Henry's Lake State Park**. For more information, contact the park headquarters 3489 Green Canyon Road, Island Park, ID 83429 (208/558-7368).

HEYBURN STATE PARK

This 7,800-acre state park surrounds Chatcolet and Benewah Lakes, along the southern end of Coeur d'Alene Lake, near the town of Plummer. The park consists of mixed conifer forest and wetlands with a boreal influence. Notable breeding birds include a great blue heron colony, osprey (often nearly 50 pairs), wood duck, ruffed grouse, red-necked grebe, northern pygmy-owl and barred owl. Summer passerines include the cordilleran flycatcher, chestnut-backed chickadee, Swainson's and varied thrushes, Townsend's solitaire, Townsend's warbler, and golden-crowned kinglet. The park is recognized as one of the state's Important Bird Areas. For more information, contact park headquarters, 1291 Chatcolet Rd., Plummer, ID (208/686-1308).

IDAHO PANHANDLE NATIONAL FOREST (including COEUR D'ALENE, KANIKSU and ST. JOE NATIONAL FORESTS)

Three national forests of northern Idaho were administratively merged in 2000. to become the 3.2-million-acre Idaho Panhandle National Forest, the boundaries of which locally extend into northeastern Washington and northwestern Montana, and include **Cabinet Mountains Wilderness** (94,000 acres). This region is probably one of the best places in the entire Rocky Mountains south of Canada to look for boreal forest birds. There are ranger stations in Avery, Bonners Ferry, Coeur d'Alene, Priest Lake, Red Ives, St, Maries, Sandpoint and Wallace. For information contact the Forest Service office, 1201 Ironwood Dr., Coeur d'Alene, ID 83814 (208/765-7221). See also the separate account (above) of the Coeur d'Alene National Forest.

KOOTENAI NATIONAL WILDLIFE REFUGE

This 2,700-acre refuge in extreme northern Idaho is located about five miles west of Bonners Ferry, and consists of montane forest and marshland habitats. The refuge checklist has 220 species of birds, including more than 90 breeders, many with distinctly boreal affinities and not easily found elsewhere in the region. The ruffed grouse, flammulated owl, and pileated and Lewis' woodpeckers are among the many elusive forest-dwelling species that nest here. Other notable breeders include the osprey, bald eagle, red-necked grebe, black tern, rufous hummingbird, varied and Swainson's thrushes, orange-crowned and MacGillivray's warblers, and Lincoln's sparrow. The refuge is identified as one of the state's "Blue Ribbon" birding sites. An on-line seasonally organized bird checklist for the refuge (including 9 owls, 8 woodpeckers and 11 finches) is available: http://www.npwrc.usgs.gov/resource/birds/ chekbird/r1/kootenai.htm. For more information, contact the refuge manager, Star Route 1, Box 160, Bonners Ferry, ID 83805 (208/267-3888). Driving north from the refuge toward Porthill one may follow the Kootenai River to the British Columbia boundary, where there are chances of seeing such boreal species as the spruce grouse, black swift, black-backed woodpecker, boreal chickadee and white-winged crossbill.

MINIDOKA NATIONAL WILDLIFE REFUGE

This refuge of 20,700 acres in south-central Idaho is located 12 miles northeast of Rupert, and extends for 25 miles along the Snake River. The reservoir behind Minidoka Dam supports Idaho's only nesting colony of American white pelicans, and is an important waterfowl migratory staging area. Other breeding marsh birds are the western grebe, double-crested cormorant, great blue heron, snowy egret, black-crowned night-heron, and white-faced ibis, plus two gulls and three terns. A total of 201 bird species have been reported, including at least 82 breeders. Recognized as a globally significant Important Bird Area, the refuge has also been identified as one of the state's "Blue Ribbon" birding sites. An on-line seasonally organized bird checklist for the refuge is available: http://www.npwrc.usgs.gov/resource/birds/chekbird/r1/mini.htm, For more information, contact the refuge manager, Rte. 4, Box 290, Rupert, ID 83350 (208/436-3580).

NEZ PERCE NATIONAL FOREST

This 987,000 acre forest in Idaho's central panhandle region includes parts of four wilderness areas and the **Wild and Scenic Salmon River**. No bird list is yet available. There are ranger stations at Elk City, Grangeville, Kooskia, Riggins and White Bird. For information contact the ranger office, Rte. 2, Box 475, Grangeville ID 83530 (208/983-1950).

OXFORD SLOUGH WATERFOWL PRODUCTION AREA

This 5,000-acre site just south of Oxford in extreme southeastern Idaho is a freshwater wetland with seasonally flooded alkali flats, supporting colonies of the white-faced ibis, Franklin's gull, Forster's and black terns, eared grebe, snowy egret, and black-crowned night-heron. The juniper uplands attract some Great Basin birds such as the juniper titmouse, black-throated gray warbler, blue-gray gnatcatcher and gray flycatcher. The site is recognized as a globally significant Important Bird Area. The northern part is open to birding year-around. The site is jointly administered by the BLM (208/766-4766), the U.S. Fish & Wildlife Service (208/237-6615) and the Idaho Dept. of Fish & Game (208/232-4703).

PAYETTE NATIONAL FOREST

This remote and highly mountainous 2.3 million-acre forest in north-central Idaho includes part of the 2.2 million-acre **Frank Church River of No Return Wilderness**, with 190 reported bird species and the largest designated wilderness outside of Alaska. Roberts (1992) described the region's birds (see Salmon–Challis National Forest description below). There are ranger stations at Connell, McCall, New Meadows and Weiser. For information contact the ranger office, P.O. Box 1025, Forest Service Bldg., McCall ID 83638 (208/634-2255).

PONDEROSA STATE PARK

This state park is located along Payette Lake, about two miles northwest of McCall, in Valley County. The park consists mostly of mature mixed coniferous forest and wetlands, and has been identified as one of the state's "Blue Ribbon" birding sites. Typical forest birds include the pileated woodpecker, olive-sided flycatcher, black-headed grosbeak, mountain chickadee, Swainson's thrush, MacGillivray's warbler, Steller's jay, and dusky flycatcher. Several cavity-nesting ducks breed here, such as the wood duck, bufflehead, hooded merganser, and probably both goldeneyes. Water-dependent birds also include the osprey, sora and many other marsh species. A printed bird list is available. For information, contact the park office, McCall, ID 83638 (208/634-2164) (208/634-2164).

ST. JOE NATIONAL FOREST

See Idaho Panhandle National Forest.

SALMON-CHALLIS NATIONAL FOREST

The Salmon-Challis National Forest covers 4.3 million acres in east-central Idaho, and includes part of the **Frank Church River of No Return Wilderness** (2.3 million acres). The Challis section includes Borah Peak, Idaho's highest mountain, at 12,662 feet, while the Salmon section includes the beautiful middle fork of the Salmon River. The **Salmon River Scenic Byway** has been identified as one of the state's "Blue Ribbon" birding sites. There are ranger stations at Challis, Clayton, Leadore, Mackay, North Fork and Salmon. Birding sites in the Payette, Salmon and Challis national forests have been described by Roberts (1992). He documented 245 regional species, including 28 waterfowl, 23 raptors, 9 woodpeckers, 13 warblers, and 11 finches. An on-line and seasonally organized bird checklist for the region lists all these species, 156 of which are summer or permanent residents: http://www.npwrc.usgs.gov/resource/birds/chekbird/r1/salmon.htm. The forest headquarters are at Highway 93 N., P.O. Box 729, Salmon, ID 83467 (208/756-2215).

SAWTOOTH NATIONAL FOREST AND SAWTOOTH NATIONAL RECREATION AREA

This national forest covers 2.1 million acres in the Salmon River's drainage of eastern Idaho's panhandle. Sawtooth Valley (in Custer County) may be visited via Idaho Highways 75 and 21, passing through spectacular mountain scenery and traversing the forest habitats of such conifer-dependent species as the northern goshawk, spruce grouse, American three-toed woodpecker, great gray owl and northern pygmy-owl (Roberts, 1992). The U.S. Forest Service published a (now-out-of-print) regional checklist of more than 140 species in 1994. There are visitor centers at Ketchum and Red Fish Lake. The Forest Service office address is 1525 Addison Ave., East Twin Falls, ID 83301 (308/733-3698). The 778,000-acre **Sawtooth National Recreation Area** lies within the Sawtooth, Challis and Boise national forests, and adjoins the **Sawtooth Wilderness** (217,000 acres, with nearly 300 miles of trails). The recreation area's headquarters are 11 miles north of Ketchum. Information on it may be obtained from its headquarters, at Star Route, Ketchum, ID 83340 (208/726-8291), or from the area's Visitors Center (208/726-5018).

SNAKE RIVER BIRDS OF PREY NATIONAL CONSERVATION AREA

This area covers an 81-mile stretch (482,000 acres) of the Snake River south of Kuna that the BLM manages for nesting raptors, including the golden eagle, red-tailed hawk, ferruginous hawk, American kestrel and prairie falcon (Bureau of Land Management. 1979). More than 700 pairs of raptors nest here, most notably about 200 pairs of prairie falcons. Other raptors using the area include the barn owl, western screech-owl, northern saw-whet owl and long-eared owl. For specific raptor information the Peregrine Fund/World Center for Birds of Prey in Boise (208/362-8687) may be of assistance. The desert-adapted black-throated sparrow may also be found in the conservation area, as well as many sage-adapted birds. The area has been identified as one of the state's Important Bird Areas and is one of the state's "Blue Ribbon" birding sites. For information, contact the BLM office, 3948 Development Ave., Boise ID 83705 (208/334-1582 or 384-3300).

TARGHEE NATIONAL FOREST

See Caribou–Targhree National Forest.

WINCHESTER LAKE STATE PARK

This 550-acre park, about 30 miles southeast of Lewiston and in the village of Winchester, is notable for its white-headed woodpeckers, which are largely limited regionally to west-central Idaho. The vegetation around the lake consists of mixed conifers, with local nesting by the ruffed grouse, barred owl, calliope hummingbird, pileated woodpecker, and red-naped sapsucker. Nesting passerines include such conifer-dependent species as the Hammond's flycatcher, Swainson's and varied thrushes, and Townsend's warbler. Resident finches include the Cassin's finch, red crossbill and evening grosbeak. For more information, contact Winchester Lake State Park, Box 186, Winchester, ID 83555 (208/924-7563).

In addition to Curlew National Grassland, the Forest Service currently manages a total of 13 national forests in Idaho. Two of Idaho's national forests (Caribou and Targhee) are partially shared with Wyoming and are part of the Greater Yellowstone ecosystem, and one (Idaho Panhandle) is shared with Washington. Idaho's national forests collectively total about 18 million acres, encompassing about a third of the state's overall area, and representing the highest percentage of national forest area relative to total state area for any of the 50 states. Collectively all of Idaho's federal lands represent nearly two-thirds of the state's overall area, which is the greatest percentage of federal land ownership among any of the four states considered here.

The Bureau of Land Management (BLM) administers 12 million acres of public lands in Idaho. The BLM manages or partly manages several wilderness areas in Idaho, among them **Big Jack Creek** (53,000 acres), **Bruneau-Jarbidge** (90,000 acres), **Hells Canyon** (214,000 acres), **Little Jack Creek** (51,000 acres), **North Fork Owyhee** (43,000 acres) and **Pole Creek** (12,000 acres). Bird lists are currently available on-line for three BLM regions:

> **Burley BLM Field Office (Twin Falls District).** Headquarters at 15 E. 200 South, Burley 83319 (208/678-5514). This region of 1.3 million acres covers much of south-central Idaho, mostly south of the Snake River and including the Snake River and Deep Creek natural resource areas. The region's non-seasonal bird checklist totals about 300 species (including 11 owls, 7 woodpeckers and 13 finches): http://www.npwrc.usgs.gov/resource/birds/chekbird/r1/burl.htm.

> **Idaho Falls BLM District.** Headquarters at 940 Lincoln Rd., Idaho Falls, ID 83401 (208/524-7505). This region of about 2.3 million acres in eastern and east-central Idaho includes the Medicine Lodge, Big Butte and Pocatello natural resource areas. The district's seasonally organized bird

checklist totals about 255 species (including 13 owls, 8 woodpeckers and 10 finches): http://www.npwrc.usgs.gov/resource/birds/chekbird/r1/falls. htm

Shoshone BLM District. Headquarters at P.O. Box 2-B, Shoshone ID 83467 (208/786-2206). This region of south-central Idaho consists of 1.7 million acres, mostly located between the Snake River and Sun Valley. Its non-seasonal bird checklist totals 278 species (including 10 owls, 10 woodpeckers and 12 finches): http://www.npwrc.usgs.gov/resource/birds/chekbird/r1/shosho.htm

The BLM's Idaho State Office address is 1397 Vinnell Way, Boise, ID 83709 (208/373-3889). There are regional field offices at Bruneau, Burley, Challis, Coeur d'Alene, Cottonwood, Four Rivers, Idaho Falls, Jarbidge, Owyhee, Pocatello, Salmon and Shoshone.

Idaho also has 29 state parks, of which at least four may be of special interest to birders. The little-developed **McCrosky State Park** (1291 Chatcolet Rd., Plummer, ID 83851; 208/686-1308) is about 25 miles north of Moscow, and consists of 4,500 acres of coniferous forest habitat. It supports the ruffed grouse, pileated woodpecker, Clark's nutcracker, gray jay, Townsend's solitaire, MacGillivray's, Nashville, Townsend's and Wilson's warblers, and evening grosbeak. The 418-acre **Priest Lake State Park** (314 Indian Creek Rd., Coolin, ID 83821 (208/443-2200) is near the town of Coolin and in the Selkirk Range. The harlequin duck has been found nesting here, as well as the dusky and spruce grouse, American three-toed and black-backed woodpeckers and Clark's nutcracker. The 200-acre **Round Lake State Park** (Box 170, Sagle, ID 83860; 208/263-3489) near Westmond has reported the pileated woodpecker, mountain bluebird and western tanager. The 960-acre **Hells Gate State Park** (5100 Hells Gate Rd., Lewiston, ID 83501; 208/743-2363) near Lewiston has 121 recorded bird species. These state parks, together with about 90 other sites and their birds, were described in the *Idaho Wildlife Viewing Guide* by Leslie Carpenter (1990).

A new (2003) edition of the *Idaho Wildlife Viewing Guide* has been written by A. L. Pope, and is available from the Idaho Dept. of Fish & Game, 600 S. Walnut, Box 25, Boise, ID 83707 (208/334-3700). The Idaho Dept. of Fish & Game also has published *The Idaho Birding Trail Guidebook* to accompany Idaho's birding trail network. This book is also available from the Dept. of Fish & Game headquarters at Boise. A more comprehensive birding guide, *A Birder' Guide to Idaho* (edited by Dan Svingen and K. Dumroese, 1997), is unfortunately now out of print, but often can be located through used-book dealers.

The Idaho Bird Observatory is headquartered in Boise, and has a bird-banding station at Lucky Peak, about 20 miles southeast, along a major raptor migration corridor. About 5,000–8,000 raptors are counted there each fall, and about 1,000 are banded. Its website is http://tingurl.com/658g9y. The Peregrine Fund/ World Center for Birds of Prey is also located in Boise (208/362-8687). The Idaho Birding Trail website (http://www.idahobirdingtrail.com) provides a field checklist with suggested site locations for finding Idaho birds, and describes 23

"Blue Ribbon" birding sites in the state. As of 2010, Idaho had designated 61 Important Bird Areas, 39 of which are located on the Idaho Birding Trail. Twenty-three Idaho sites have been identified as "Blue Ribbon" locations that provide the best birding in the state. A comprehensive state checklist of 409 species, plus a digital atlas of biological data for each species is available on-line: http://imnl.isu.edu/DIGITALATLAS/bio/birds.htm. General information on Idaho's state parks and other tourism sites may be obtained from Tourism Information, Statehouse, Room 108, Boise ID 83720 (208/334-2470).

Montana

BEARTOOTH SCENIC BYWAY AND PASS

This alpine highway (U.S. 212) extends from Red Lodge to Cooke City, Montana, with about half its length within Wyoming. The road has several high-altitude passes, including Daisy, Lulu and Coulter, ranging from about 8,000 to 10,000 feet of elevation, as well as the Beartooth Pass, at 10,947 feet. During summer there is a good chance of seeing the black rosy-finch and American pipit along these passes. At somewhat lower timberline elevations one may encounter the golden eagle, prairie falcon, Clark's nutcracker and white-crowned sparrow. Still lower and taller conifer forests support the dusky grouse, red crossbill, pine grosbeak, mountain chickadee, as well as several montane warblers and sparrows. Tourist or road-condition information may be obtained from the Beartooth Ranger Station, Box 4300, Red Lodge, MT 59068 (406/446-2102). The Yellowstone National Park office (307/344-7381) might also be able to provide current driving information on this spectacular road that traverses some of America's most stunning alpine vistas.

BEAVERHEAD–DEERLODGE NATIONAL FOREST

This gigantic national forest of 3.25 million acres straddles the Continental Divide, and has altitudes ranging from 4,075 to 10,950 feet. The forest includes part of one wilderness area (**Lee Metcalf**, 249,000 acres), and part of it extends into the northwestern section of the Greater Yellowstone ecosystem. Its typical birds include the dusky and spruce grouse, calliope and rufous hummingbirds, great gray, boreal and northern saw-whet owls, several woodpeckers, at least four breeding warblers, and such finches as the red crossbill, evening grosbeak and gray-crowned rosy-finch. There are ranger stations for the Deerlodge section at Deer Lodge, Dillon, Philipsburg and Whitehall, and for the Beaverhead section at Dillon, Ennis, Sheridan, Wisdom and Wise River. The Deerlodge bird list includes about 260 species and might be available from the Deerlodge National Forest, P.O. Box 400, Federal Building, 400 North Main, Butte, MT 59703 (406/496-3400). The address for the Beaverhead National Forest is 610 N. Montana St., Dillon, MT 59725 (406/683-3900)

BEAVER CREEK PARK

This 10,000-acre county park is located ten miles south of Havre on Beaver Creek Road, in Montana's Bears Paw Mountains. Its habitats range from riparian willow and alder thickets to Douglas-fir montane forest, and its notable birds include the golden eagle, prairie falcon, MacGillivray's warbler, black-headed grosbeak and western tanager. A 3.5-mile nature trail is present. Information may be obtained from Beaver Creek Park, Shambo Road, Box 368, Havre, MT 59501 (406/395-4965).

BENTON LAKE NATIONAL WILDLIFE REFUGE

This refuge of more than 12,000 acres is located about ten miles north of Great Falls. It consists mostly of mixed-grass grasslands and associated wetlands, and its bird life includes a variety of shorebirds, waterfowl, and marsh birds, as well as a considerable diversity of raptors. Breeding colonies of eared grebes and Franklin's gulls are among the conspicuous marsh birds, as well as breeding by the sora, black tern, American avocet and Wilson's phalarope. Refuge specialties include the white-faced ibis, long-billed curlew and chestnut-collared longspur. The refuge list totals at least 199 species, including 59 breeders. An on-line bird checklist for the refuge is available: http://www.npwrc.usgs.gov/resource/birds/chekbird/r6/benton.htm. For more information, contact the refuge manager, Benton Lake National Wildlife Refuge, P.O. Box 450, Black Eagle, MT 59414 (406/727-7400).

BIGHORN CANYON NATIONAL RECREATION AREA AND PRYOR MOUNTAINS

This national recreation area (120,000 acres) includes part of adjacent Wyoming. The area consists of a long reservoir formed by the Yellowtail Dam at Fort Smith, plus two administrative districts. The northern district is headquartered at Fort Smith, and is largely wooded. In contrast, the southern district extends into Wyoming, and is typical canyon and range country, with associated species such as the peregrine, prairie falcon, golden eagle, rock and canyon wrens and pinyon jay. The southern district is headquartered at Lovell, Wyoming (307/548-2251) (see Wyoming section for description). An on-line monthly-organized bird checklist of more than 200 species is available: http://www.npwrc.usgs.gov/resource/birds/chekbird/r6/pryor.htm. For more information, contact the Area headquarters, P.O. Box 458, Fort Smith, MT 59035 (406/666-2412).

BITTERROOT NATIONAL FOREST

This national forest of 1.6 million acres extends into Idaho. Its habitats range from grasslands at about 3,500 feet of elevation to alpine meadows at more than 10,000 feet, and includes parts of three wilderness areas: **Selway-Bitterroot**

(1.3 million acres, in the upper Selway River valley), the **Frank Church River of No Return** (2.4 million acres, in the Salmon River drainage), and the **Anaconda-Pintler** (158,000 acres, along the crest of the Anaconda Range). The forest's birds include the ruffed and spruce grouse, seven owls (including flammulated, boreal and northern pygmy-owl), Lewis's, black-backed and pileated woodpeckers, Vaux's swift, western and mountain bluebirds, Swainson's, hermit and varied thrushes, and both the red and white-winged crossbills. There are ranger stations at Darby, Stevensville and Sula. The total bird list includes over 170 species and is available on-line at: http://www.npwrc.usgs.gov/resource/birds/chekbird/r6/bitterrt.htm. For more information contact the Bitterroot National Forest, 1801 First St. North, Hamilton, MT (406/363-7117).

BOWDOIN NATIONAL WILDLIFE REFUGE

This prairie refuge of some 15,000 acres is similar to the preceding one in that it consists mostly of mixed-grass grasslands and associated wetland habitats. The refuge is located about seven miles east of Malta, Montana. Colonial nesting marsh birds include American white pelican, double-crested cormorant, eared grebe, common and black terns, and California, Franklin's, and ring-billed gulls. Other nesting marsh birds include the willet, American avocet, Wilson's phalarope, sora and marsh wren. Upland areas attract the sharp-tailed grouse, greater sage-grouse, long-billed curlew and marbled godwit. The refuge list totals at least 248 species, at least 105 of which are breeders. A bird checklist for the refuge is available on-line: http://www.npwrc.usgs.gov/resource/birds/chekbird/r6/bowdoin.htm. For more information, contact the refuge manager, P.O. Box J, Malta, MT 59538 (406/654-2863).

CANYON FERRY WILDLIFE MANAGEMENT AREA

Canyon Ferry Wildlife Management Area (3,500 acres) is located at the south end of Canyon Ferry Lake near Townsend, not far from where Meriwether Lewis collected the type specimen of the Lewis's woodpecker. Waterfowl using the area include migrating tundra and trumpeter swans, the common loon, Canada goose and many species of ducks. During the summer American white pelicans, ospreys, and Caspian terns are likely to be seen, as well as various shorebirds. From early November to mid-December large numbers of bald eagles may be observed from the eagle-viewing area below Canyon Ferry Dam at Riverside Campground. For more information contact the Game & Parks office at 1400 19th Ave., Bozeman MT 59715 (406/994-4947, or 406/444-2535).

CHARLES M. RUSSELL NATIONAL WILDLIFE REFUGE

This refuge consists of some 1,700 square miles of grassland habitats, surrounding Fort Peck Reservoir. Identified as a globally significant Important Bird Area, the refuge attracts at least 240 species, and includes such grassland- and rim-

rock-breeders as the ferruginous hawk, golden eagle, prairie falcon, piping plover and mountain plover. An on-line and seasonally organized bird checklist for
the refuge is available: http://www.npwrc.usgs.gov/resource/birds/chekbird/r6/
cmrussel.htm. For more information, contact the refuge manager, P.O. Box 110,
Lewistown, MT 59457 (406/538-8706).

CUSTER NATIONAL FOREST

This national forest of 1.2 million acres extends partly into South Dakota, and
has several mountains exceeding 12,000 feet elevation, including Granite Peak,
the state's highest peak at 12,850 feet. There is one wilderness area (**Absaroka-Beartooth**, 944,000 acres, including Granite Peak, Montana's highest
peak at 12,799 feet), plus the Beartooth National Scenic Byway (see description above). There is no bird list, but part of the forest is located just northeast
of Yellowstone Park, so that park's list should be useful here. There are ranger
stations at Ashland, Camp Creek, Dickinson, Lemmon, Lisbon, Red Lodge and
Watford City. For more information, contact the forest office at P.O. 2556, Billings, MT 59103 (406/657-6361).

DEERLODGE NATIONAL FOREST

See Beaverhead-Deerlodge National Forest.

FLATHEAD NATIONAL FOREST

This national forest of 3.6 million acres lies in the heart of the northern Rocky
Mountains, along the western slope of the Continental Divide and adjacent to
Glacier National Park. Nearly half of it is designated as three wilderness areas: **Mission Mountains** (74,000 acres), **Great Bear** (286,000 acres) and the
enormous **Bob Marshall** (1.0 million acres, encompassing three river drainages and several mountain ranges). There are ranger stations at Bigfork, Columbia Falls, Hungry Horse and Whitefish. Virtually all the Rocky Mountains
bird species have been reported here, including the northern saw-whet and
boreal owls, American three-toed and black-backed woodpeckers, dusky and
spruce grouse, Vaux's swift, Swainson's and varied thrushes, and white-tailed
ptarmigan. The forest's bird list includes over 200 species and is available
from the Flathead National Forest, 1935 Third Ave. East, Kalispell, MT 59901
(406/755-5401).

FREEZEOUT LAKE WILDLIFE MANAGEMENT AREA

This 12,000-acre lake is located just north of Fairfield, and is notable for its vast
numbers of migratory waterfowl in spring, as well as other migrating birds.
Summer access is somewhat restricted. Many waterfowl nest here, as well as
the black-necked stilt, American avocet, long-billed curlew, marbled god-

wit, Wilson's phalarope, plus three gulls and three terns. For information contact the Montana Dept. of Fish, Wildlife & Parks, PO. Box 6610, Great Falls, MT 59405 (406/454-3441 or 406/467-2646).

GALLATIN NATIONAL FOREST

This 1.7 million-acre national forest bordering northwestern Wyoming encompasses parts of two wilderness areas: **Lee-Metcalf** and **Absaroka-Beartooth.** It borders the north boundary of Yellowstone National Park, so the park's bird list probably includes all of the species found in these forests. There are ranger stations at Big Timber, Gardiner, Livingston and West Yellowstone. For information contact the Forest Service office, Federal Building, Box 130, Bozeman, MT 59715 (406/587-5271).

GLACIER NATIONAL PARK

This park consists of nearly 1,600 square miles in northwestern Montana, and includes some of the most spectacular glacial topography to be seen anywhere south of Canada. Logan Pass at 6,664 feet lies at the crest of the Continental Divide. The park ranges in altitude from slightly over 3,000 feet at its eastern boundary to several peaks exceeding 10,000 feet; Mt. Cleveland reaches about 10,450 feet. About a third of the park lies in the alpine zone. Below 7,000-8,000 feet the park is largely covered by montane coniferous forest, especially lodgepole pine, Engelmann spruce and subalpine fir. On the lower western slopes there is a moister plant community dominated by western red-cedar and western hemlock, and more bird species associated with the Pacific Northwest. There are over 200 lakes and, at least until recently, over 50 glaciers, although these are now rapidly shrinking or even disappearing. Terry McEneaney's 1993 book *The Birder's Guide to Montana* has a comprehensive account of Glacier National Park's birds. The park has been identified as a globally significant Important Bird Area, and its bird list includes at least 223 species, of which at least 124 are known breeders. Park specialties include the hooded merganser, spruce grouse, Vaux's swift, black swift, chestnut-backed chickadee, and even a few grassland birds such as LeConte's sparrow and chestnut-collared longspur. A major bird attraction is provided by the concentrations of several hundred bald eagles at McDonald Creek each October. A checklist of Glacier National Park birds is available from the Glacier Natural History Association. For more information, contact Glacier National Park, West Glacier, MT 59936 (406/888-5441).

HELENA NATIONAL FOREST

This 975,000-acre national forest includes **Gates of the Mountains** and **Scapegoat** (240,000 acres) wilderness areas, and 1,600 miles of forest roads. It is mostly located along the Continental Divide, but it also includes the Elkhorn and Big Belt Mountains. No bird list is yet available. There are ranger stations as Helena, Townsend and Lincoln. For information, contact the forest office at the Federal Bldg., Drawer 10014, Helena, MT 59601 (406/449-5201).

KANIKAU NATIONAL FOREST

This 1.4 million-acre national forest is mostly located in Idaho, as part of the Idaho Panhandle National Forest (see Idaho listing), but 468,000 acres are in northwestern Montana.

KOOTENAI NATIONAL FOREST

This 2.2 million-acre national forest is located in the northwestern corner of Montana and adjacent Idaho. It includes the **Cabinet Mountains Wilderness** (94,000 acres), with some peaks reaching more than 8,600 feet. No bird list is yet available, but one is available for the Kootenai National Wildlife Refuge (see Idaho listing). There are ranger stations at Eureka, Fortine, Libby, Trout Creek and Troy. For information contact the Forest Service office at P.O. Box AS, W. Highway 2, Libby, MT 59923 (406/293-6211).

LEE METCALF NATIONAL WILDLIFE REFUGE

This 2,700-acre refuge is located in the Bitterroot Mountains of Montana, about 25 miles south of Missoula. The refuge is centered along the Bitterroot River, at about 3,000 feet elevation and includes coniferous woodland as well as open field and aquatic habitats. The refuge attracts at least 235 species, including 106 breeders. Woodland species include at least six woodpeckers, including the pileated and Lewis's, as well as the hooded merganser, wood duck and rufous hummingbird. An on-line seasonally organized bird checklist is available: http://www.npwrc.usgs.gov/resource/birds/chekbird/r6/metcalf.htm. For more information, contact the refuge manager, P.O. Box 257, Stevensville, MT 59870 (406/777-5181).

LEWIS AND CLARK NATIONAL FOREST

This 1.8 million-acre national forest is located on the eastern slopes of the Continental Divide, south of Glacier National Park. It includes part of two wilderness areas: **Bob Marshall** (1.0 million acres) and **Scapegoat** (240,000 acres). No bird lists are yet available for the forest. There are information stations at Augusta and Niehart, and ranger stations at Choteau, Harlowtown, Stanford and White Sulfur Springs. For more information contact the Forest Service office, Box 871, 1601 2nd Ave. N., Great Falls, MT 59403 (406/727-0901).

LOLO NATIONAL FOREST

This national forest of 2.1 million acres lies in the heart of the northern Rocky Mountains, and includes the Lolo Pass Trail of Lewis and Clark fame, parts or all of three wilderness areas; **Scapegoat** (240,000 acres), **Selway-Bitterroot** (1.3 million acres), and **Welcome Creek** (28,000 acres), in addition to a 60,000-acre **Rattlesnake National Recreation Area**. There are district ranger stations at Huson, Missoula, North Seeley Lake, Plains, and Superior. The forest's birds include the harlequin duck, ruffed grouse, sandhill crane, great gray and boreal owls, pileated woodpecker, red-naped and Williamson's sapsuckers, black-backed and American three-toed woodpeckers, willow, least and cordilleran flycatchers, and several western warblers. The total bird list includes over 200 species and is available from the Lolo National Forest, Building 24, Fort Missoula, Missoula, MT 59801 (406/329-3750).

MEDICINE LAKE NATIONAL WILDLIFE REFUGE

This 31,000-acre Great Plains refuge in northeastern Montana consists of marshes and grasslands, with many typical prairie nesting species. The refuge list totals at least 228 species, including 106 breeders, and such typical native prairie species as the Sprague's pipit, Baird's sparrow, LeConte's sparrow, and chestnut-collared longspur. The many marsh nesters include one of the largest nesting colonies of the American white pelican south of Canada. Also nesting are the double-crested cormorant, California and ring-billed gulls, eared and western grebes, American bittern, black-crowned night-heron, sora, American avocet, Wilson's phalarope and common tern. An on-line bird checklist is available: http://www.npwrc.usgs. gov/resource/birds/chekbird/r6/medicine.htm. For more information, contact the refuge manager, Medicine Lake, MT 59247 (406/789-2303).

MISSOURI RIVER HEADWATERS STATE PARK

This 527-acre state park is located north of the town of Three Forks, where Lewis and Clark found the sources of the Missouri River, at the point where three mountain-fed rivers coalesce. The birds are mostly those of riparian woodlands, although some rimrock and steep cliffs are also present. Not far from here Meriwether Lewis discovered and collected the first known Lewis's woodpecker. To the west lies the Rocky Mountains, while to the east the Great Plains may be seen on the far horizon, at this transition zone of eastern and Rocky Mountain faunas. No bird list is yet available. For information contact the park office at Three Forks, MT 59752 (406/285-3610).

NATIONAL BISON RANGE

This 19,000-acre area lies southwest of Glacier National Park in the Flathead Valley, and consists of nearly 30 square miles of grassland and forest habitats. In addition to its bison herd and other large mammals, the refuge attracts at least 187 bird species, including breeding golden eagles, wood ducks, and hooded mergansers. An on-line seasonally organized bird checklist is available: http:// www.npwrc.usgs.gov/resource/birds/chekbird/r6/bison.htm. For more information, contact the refuge manager, Moiese, MT 59824 (406/644-2211).

NINEPIPE AND PABLO NATIONAL WILDLIFE REFUGES

These refuges are northeast of the National Bison Range, and are managed from it. They collectively consist of about seven square miles of water, marsh and up-land grasslands. These two refuges, each of about 2,000 acres, attract at least 188 bird species, including 74 breeders. At least 16 species of nesting waterfowl are present, plus five nesting grebes, two gulls, two terns, and three blackbirds. An on-line bird checklist is available: http://www.npwrc.usgs.gov/resource/ birds/chekbird/r6/ninepipe.htm. For more information, contact the refuge man-ager, c/o National Bison Range, Moiese, MT 59824 (406/644-2211).

PINE BUTTE SWAMP & PRESERVE

This large (18,000-acre) Nature Conservancy preserve is located west of Choteau on Teton River Road. The preserve requires permission to enter, but the asso-ciated public road parallels the Teton River, where many riparian birds may be seen. The South Fork Road passes through the corner of the preserve, and leads into increasingly more mountainous terrain. Ruffed grouse, northern goshawks, Clark's nutcrackers, and common ravens might be encountered here. Beyond the preserve is the Lewis and Clark National Forest, and trails leading to the Bob Marshall Wilderness. For information on the preserve, contact The Nature Conservancy, Pine Butte Swamp Preserve, Star Rte. 34B, Choteau, MT 59422 (406/466-5526).

RED ROCK LAKES NATIONAL WILDLIFE REFUGE

This refuge, famous for helping preserve the trumpeter swan when it was listed as federally endangered, is located in the Centennial Valley of southwestern Montana, directly west of Yellowstone National Park. The refuge encompasses some 42,000 acres and ranges in elevation from 6,000 to nearly 10,000 feet. Its centerpiece is a 12,000-acre marsh that is a primary breeding area for trumpeter swans, as well as for more than 20 other species of waterfowl. Eight species of dabbling ducks, seven diving ducks (including both goldeneyes and the buffle-head), three mergansers and the ruddy duck all nest here. The refuge attracts over 230 bird species, including at least 152 breeders, which represents a no-tably high species diversity. An on-line bird checklist is available: http://www. npwrc.usgs.gov/resource/birds/chekbird/r6/redrock.htm. For more information, contact the refuge manager, Box 15, Lima, MT 59739 (406/276-3347).

SWAN RIVER NATIONAL WILDLIFE REFUGE

This 1,600-acre refuge is located about 40 miles southeast of Kalispell, and includes a large marsh plus surrounding mature coniferous forests. The refuge bird list includes 171 species, with nesting by at least 78 species, including 16 waterfowl, four grebes, and three hummingbirds. There are also breeding populations of the bald eagle, ruffed grouse, chestnut-backed chickadee and varied thrush. An on-line bird checklist is available: http://www.npwrc.usgs.gov/resource/birds/chekbird/r6/swan.htm. For more information, contact the refuge manager, c/o Northwest Montana W.M.S., 700 Creston Hatchery Road, Kalispell, MT 59739 (406/755-7870).

TERRY BADLANDS

This little-visited badlands region of 44,000 acres is managed by the BLM, and is located along the Yellowstone River northwest of Terry. To reach it one exits I-94 at Terry and takes Highway 253 north. After two miles, turn west onto an unimproved road that passes through the badland. It is six more miles to a scenic overlook, revealing a mixture of shrubs and small trees, including an eastern outlier population of limber pines. Various grassland birds might be seen, including the golden eagle, prairie falcon, upland sandpiper, long-billed curlew and pinyon jay. With great luck the highly elusive and grassland-endemic Sprague's pipit might be seen. The area's birds are probably similar to those of Makoshika and Medicine Rocks state parks (see below). For information, and to verify road conditions, contact the local BLM office in Terry (406/232-7000).

WARM SPRINGS PONDS

This is a small Montana Fish, Wildlife and Parks property located near Anaconda. Its non-seasonal on-line checklist includes about 140 species: http://www.npwrc.usgs.gov/resource/birds/chekbird/r6/warmspri.htm.

Montana's boundaries encompass part or all of ten national forests. These forests plus Montana's BLM lands and national park lands total more than 37 million acres, or about 40 percent of the state's total area. Information on all the national forests of Montana may be obtained from the Forest Service's Northern Region Office, 200 E. Broadway, P.O. Box 7669, Missoula, MT 59801 (406/329-3511).

The BLM administers 8.1 million acres of public lands in Montana. The state has three BLM Districts, organized into 11 Resource Areas, which can be contacted for regional information. Some BLM sites of probable interest to birders include the **Humbug Spires Special Recreation Management Area**, 11,000 acres of highly eroded rock formations 30 miles south of Butte off I-15, exit 99 (information at BLM offices in Dillon and Butte). There is also the **Up-**

per Missouri National Wild and Scenic River, a 149-mile stretch of undeveloped Missouri River, mostly accessible only by boat or canoe from Fort Benton. The BLM's state office address is 222 N. 32nd St., P.O. Box 36800, Missoula, MT 59107 (406/329-3511). Local BLM resource area offices are located at Big Dry (406/232-7000), Billings (406/657-6262), Dillon (406/683-2337), Garnet (406/329-3914), Great Falls (406/727-0503), Havre (406/265-5891), Headwaters (406/265-5891), Judith (406/538-7461), Phillips (406/654-1240), Powder River (406/232-7000) and Valley (406/228-4316).

The Montana Department of Fish, Wildlife and Parks administers 36 state parks. Parks of possible interest to birders include the 2,165-acre **Wild Horse Island State Park** (Big Arm, MT; 406/752-5501), an island in Flathead Lake with nesting bald eagles, ospreys and northern goshawks. The 50-acre **Lost Creek State Park** (Anaconda, MT 59711; 406/542-5500) near Anaconda has a steep canyon with 1,200-foot cliffs, which should host some cliff-nesting species. The 260-acre **Giant Springs State Park** (Great Falls, MT 59405; 406/454-3441) has a half-mile nature trail, and a bird list of more than 150 species. The 8,123-acre **Makoshika State Park** (1301 Snyder St., Glendive, MT 59330; 406/365-6256) is a badlands area that has golden eagles, prairie falcons, rock wrens, and grassland sparrows such as the Brewer's, vesper and lark. Park road conditions should be verified in advance from the visitor center. The 316-acre **Medicine Rocks State Park** (Ekalaka, MT; 406/232-4365) near Ekalaka is similarly badland-like, with nesting golden eagles, ferruginous hawks, merlins and prairie falcons, as well as many grassland birds such as sharp-tailed grouse. Carol and Hank Fischer (1995) have described all these sites and their birds.

In addition to its ten national forests, Montana also has seven state forests, including **Clearwater** (43,000 acres, near Bonner, MT 59823), **Coal Creek** (93,000 acres, near Polebridge, MT 59928), **Lincoln** (12,000 acres, near Lincoln, MT 59639), **Stillwater** (90,000 acres, between Eureka and Kalispell), **Sula** (13,000 acres, near Conner, MT 59827), **Swan River** (40,000 acres, near Bigfork, MT 59911), and **Thompson River** (34,000 acres, near Darby, MT 59829). Most of these forests have no direct phone numbers, and bird lists are not currently available. For more information on state forests and parks, contact the Fish, Wildlife and Parks Department's State Headquarters, 1420 E. Sixth Ave., Helena, MT 59620 (406/444-3750). General information may be obtained from Montana Tourist Information, 1424 9th Ave., Helena MT 59620 (406/444-2654 or 800/541-1447).

Montana birding information can also be obtained at http://montanabirdingtrail.org. There one may find information on the state's first two birding trails, the Bitterroot Birding and Nature Trail, and the Northeastern Plains Birding Trail. Local checklists, maps, and other birding information can be downloaded at http://birding.com/wheretobird/Montana.asp. The address of the Montana Audubon Council is P.O. Box 595, Helena, MT 59624. There are local Audubon chapters at Bigfork, Billings, Bozeman, Deer Lodge, Great Falls, Hamilton, Helena and Miles City. The Montana chapter of the National Audubon Society has a useful website at http://mtaudbon.org, and Montana's Outdoor Birding Group blog can be accessed at http://groups.yahoo.com/group/MOB-Montana. A non-annotated collective state checklist including about 390 species is available on-line: http://www.npwrc.usgs.gov/resource/birds/chekbird/r6/mon.htm.

Wyoming

BIGHORN CANYON NATIONAL RECREATION AREA

The Wyoming district of this enormous (120,000-acre) two-state national recreation area covers about 30,000 acres. Lovell is the headquarters for this section, and has a Forest Service Ranger Station located at the eastern end of town. The Pryor Mountain Wild Horse Range is located a few miles north of town, as is a wildlife habitat area jointly operated by several private, state and federal agencies. Highway U.S. 37 passes through sagebrush–juniper habitats and leads to sites overlooking a 1,000-foot canyon, where golden eagles, turkey vultures and various hawks and falcons can often be seen soaring above Bighorn Lake. The Montana district is much larger, and is headquartered at Fort Smith (see description in Montana listing above). The recreation area's on-line bird checklist has more than 200 species: http://www.npwrc.usgs.gov/resource/birds/chekbird/r6/bighorn.htm.

BIGHORN NATIONAL FOREST

This 1.1 million-acre national forest in the Big Horn Mountains includes a wilderness area, **Cloud Peak**, of 195,000 acres. The region's highest peak, Cloud Peak, rises to 13,165 feet elevation, and is exceeded only by a few peaks in the Wind River range. There are 300 lakes in the forest, as well as glacial fields and alpine tundra. There are also ranger stations at Buffalo, Greybull, Lovell, and Worland. Helen Downing's (1990) survey of north-central Wyoming birds included Bighorn National Forest. Long out-of-print, it is currently being updated by Jacqueline Canterbury and Paul Johnsgard. The forest is headquartered at 2013 Eastside 2nd St., Sheridan, WY 82801 (307/674-2600).

BLACK HILLS NATIONAL FOREST

This 1.2 million-acre national forest is mostly located in South Dakota, but extends into northeastern Wyoming. The forest's highest peak (in South Dakota) is at 7,242 feet of elevation, well below the subalpine zone. Much of the forest consists of ponderosa pines, which are now under severe attack by pine bark beetles. Some of the Black Hills' bird life is comprised of eastern-oriented birds, such as the blue jay, red-headed woodpecker, eastern bluebird, brown thrasher and indigo bunting. However, the forest also includes several typical Rocky Mountain species, such as the red-naped sapsucker, and the dusky and cordilleran flycatchers. There are also many species with general western orientations, such as the white-throated swift, Townsend's solitaire and MacGillivray's warbler. The Black Hills region is also home to an endemic "white-winged" race of the dark-eyed junco. A comprehensive *Birds of the Black Hills* (Pettingill and Whitney, 1965) is now out-of-print, but an updated bird list of 195 species reported for the forest is available from the Black Hills National Forest, R.R. 2, Box 200, Custer, SD 57730 (605/673-2251).

BRIDGER-TETON NATIONAL FOREST

This enormous 3.4-million acre national forest is an important part of the Greater Yellowstone ecosystem. The forest encompasses 1.2 million acres of wilderness, including the **Teton** (575,000 acres), **Gros Ventre** (287,000 acres), **Winegar Hole** (14,000 acres) and **Bridger** (428,000 acres) wilderness areas, and ranges in altitude from about 6,000 to 13,785 feet (Gannet Peak, Wyoming's highest peak). Its vegetation varies from sagebrush and other lowland arid brush to alpine meadows and glaciers. Species such as the greater sage-grouse occupy sagebrush, while the harlequin duck, common merganser and American dipper inhabit clear and swift mountain streams. Water and marsh birds such as the trumpeter swan, ring-necked duck, Barrow's goldeneye, bufflehead and sandhill crane nest around beaver ponds. Several species designated by the Forest Service as rare or sensitive also occur here, such as the northern goshawk, great gray owl, boreal owl, American three-toed woodpecker and Hammond's flycatcher. The Forest Service's remarkably large list of 355 birds represents over 90 percent of the species that occur regularly in Wyoming, and includes at least 157 summer residents, 80 permanent residents, 68 migrants, and 14 winter visitors. A "Birds of Jackson Hole" checklist published by the Wyoming Game and Fish Dept. contains 340 species and covers most of the Bridger-Teton National Forest. It is available from the Wyoming Game & Fish Dept., 260 Buena Vista Dr., Lander, WY 82520-9902, and probably also at the Bridger-Teton National Forest headquarters, P.O. Box 1888, Jackson, WY 83001 (307/733-2752, or 307/739-5500). Information is also available at ranger stations in Afton, Big Piney, Kemmerer, Moran and Pinedale. See also the Grand Teton National Park and Jackson Hole account.

COKEVILLE MEADOWS NATIONAL WILDLIFE REFUGE

This newly established refuge is located about 20 miles south of Cokeville along Bear River, in extreme southwestern Wyoming. The refuge was under development and was still acquiring land as of 2010, but should provide excellent birding for waterfowl and marsh birds when it is opened to the public. It is managed out of Seedskadee National Wildlife Refuge (see contact information below).

DEVILS TOWER NATIONAL MONUMENT

This national monument covers 1,346 acres in northeastern Wyoming near the western limit of the Black Hills. Typical western-oriented species include the white-throated swift, violet-green swallow, black-headed grosbeak, spotted towhee and Bullock's oriole. The monument's on-line bird checklist has about 160 species: http://www.npwrc.usgs.gov/resource/birds/chekbird/r6/devil.htm. For more information contact the headquarters at P.O. Box 10, Gillette, Devils Tower, WY 82714 (307/467-5283).

FLAMING GORGE NATIONAL RECREATION AREA

This large area in southwestern Wyoming and adjacent Utah is centered on the 90-mile-long Flaming Gorge Reservoir, and includes part of Ashley National Forest (which totals 1.38 million acres, mostly in Utah). The area's bird list includes 259 species, with notable numbers of raptors (15 species), waterfowl (24 species) and especially shorebirds and gulls (31 species). A checklist is available from Red Canyon Lodge, 2450 Red Canyon Lodge, Dutch John, UT 84023 (435-889-3759), or the Ashley National Forest, Flaming Gorge Ranger District, P.O. Box 279, Manila, UT 84046 (435/784-3445).

FORT LARAMIE NATIONAL HISTORIC SITE

The seasonally organized bird checklist for this small historic fort in the North Platte valley has about 160 species: http://www.npwrc.usgs.gov/resource/birds/ chekbird/r6/flaramie.htm. For more information contact the site headquarters at 965 Gray Rocks Road, Fort Laramie WY 82212 (307/837-2221).

FOSSIL BUTTE NATIONAL MONUMENT

This 8,000-acre site is a 50-million year-old Eocene lake bed in southwestern Wyoming that is world-famous for its remarkable fish fossils, but also has produced some notable bird fossils. The non-seasonal on-line bird checklist for this arid and scrub-dominated site has about 100 species: http://www.npwrc.usgs. gov/resource/birds/chekbird/r6/fossil.htm. For more information contact the headquarters at P. O. Box 592, Kemmerer, WY 83101 (307/877-4450).

GRAND TETON NATIONAL PARK AND JACKSON HOLE

This national park includes the adjacent Teton-Range and part of the valley to the south (Jackson Hole). The region consists of more than 480 square miles of high mountains (with a maximum elevation of 13,766 feet), coniferous forests, and sage-covered plains and riparian valleys between 6,000 and 7,000 feet. All the roads are limited to lower elevations, but many hiking trails extend into the mountain forests and alpine zone. The park is recognized as one of Wyoming's

Important Bird Areas; park specialties include breeding trumpeter swans, sand-hill cranes, greater sage-grouse, and black rosy-finches. Follett (1986) provided a seasonally organized list of 226 Grand Teton National Park birds, 15 of which he considered accidentals. Of the regularly occurring species, 170 were reported during summer, and most of these might be considered as prospective if not proven breeders. The nearby Jackson Hole valley, National Elk Refuge, and the adjacent Bridger-Teton National Forest (see descriptions above) are all part of the Greater Yellowstone ecosystem, and support many additional species and habitats. Bert Raynes (1984) provided individual accounts of 68 species of greater Jackson Hole birds, and a seasonally organized checklist of 293 species. His "pocket guide" (2000) list provides relative seasonal abundance and habitat information on 301 species. A checklist of 305 species for the region, with accompanying information on local birding sites, is available on-line: http://www.npwrc.usgs.gov/resource/birds/chekbird/r6/jackhole.htm. An updated hard-copy checklist for the greater Jackson Hole region has 340 species (Raynes and Raynes, 2008), and is available from the Wyoming Game & Fish Dept., 260 Buena Vista Dr., Lander WY 82520). For information on Grand Teton National Park contact the park headquarters, Moose, WY 83012 (307/733-2880).

HUTTON LAKE NATIONAL WILDLIFE REFUGE

This 1,900-acre refuge is located 12 miles south of Laramie, at an elevation of 7,150 feet. The refuge consists of five small lakes, a marsh and uplands, and attracts at least 153 species, including 61 breeders. Nesting marsh birds include the eared grebe, American avocet, Wilson's phalarope, Forster's tern and black-crowned night-heron. Notable land birds include the golden eagle, McCown's longspur, and sage thrasher. Hutton Lake is managed out of Colorado's Arapaho National Wildlife Refuge (303/723-8202). The nearby **Bamforth National Wildlife Refuge** was established to protect the endangered Wyoming toad, and is off-limits to the public.

MEDICINE BOW NATIONAL FOREST

This 1.6-million acre national forest in south-central Wyoming consists of three rather small but beautiful mountain ranges, the Medicine Bow, Snowy and Sierra Madre. The forest includes **Encampment River** (10,400 acres), **Houston Park** (31,000 acres), **Platte River** (23,000 acres) and **Savage Run** (15,000 acres) wilderness areas, and reaches a maximum elevation of 12,013 feet (Medicine Bow Peak). Two high-elevation passes are the Snowy Range Pass at 10,846 feet, and Battle Pass, at 9,955 feet. No bird list for this forest is yet available, but Brooklyn Lake and Lewis Lake in the Snowy Range are good places for finding alpine species such as American pipits and brown-capped rosy finches. The white-tailed ptarmigan has now apparently vanished from these mountains. Battle Creek Campground in the Sierra Madre Range is at a lower altitude, in mixed coniferous-deciduous forest. The campground has some scrub oaks that at times have attracted band-tailed pigeons, and support an assortment of western flycatchers, towhees and jays (Scott, 1993). There are ranger stations at Douglas, Encampment, Laramie and Saratoga. For information contact the Forest Service office, 2468 Jackson St., Laramie, WY 82070 (307/745-2300).

NATIONAL ELK REFUGE

This 25,000-acre refuge is located just north of Jackson and is part of the Greater Yellowstone ecosystem. Although primarily dedicated to providing winter habitat for up to nearly 10,000 elk, the largely grassland- and sage-covered refuge has attracted at least 219 bird species, including 125 breeders. The greater sage-grouse, golden eagle, sandhill crane, trumpeter swan and long-billed curlew are among the refuge's notable nesting species. A seasonally organized bird list can be found on-line at: http://www.npwrc.usgs.gov/resource/birds/chekbird/r6/elkref.htm. For more information, contact the refuge manager, P.O. Box C, Jackson, WY 83001 (307/733-9212).

SEEDSKADEE NATIONAL WILDLIFE REFUGE

This refuge of 14,455 acres in southwestern Wyoming lies at about 6,000 feet elevation, in the Green River valley. The refuge is dominated by sagebrush and other arid-adapted plants, and has an associated bird fauna, together with bluff-associated and riparian woodland birds. A graveled auto tour route parallels the Green River and provides many overlook views. The refuge attracts at least 227 species, including 120 breeders. Breeding raptors include the osprey, golden eagle, prairie falcon, merlin and short-eared owl. Breeding marsh birds include the white-faced ibis, sora, Virginia rail, Forster's tern and great blue heron. Landbirds include the western scrub-jay, pinyon jay, blue-gray gnatcatcher, western bluebird and green-tailed towhee. An on-line and seasonally organized bird checklist for the refuge is available: http://www.npwrc.usgs.gov/resource/birds/chekbird/r6/seedskad.htm. Further information can be obtained from the refuge manager, P.O. Box 67, Green River, WY 82935 (307/875-2187). About 50 miles to the south is the 207,000-acre **Flaming Gorge National Recreation Area** (201,000 acres), where scrubby juniper habitats attract the juniper titmouse, gray flycatcher, Bewick's wren and black-throated gray warbler, all of which are relatively uncommon species in Wyoming. The Flaming Gorge National Recreation Area is partly in Utah and is part of the Ashley National Forest, headquartered at 355 N. Vernal Ave., Vernal, UT 84078 (801-789-1181).

SHOSHONE NATIONAL FOREST

This 2.4 million-acre forest extends over much of central and southern Wyoming and includes five wilderness areas: **Absaroka** (24,000 acres), **Fitzpatrick** (199,000 acres), **North Absaroka** (360,000 acres, with peaks to 11,700 feet), **Popo Agie** (101,000 acres, with 20 peaks exceeding 12,000 feet) and

Washakie (714,000 acres, with peaks exceeding 13,000 feet). There are ranger stations at Cody, Dubois, Lander, Meeteetse and Powell. No bird list for this forest is yet available, but the Jackson Hole list may be applicable, For information contact the Forest Service office, P.O. Box 2140, W. Yellowstone Hwy., Cody, W 82414 (207/527-6241).

TARGHEE NATIONAL FOREST

See Caribou–Targhree National Forest (in Idaho section).

THUNDER BASIN NATIONAL GRASSLAND

This national grassland covers 572,000 acres in northeastern Wyoming, and is extensively interspersed with BLM, state and private lands. It consists mostly of shortgrass prairie, with associated grassland birds. The site's on-line bird checklist has over 200 species: http://www.npwrc.usgs.gov/resource/birds/chekbird/ r6/thunder.htm. Reputed breeders include 13 raptors, six owls and five woodpeckers, of which many are forest-dependent. However, this list is unreliable as to the breeding status of some species, such as including several shorebirds that nest only in the high arctic. The grassland is managed by the Forest Service, 2468 Jackson St., Laramie, WY 82070 (307/745-2300).

WASATCH-CACHE NATIONAL FOREST

This 1.2 million-acre national forest is mostly located in Utah, but extends into southwestern Wyoming along the northern edge of the Uinta Range. The forest's maximum elevation is 13,528 feet, and includes a 456,000-acre **High Uintas Wilderness**. Some of the birds of special interest are the northern goshawk, golden eagle, dusky grouse, white-tailed ptarmigan, American three-toed woodpecker and northern saw-whet owl. Notable finches include the black rosy-finch, pine grosbeak and red crossbill. The bird list contains over 200 species and is available from the National Forest office, 8230 Federal Building. 125 South St., Salt Lake City, UT 84138 (801/524-5030).

YELLOWSTONE NATIONAL PARK

This, the oldest U.S. national park, is also the largest south of Canada, covering more than 3,400 square miles. Most of the roads are at elevations in excess of 7,000 feet, with passes as high as 8,850 feet. The park's highest point (Eagle Peak) is 11,358 feet. Mount Washburn is lower (at 10,243 feet), but is fairly accessible by trail and supports alpine birds such as American pipits and black rosy-finches. The majority of the land area of the park is covered with lodgepole pine forests that are now regenerating from massive 1988 forest fires that blackened 36 percent of the park's land area. Douglas-fir, Engelmann spruce and subalpine fir occur at the higher elevations. Some fairly extensive areas of sage-dominated grasslands exist between 5,000 and 7,500 feet, mostly in the northern parts of the park, which is an important year-around habitat for large grazing mammals and their predators. Yellowstone Lake has the only nesting colony of the American white pelican in any national park, and the park is also notable for its nesting populations of the bald eagle, osprey, sandhill crane and trumpeter swan. Park rarities include the Caspian tern, harlequin duck, pinyon jay, and sage thrasher.

In their survey of Montana wildlife-watching sites, Carol and Hank Fischer (1990) described 12 outstanding birding areas in the park, which has been recognized as one of Wyoming's Important Bird Areas. Terry McEneaney's 1988 book, *Birds of Yellowstone*, is a comprehensive birding reference, and includes a great deal of information on local birds and bird-finding. McEneaney has recently (2004) provided an on-line park checklist of 318 species (http://www. nps.gov/yell/naturescience/upload/bird2004.pdf). He also has produced (2008a) a printed park checklist of 323 species, including nearly 150 breeders. This checklist is available at nominal cost from the Yellowstone Association, at Mammoth Hot Springs, P.O. Box 117, Yellowstone National Park WY 82190 (406/848-2400). The Yellowstone Association (www.Yellowstone Association.org) also operates the Yellowstone Institute, which offers short courses on regional birds and other nature-oriented subjects (866/439-7375), as well as holding field seminars and directing private nature tours (406/848-2400). The address of Yellowstoe Park headquarters is P.O. Box 168, Yellowstone National Park, WY 82190 (307/344-7381).

Besides the Bridger-Teton National Forest to the south of Yellowstone National Park, several other Wyoming national forests are also part of the Greater Yellowstone ecosystem. Part of the Caribou-Targhee National Forest borders the western boundary of Yellowstone and Grand Teton parks, and the Shoshone National Forest borders the eastern edge of Yellowstone National Park. The Gallatin National Forest borders the northwestern and northern parts of Yellowstone Park, and the Custer National Forest barely reaches its northeasternmost corner. Collectively these seven national forests, three national wildlife refuges, two national parks, and other regional federal and state lands total about 18,000 square miles (an area equal to that of Vermont plus New Hampshire), and comprise the greater Yellowstone ecosystem, the largest nationally preserved and intact ecosystem south of Canada. Information on Wyoming's national forests can be obtained from the Forest Service's regional headquarters, Rocky Mountain Region, P.O. Box 25127, Lakewood, CO 80225 (305/275-5350).

Wyoming also has nearly 18 million acres of public-access land under management by the Bureau of Land Management (BLM). Several sites among the many BLM properties are of possible interest to birders. **Muddy Mountain Environmental Education Area** (12,000 acres, 10 mi. S of Casper) includes peaks exceeding 8,000 feet, with golden eagles and other raptors. Another BLM sites of potential interest to birders is the **Middlefork Recreation**

Area (48,239 acres, 11 mi. west of Kaycee) with steep 1,000-foot canyon walls, above the Powder River, with associated raptors and greater sage-grouse on the uplands. Other BLM sites of less certain interest include **Goldeneye Wildlife and Recreation Area** (1,153 acres, including the 500-acre Goldeneye Reservoir, 20 mi. west of Casper), and **West Slope** (448,300 acres, 15 mi. east of Lovell, on the western slope of the Bighorn Mountains).

Many of Wyoming's BLM properties are of special importance to the rapidly declining greater sage-grouse and other sage-adapted species, such as **Shirley Basin** (Carbon & Natrona counties) and the **Red Desert** (Fremont and Sweetwater counties), both of which have been designated as state-level Important Bird Areas. Two other BLM locations in the Red Desert region that are also state-level Important Bird Areas and that support many sage-grouse leks are **Ninemile Draw** and **Little Sandy Landscape**. This species has been proposed for federal listing as threatened, and Wyoming offers the best opportunities for preserving it.

Information on the BLM lands mentioned above can be obtained from the Wyoming State BLM Office, P.O. Box 1828, Cheyenne, WY 82003 (307/775-6256). Local information can also be obtained from regional field offices at Buffalo (307/684-1100), Casper (307/261-7600), Cody (307/578-5900), Kemmerer (307/828-4500), Lander (307/332-8400), Newcastle (307/746-6600), Pinedale (307/367-5300), Rawlins (307/328-4200), Rock Springs (307/352-0256) and Worland (307/347-5100).

On-line bird lists are available for the two following BLM districts:

Northeastern Wyoming, Casper District (office at Casper). This on-line but non-seasonal and non-annotated bird checklist has about 375 species: http://www.npwrc.usgs.gov/resource/birds/chekbird/r6/new.htm.

Southeastern Wyoming, Rawlins District (office at Rawlins). This on-line bird checklist (seasonally organized and slightly annotated) has about 350 species: http://www.npwrc.usgs.gov/resource/birds/chekbird/r6/rawlins.htm.

Wyoming has 12 state parks: **Bear River** (Evanston, 290 acres, 307/789-6540); **Boysen** (Shoshoni, 34,705 acres, 307/876-2796), **Buffalo Bill**, (W of Cody, 12,000 acres, 307/587-9227), **Curt Gowdy** (Cheyenne, 1,960 acres, 307/632-7946), **Edness K. Wilkins** (Casper, 319 acres, 307/577-5150), **Glendo** (Glendo, 19,000 acres), **Guernsey** (Guernsey, 8,638 acres, 307/836-1942), **Hawk Springs** (Hawk Springs, 2,000 acres, 307/836-2334), **Hot Springs** (Thermopolis, 1,029 acres, 307/864-2176), **Keyhole** (Moorcroft, 15,674 acres, 307/322-2220), **Seminoe** (north of Sinclair, 21,741 acres, 307/320-3013), and **Sinks Canyon** (southwest of Lander, 600 acres, 307/332-6333). Bird lists are not yet available for these parks. General park information may be obtained from the Wyoming Travel Commission, I-25 at College Drive, Cheyenne, WY 82002 (307/777-7777, or 800/225-5996), or from the Information Section, Wyoming Game & Fish Dept., 5400 Bishop Blvd., Cheyenne, WY 82006 (207/777-4600).

There is an on-line, non-annotated, collective checklist of Wyoming birds that includes 394 of the 426 species documented for Wyoming as of 2010: http:// www.npwrc.usgs.gov/resource/birds/chekbird/r6/wyor.htm. There are six local Audubon Society chapters in Wyoming, at Casper, Cody, Jackson, Lander, Laramie, and Sheridan. The state Audubon Society headquarters address is 358 N. 5th St., Unit A, Laramie, WY 82072 (www.audubonwyoming.org). As of 2010 Wyoming's Audubon Society had identified 44 Important Bird Areas, which are listed on their website: (www.audubonwyoming.org//BirdSci_IBA.html).

Canadian National Parks in the Rocky Mountain Region

Alberta

BANFF NATIONAL PARK

This internationally famous national park encompasses some 1.64 million acres along the western border of Alberta, and is largely a montane park, with a maximum elevation of about 11,800 feet (Mt. Forbes). The **Bighorn Wildland and Recreation Area** and **Rocky Mountain Forest Reserve** connect Banff and Jasper Parks, and collectively include the **White Goat, Siffleur** and **Ghost River** wilderness areas. Much of the park is above timberline, and several large glaciers are within its boundaries. There are several hundred miles of trails, and a 143-mile scenic drive. Its bird list includes over 260 species. The tundra area at Bow Summit is a good place to look for alpine birds such as the white-tailed ptarmigan, American pipit and gray-crowned rosy-finch. Black swifts might be seen in Johnson Canyon, and other park specialties include some boreal breeders such as the alder flycatcher, white-throated sparrow and purple finch. A bilingual bird checklist covering both Banff and Jasper national parks is available. For more information, contact Banff National Park, Box 900, Banff, AB, TOl OCO, Canada (403/762-1550), or Parks Canada, 25-7-N Eddy St., Gatineau, Quebec KIA OMS (888-773-8888 in Canada, or internationally, 613-860-1251).

JASPER NATIONAL PARK

This national park consists of some 2.68 million acres of mountain and alpine scenery along the western border of Alberta, directly north of Banff National Park. Its maximum elevation is 12,294 feet at Mt. Columbia, the highest point in Alberta. There are many glaciers, including the famous Columbia Icefield. In contrast, some of the river valleys are only slightly above 3,000 feet, providing an enormous vertical habitat range. Timberline here occurs are about 6,000 feet. Adjoining Jasper Park to the north is **Willmore Wilderness,** of 1.1 million acres, with 160 miles of trails and connecting with Jasper's trail network. The park's bird life includes at least 277 species, including many boreal and sub-arctic species such as the greater yellowlegs, willow ptarmigan. palm warbler and rusty blackbird. A checklist of park birds is available. For more information, contact Jasper National Park, Box 10, Jasper, AB, TOE lEO, Canada (780/652-6176), or Parks Canada, 25-7-N Eddy St., Gatineau, Quebec KIA OMS (888-773-8888 in Canada; internationally 613-860-1251).

WATERTON LAKES NATIONAL PARK

This national park is contiguous with the U.S.'s Glacier National Park, and shares many of the same habitat types and topographic characteristics. It covers 140,000 acres, and has a maximum elevation of 8,833 feet. A bilingual bird checklist is available. The park's bird list includes at least 255 species, of which at least 37 are residents, and 149 are known or probable breeders. Park specialties are largely eastern-oriented birds, and include the black-billed cuckoo, ruby-throated hummingbird, brown thrasher, ovenbird, indigo bunting and LeConte's sparrow. For more information, contact Watertown Lakes National Park, AB, TOK 2MO, Canada (403/859-2150) or Parks Canada, 25-7-N Eddy St., Gatineau, Quebec KIA OMS (888-773-8888 in Canada, or internationally, 613-860-1251).

British Columbia

BOWRON LAKE PROVINCIAL PARK

This provincial park of about 300,000 acres is in the Cariboo Mountains of southeastern British Columbia, It is notable for its large number of beautiful lakes and rivers, and can be traversed only by canoe or kayak. No bird-related information is yet available. For general information, contact Tourism British Columbia (800/435-5622).

GLACIER NATIONAL PARK

Glacier National Park (328,000 acres) has over 100 glaciers, and more than 100 miles of trails. Rogers Pass is at 4,225 feet, and its highest point, Mount Dawson, exceeds 11,000 feet. Its topography and bird life are very similar to those of Yoho and Kootenay parks. A bird checklist is available. For more information contact Parks Canada, 25-7-N Eddy St., Gatineau, Quebec KIA OMS (888-773-8888 in Canada, or internationally, 613-860-1251).

HAMBER PROVINCIAL PARK

This provincial park of 60,000 acres is on the western side of the Continental Divide in southeastern British Columbia, and is connected to adjacent Jasper National Park via trails. This is a wilderness park with no road access, and no bird-related information is yet available. For general information, contact Tourism British Columbia (800/435-5622).

KOOTENAY NATIONAL PARK

This national park encompasses 347,000 acres, and is contiguous with Yoho National Park on the north, and with Banff National Park on the east. Its highest elevation of 10,511 feet is located on the western edge of the park, but much of the highway is below 5,000 feet. A checklist of the park's birds is available. For more information contact Kootenay National Park, Radium Hot Springs, BC, VOA 1MO, Canada (250/347-9505), or Parks Canada, 25-7-N Eddy St., Gatineau, Quebec KIA OMS (888-773-8888 in Canada, or internationally, 613-860-1251).

MOUNT ASSINIBOINE PROVINCIAL PARK

This provincial park of 94,000 acres is on the western side of the Continental Divide in southeastern British Columbia, and south of Kootenay National Park. Its peaks include Mount Assiniboine, at 11,870 feet. At least eighty-four species of birds have been reported. For information, contact Tourism British Columbia (800/435-5622).

MOUNT REVELSTOKE NATIONAL PARK

This relatively small national park (64,000 acres) is located west of the main chain of mountains forming the Continental Divide, and its highest point is Mount Revelstoke, at 6,358 feet. The interpretive trails pass through ancient lowland rainforests and wet montane forests, and a paved parkway leads to subalpine meadows. No bird lists are available. For more information contact Mount Revelstoke National Park, Box 350, Revelstoke, BC, VOE 2S0, Canada (350/837-7500), or Parks Canada, 25-7-N Eddy St., Gatineau, Quebec KIA OMS (888-773-8888 in Canada, or internationally, 613-860-1251).

MOUNT ROBSON PROVINCIAL PARK

This provincial park of 536,000 acres is on the western side of the Continental Divide in southeastern British Columbia, and adjoining Jasper National Park. Its peaks includes Mount Robson, at 12,972 feet the highest peak in the Canadian Rockies. At least 182 bird species have been reported in the park. For general information, contact Tourism British Columbia (800/435-5622).

WELLS GRAY PROVINCIAL PARK

This provincial park of 1.8 million acres is in the Caribou Mountains of southeastern British Columbia, and is notable for its many lakes rivers, waterfalls, and glaciers. At least 219 bird species have been reported in the park. In the same region are **Caribou Mts.** and **Bowron Lake** provincial parks. For general information, contact Tourism British Columbia (800/435-5622).

YOHO NATIONAL PARK

This national park encompasses 324,000 acres of montane habitats, and has a maximum elevation of about 11,000 feet (Mt. Goodsir). It has 28 mountains exceeding 9,800 feet, and 26 alpine lakes. It lies along the western border of Banff National Park, and the two parks share many bird species. A checklist of park birds is available. For more information, contact Yoho National Park, Field, BC, VOA 100, Canada (250/343-6221), or Parks Canada, 25-7-N Eddy St., Gatineau, Quebec KIA OMS (888-773-8888 in Canada, or internationally, 613-860-1251).

Trumpeter Swan, adult.
Drawing by author.

Chapter 2

Birds of the Central and Northern
Rocky Mountains

Ducks, Geese & Swans

Greater White-fronted Goose (*Anser albifrons*)

Status: A local and relatively rare migrant in the region, becoming more common eastwardly on the Great Plains. Only vagrant migrants appear in the mountain parks.

Habitats and Ecology: While on migration these birds sometimes fly in company with Canada geese, especially in the case of birds separated from their own flocks. Like Canada geese, they field-forage in grainfields and other croplands, and usually roost in large, open marshes. Breeding is done on open arctic tundra, and in general the birds avoid forested areas.

Suggested Viewing Locations: Wetlands and agricultural fields east of the Rockies attract these birds during migration. There are few Wyoming spring records, but several have been reported from the Goshen Hole area (Goshen County). They are more common during October and November (Scott, 1993).

Population: National Breeding Bird Survey trend data are not available. The North American 2009 Pacific Coast winter population was estimated at about 537,000 birds, or 14% below the 2000 estimate, and the mid-continent fall population was about 752,000, or well below the 2000 estimate (U.S.F.W.S., 2009a).

Snow Goose (*Chen caerulescens*)

Status: An uncommon to rare migrant throughout the region, becoming more common east of the Rockies, especially to the north. Increasingly common nationally in recent years.

Habitats and Ecology: Generally associated with large marsh and wetland habitats; feeding in dry fields is done less frequently than in Canada geese, and rootstalks and tubers of marshland plants are more regularly eaten.

Suggested Viewing Locations: More common during fall than spring, wetlands and agricultural fields east of the Rockies attract these birds during migration. In Wyoming, locations such as Yellowstone Lake, Table Mountain Wildlife Unit (Goshen County), Bump Sullivan Reservoir (near Yoder), and Cliff Graham Reservoir (Uinta County) are good choices (Dorn and Dorn, 1990). Ocean Lake, and Lake DeSmet and Table Mountain are also good viewing locations (Scott, 1993). The Snake River corridor also attracts many birds (Faulkner, 2010).

Population: National Breeding Bird Survey trend data are not available. Winter or spring 2009 national population estimates include about 1.4 million greater snow geese and nearly four million lesser snow geese (U.S.F.W.S., 2009a).

Ross's Goose (*Chen rossii*)

Status: A localized and generally uncommon migrant in the region. However, large flocks of snow geese often have Ross's geese among them, especially in more northerly areas.

Habitats and Ecology; Migrating birds have similar habitat needs to those of snow geese, and often mingle with them, especially in spring. During that time both species often forage in grain stubble, where the shorter bill of the Ross's goose presumably allows for closer cropping of grasses or other vegetation.

Suggested Viewing Locations: Most migrants pass well to the east of the Rockies, along the eastern edge of the region. The Wyoming areas listed for snow geese are also favorable for seeing Ross's geese, such as Table Mountain. Most are seen along the eastern tier of counties.

Population: National Breeding Bird Survey trend data are not available. Because of difficulties in field separation from snow geese, no attempts are made to specifically identify and inventory Ross's geese. One enormous nesting colony (Karrak Lake, in the tundra lowlands of arctic Canada's Queen Maud Gulf) had 726,000 birds in 2008, and comprised a substantial percentage of this species' gradually increasing population (U.S.F.W.S., 2009a).

Canada Goose (*Branta canadensis*)

Status: The Canada goose is a widespread, virtually pandemic breeder throughout the region, while the cackling goose is an arctic-breeding migrant, especially in the plains region.

Habitats and Ecology: The extremely adaptable Canada goose sometimes nests within the city limits of large cities, but also occurs on prairie marshes, beaver ponds, and forest-edged mountain lakes. Beaver lodges or muskrat houses provide safe and favored nest sites in many areas.

Suggested Viewing Locations: Canada geese occupy the majority of wetlands in the region, especially protected wetlands and larger lakes. Table Mountain and Springer Lake are especially favored Wyoming sites (Scott, 1993). The densest regional breeding concentrations are probably in southwestern Montana, northwestern Wyoming, and eastern Idaho.

Population: National Breeding Bird Survey trend data indicate a significant annual population increase of 7.4%. Trend estimates for the Rocky Mountains & Plains States region (U.S.F.W.S. Region 6) indicate a significant increase of 6.0%. There may have been more than five million Canada geese in North America by 2009 (U.S.F.W.S., 2009a).

Cackling Goose (*Branta hutchsinii*)

Status: Probably a more common migrant through the region than is generally realized.

Suggested Viewing Locations: Cackling geese are most likely seen in wetlands east of the Rockies during migration in March–April and October–November. Because the latter species was only recently recognized as specifically distinct from the Canada goose, regional records for it are still quite limited.

Suggested Viewing Locations: This tundra breeder is most often seen on the eastern plains of the region. A few sight records are from Johnson and Natrona counties.

Population: Various recent U.S. Fish & Wildlife Service reports suggest that the total North American population of cackling geese might be at least 750,000 birds.

Trumpeter Swan (*Cygnus buccinator*)

Status: An uncommon to rare permanent resident in the vicinity of Yellowstone and Grand Teton parks, generally rare or accidental elsewhere. In the late 1970's Yellowstone Park supported about 20 nesting pairs in or very near the park, while Red Rock Lakes Refuge to the west of Yellowstone contained about 27 nesting pairs. McEneaney (1988) judged that park's population to be less than 15 pairs during the late 1980's, with about 40 individuals typically present during summer months. Grand Teton National Park has long had nesting trumpeter swans. Johnsgard (1986) judged the Grand Teton population to be about six pairs during the early 1980's. The Greater Yellowstone region population has been in serious decline, and by 2009 fewer than 400 birds were present. Up to a half dozen pairs still nest in Grand Teton Park and the National Elk Refuge, usually on isolated lakes or beaver ponds. Only local nestings have occurred outside this region, including ones at Colony and Star Valley (Scott, 1993).

Habitats and Ecology: In the Rocky Mountain region this species is mostly limited to fairly large (usually over 30-acre) ponds having considerable aquatic vegetation and relative seclusion from disturbance by humans. Beaver ponds are most often used in the Jackson Hole area, and nests are sometimes built on their lodges.

Suggested Viewing Locations: The largest local population in the Rocky Mountains region is at Red Rock Lakes National Wildlife Refuge. The Madison River is an excellent locations for finding these rare swans in Yellowstone National Park (McEneaney, 1988). Trumpeter swans are also usually easily visible at the National Elk Refuge near Jackson.

Population: National Breeding Bird Survey trend data are not available, but the North American population probably exceeded 30,000 birds as of 2009. The Rocky Mountain population (estimated at 3,700 birds in 2000) extends from Canada's Yukon region southeast to central Alberta. There is also a resident population in eastern Oregon, and one in the Greater Yellowstone region, including Red Rock Lakes National Wildlife Refuge. The latter population has been in serious decline, and by 2009 fewer than 400 birds were present in the Greater Yellowstone region. Owing to its overall current status, the species has been removed from the national list of threatened and endangered species.

Tundra (Whistling) Swan (*Cygnus columbianus*)

Status: A migrant throughout, ranging in abundance from rare to fairly common. Probably more common in northern areas, and especially in prairie marshes rather than in forested areas, where trumpeter swans are more likely to occur.

Habitats and Ecology: On migration, tundra swans frequent favored stopover points that are used every year on their way to and from arctic nesting grounds. These are usually shallow marshes rich in submerged vegetation, which is the major food source for these birds. Field-feeding on dry land has also been observed rarely in migrating birds.

Suggested Viewing Locations: Good Wyoming localities for seeing this species during migration include Yellowstone Park, Ocean Lake in Fremont County, and the Wheatland area (Dorn and Dorn, 1990). Most Wyoming records are from the northwestern corner of the state (Faulkner, 2010).

Population: National Breeding Bird Survey trend data are not available. By 2009 the Atlantic and Pacific flyway populations were estimated to total about 100,000 birds each (U.S.F.W.S., 2009a).

Wood Duck (*Aix sponsa*)

Status: A local summer resident in much of the region, at least from Yellowstone Park northward. To the south it is a local migrant, with rare breeding in eastern Colorado.

Habitats and Ecology: During the breeding season these birds are found among woodlands having fairly large trees offering nesting holes, and frequently also those having acorns or similar nutlike foods in abundance. Even outside the breeding season the birds are usually associated with flooded woodlands rather than open marshes. In the interior Northwest, wood ducks begin breeding in mixed coniferous forests during the mature forest stage of succession (Sanderson, Bull and Edgerton, 1980).

Suggested Viewing Locations: Wooded streams such as the Laramie, North Platte, Belle Fourche and Shoshone Rivers in Wyoming attract wood ducks (Dorn and Dorn, 1990), as do similar Plains rivers in Montana and Colorado. In Wyoming they are common to uncommon breeders along the eastern flank of the Bighorns, the Wind River and North Platte, and on streams near the Black Hills (Faulkner, 2010). The densest regional breeding concentrations are probably in Montana (*e.g.*, Bowdoin, Les Metcalf and Medicine Lake national wildlife refuges). and Idaho (Deer Flat and Kootenai national wildlife refuges).

Population: National Breeding Bird Survey trend data indicate a significant annual population increase of 3.4%. Trend estimates for the Rocky Mountains

& Plains States region (U.S.F.W.S. Region 6) indicate a significant annual increase of 6.7%. Recent population estimates are of about 2,800,000 wood ducks in eastern regions of North America, 665,000 in central regions, and 66,000 in western regions.

Gadwall (*Anas strepera*)

Status: A widespread but only occasional nester in the montane parks, becoming much more common on the prairie marshes to the east. It is relatively rare in the Canadian mountains.

Habitats and Ecology: This prairie-adapted dabbling duck prefers shallow marshes with grassy or weedy nesting cover, especially where islands are present.

Suggested Viewing Locations: This species is widespread throughout the region during migration. The densest regional breeding concentrations are probably in the plains of northern Montana (*e.g.*, Bowdoin and Medicine Lake national wildlife refuges) and southeastern Idaho (Bear Lake and Camas national wildlife refuges).

Population: National Breeding Bird Survey trend data indicate a significant annual population increase of 3.2%. Trend estimates for the Rocky Mountains & Plains States region (U.S.F.W.S. Region 6) indicate a significant increase of 5.0%. North American breeding grounds surveys in 2009 indicated a total population of 3.05 million birds, 71% above the long-term average (U.S.F.W.S., 2009a).

American Wigeon (*Anas americana*)

Status: A common to rare summer visitor throughout the region, but less common in the montane parks than on the plains, and relatively rare in the Canadian parks, at least in summer.

Habitats and Ecology: Associated with relatively open marshes and lakes having abundant aquatic vegetation at or near the surface, and in the breeding season favoring areas with sedge meadows or with shrubby or partially wooded habitats nearby. Wigeons are strongly vegetarian, and spend more time grazing grassy vegetation than do most ducks.

Suggested Viewing Locations: This species is widespread throughout the region during migration, and nests throughout Wyoming. The densest regional breeding concentrations are probably in the plains of northern Montana (*e.g.*, Bowdoin and Medicine Lake national wildlife refuges) and southern Alberta.

Population: National Breeding Bird Survey trend data indicate a nonsignificant annual population decline of 0.6%. North American breeding grounds surveys in 2009 indicated a total population of 2.47 million birds, five percent below the long-term average (U.S.F.W.S., 2009a).

Mallard (*Anas platyrhynchos*)

Status: An abundant permanent resident throughout the region, breeding nearly everywhere ponds or marshes occur, and in all of the montane parks.

Habitats and Ecology: This highly adaptable species nests on nearly aquatic habitats, but prefers non-forested areas over forested ones, and shallow waters over deeper ones. Mallards quickly locate and utilize protected areas, even when close to human activities, and thus remain common in spite of intensive hunting pressures on them.

Suggested Viewing Locations: This species can be found on regional wetlands almost anywhere throughout the year. The densest regional breeding concentrations are probably in the plains of northern Montana (*e.g.,* Bowdoin and Medicine Lake national wildlife refuges) and southern Alberta.

Population: National Breeding Bird Survey trend data indicate a significant annual population increase of 0.5%. Trend estimates for the Rocky Mountains & Plains States region (U.S.F.W.S. Region 6) indicate a significant increase of 2.5%. The 2009 North American breeding population was estimated at 8.5 million birds, 13% above the long-term average (U.S.F.W.S., 2009a).

Blue-winged Teal (*Anas discors*)

Status: A summer resident throughout the region, breeding in most if not all the montane parks. It is common on the grasslands and foothills, especially in the prairie pothole country.

Habitats and Ecology: This species favors relatively small, shallow marshes over larger and deeper ones, especially those that are surrounded by grass or sedge meadows. Migration in spring occurs fairly late, as does pair formation, but nonetheless renesting efforts are fairly common following nest failure.

Suggested Viewing Locations: This species is widespread throughout the Rocky Mountains region during migration. The densest regional breeding concentrations are probably in the plains of southern Alberta.

Population: National Breeding Bird Survey trend data indicate a nonsignificant annual population decline of 0.7%. North American breeding grounds surveys in 2009 indicated a total population of 7.4 million birds, 60% above the long-term average (U.S.F.W.S., 2009a).

Cinnamon Teal (*Anas cyanoptera*)

Status: A summer resident virtually throughout the entire region, except in the northernmost areas. In Alberta, nesting occurs north to the Tofield area, but apparently not to Banff.

Habitats and Ecology: Associated with small, shallow and often somewhat alkaline marshes of western North America, overlapping with but largely replacing the blue-winged teal in drier regions.

Suggested Viewing Locations: This species is widespread throughout the Rocky Mountains region during migration, especially westwardly. The densest regional breeding concentrations are probably in southeastern Idaho (*e.g.,* Camas and Grays Lake national wildlife refuges).

Population: National Breeding Bird Survey trend data indicate a nonsignificant

annual population decline of 1.0%. The North American population has
been estimated as 260,000 birds (Wetlands International, 2002).

Northern Shoveler *(Anas clypeata)*

Status: A summer resident essentially throughout the entire region, although
  uncommon to rare in the montane parks, and common only on wetlands of
  the plains and foothills.

Habitats and Ecology: The specialized bill of this species allows for filter-feed-
  ing of surface organisms, and submerged plants sometimes also provide a
  supply of organisms that can be reached from the surface.

Suggested Viewing Locations: This species is widespread throughout the Rocky
  Mountains region during migration. The densest regional breeding con-
  centrations are probably in the plains of northern Montana (*e.g.*, Benton
  Lake, Bowdoin and Medicine Lake national wildlife refuges) and southern
  Alberta.

Population: National Breeding Bird Survey trend data indicate a significant annual population increase of 2.3%. Trend estimates for the Rocky Mountains
  & Plains States region (U.S.F.W.S. Region 6) indicate a significant increase
  of 4.8%. North American breeding grounds surveys in 2009 indicated a
  total population of 4.38 million birds, 92% above the long-term average
  (U.S.F.W.S., 2009a).

Northern Pintail *(Anas acuta)*

Status: A year-round or summer resident throughout the region, breeding in
  suitable habitats in most and perhaps all the montane parks.

Habitats and Ecology: This is a tundra- and prairie-adapted breeding species,
  and it is rarely found in heavily wooded wetlands. It can breed on small
  and temporary ponds as well as permanent marshes, and frequently nests
  on dry land in extremely exposed situations well away from water.

Suggested Viewing Locations: This species is widespread on wetlands throughout the Rocky Mountains region during migration. The densest regional
  breeding concentrations are probably in Montana (*e.g.*, Bowdoin, Medicine
  Lake and Red Rock Lakes national wildlife refuges) and adjacent Alberta.

Population: National Breeding Bird Survey trend data indicate a nonsignificant
  annual population decline of 2.6%. However, trend estimates for the Rocky
  Mountains & Plains States region (U.S.F.W.S. Region 6) indicate a significant increase of 1.9%. North American breeding grounds surveys in 2009
  indicated a total population of 3.22 million birds, 20% below the long-term
  average (U.S.F.W.S., 2009a).

Green-winged Teal *(Anas crecca)*

Status: A summer resident over nearly the entire region, probably breeding in
  all the montane parks except perhaps Rocky Mountain, where it is rare in
  summer. It is very common on the prairie marshes and foothill areas.

Suggested Viewing Locations: Occurs throughout the Rocky Mountains region

during migration. The densest regional breeding concentrations are found
east of the Rocky Mountains in Alberta.

Population: National Breeding Bird Survey trend data indicate a nonsignificant
annual population increase of 0.9%. North American breeding grounds surveys in 2009 indicated a total population of 3.44 million green winged teal,
79% above the long-term average (U.S.F.W.S., 2009a).

Canvasback *(Aythya valisineria)*

Status: A local and uncommon to rare summer resident over much of the region, being relatively rare in the montane parks (and not known to breed in
any), and most abundant in prairie potholes and marshes.

Habitats and Ecology: In the breeding season, canvasbacks are found on shallow prairie marshes with abundant growths of emergent vegetation and
also open water areas that frequently are rich in aquatic plants such as
pondweeds.

Suggested Viewing Locations: Deeper and larger marshes, lakes and reservoirs attract this species on migration. Nesting in Wyoming is local although widespread in south-central and western areas, and probably includes Cokeville Meadows (Scott, 1993; Faulkner, 2010). The densest regional breeding
concentrations are probably in the plains of northern Montana (*e.g.*, Benton Lake, Bowdoin and Medicine Lake national wildlife refuges) and southern Alberta.

Population: National Breeding Bird Survey trend data indicate a nonsignificant
annual population decline of 0.2%. North American breeding grounds surveys in 2009 indicated a total population of about 700,000 birds, or 16%
above the long-term average (U.S.F.W.S., 2009a).

Redhead *(Aythya americana)*

Status: A summer resident over most of the region, and a locally uncommon
to rare breeder; rare in the montane parks but fairly common in prairie
marshes.

Habitats and Ecology: Breeding habitats consist of non-forested country with
water areas sufficiently deep to provide permanent, fairly dense emergent
vegetation as nesting cover. Water areas at least an acre in size are preferred
for nesting, with substantial areas of open water for taking off and landing.

Suggested Viewing Locations: This species is widespread on larger marshes and
lakes. throughout the Rocky Mountains region during migration. In Wyoming, Hutton Lake National Wildlife Refuge attracts migrating birds (Dorn
and Dorn, 1990). Nesting in Wyoming is widespread except for the eastern plains (Faulkner, 2010), and extends from Cokeville to Table Mountain
(Scott, 1993). The densest regional breeding concentrations are probably in
the plains of southeastern Idaho (*e.g.*, Bear Lake and Camas national wildlife refuges), northern Montana (Bowdoin, Lee Metcalf and Ninepipe national wildlife refuges), and southern Alberta.

Population: National Breeding Bird Survey trend data indicate a nomsignificant annual population decline of 0.9%. North American breeding grounds surveys in 2009 indicated a total population of 1.04 million birds, or 62% above the long-term average (U.S.F.W.S., 2009a).

Ring-necked Duck (*Aythya collaris*)

Status: Relatively common in woodland ponds from the Tetons northward, but rare or absent from prairie marshes during the breeding period.

Habitats and Ecology: Unlike any of its near relatives, the ring-necked duck is strongly associated with beaver ponds and other forest wetlands, where it is often among the commonest of breeding ducks. Sedge-meadow marshes and boggy areas are preferred for nesting, and the presence of water lilies and associated heather cover seem to be an important part of breeding habitats.

Suggested Viewing Locations: This species is widespread on larger marshes and lakes throughout the region during migration. It is the most common nesting duck in the Jackson Hole area of northwestern Wyoming (Scott, 1993), and breeds in the northwestern and south-central parts of that state (Faulkner, 2010). The densest regional breeding concentrations are probably in the wetlands of Montana (*e.g.*, Medicine Lake National Wildlife Refuge) and Idaho (Camas and Kootenai national wildlife refuges).

Population: National Breeding Bird Survey trend data indicate a significant annual population increase of 2.5%. Breeding grounds surveys in 2009 indicated a total population of 551,000 birds in eastern North America (U.S.F.W.S., 2009a). No estimates for the Rocky Mountain region are available.

Greater Scaup (*Aythya marila*)

Status: A relatively rare wintering migrant or vagrant throughout the region.

Habitats and & Ecology: To be expected on rather large water areas, such as lakes and deeper marshes.

Suggested Viewing Locations: This species is widespread on larger marshes and lakes throughout the Rocky Mountains region during migration.

Population: National Breeding Bird Survey trend data are not available. No continental population estimates are available for this scaup, which is not separated from lesser scaup during national surveys (see below). Winter U.S. estimates of greater scaups from 1961–2000 have ranged from 140,000–699,000,and show a consistent long-term declining trend (Kessel, Rocque and Barclay, 2002).

Lesser Scaup (*Aythya affinis*)

Status: A regular migrant and local summer resident through the region, except in the most montane or driest areas. In the montane parks, common only in Yellowstone as a breeder.

Habitats and Ecology: This is largely a prairie-adapted breeder, and is associated also with ponds in the foothill woodlands, especially those supporting good populations of amphipods and other aquatic invertebrates.

Suggested Viewing Locations: This species is widespread on larger marshes and lakes throughout the region during migration. Nesting is widespread in Wyoming except in the northeast (Faulkner, 2010), at sites such as Cokeville, Seedskadee and Table Mountain (Scott 1993). The densest regional breeding concentrations are probably in the plains of southern Idaho (*e.g.,* Camas and Minidoka national wildlife refuges), northern Montana (Benton Lake and Medicine Lake national wildlife refuges), northwestern Wyoming (Yellowstone National Park) and southern Alberta.

Population: National Breeding Bird Survey trend data indicate a nonsignificant annual population increase of 0.7%. Breeding grounds surveys in 2009 indicated a total population of 4.2 million scaups of both species, or 18% below the long-term average (U.S.F.W.S., 2009a).

Harlequin Duck (*Histrionicus histrionicus*)

Status: A local summer resident on mountain streams from the Wind River Range northward, becoming commoner to the north. An infrequent and an apparently rare breeder in the Tetons, but local on the Yellowstone River in Yellowstone National Park. Common at Glacier (Avalanche and Two Medicine Lakes, Roes Creek, Fish Creek campground) and Watertown (Watertown River) and in the rapid streams of the other Canadian parks.

Habitats and Ecology: Associated with clear, rapidly flowing streams, where aquatic insects such as caddis larvae abound; often found where dippers also occur.

Suggested Viewing Locations: McEneaney (1988) judged the Yellowstone National Park population of this beautiful diving duck at less than 20 pairs during the late 1980's. The Yellowstone River provides some traditional locations for finding these torrent-loving birds in Yellowstone National Park, such as just east of Tower Junction (but they now occur only infrequently below Fishing Bridge at LeHardy Rapids). Other reported Wyoming sites include Snake River tributaries in Grand Teton National Park, the Middle Fork (Red Fork) of the Powder River, Shell Creek in the Bighorn Mountains, and Dinwoody Creek in the Wind River Range. Harlequin ducks are easily found in various parts of northern Montana, Idaho and adjacent Canada, such as at Glacier National Park and all of the regional Canadian national parks. In Idaho the best place to see harlequins ducks is at Wilderness Gateway Campground, 40 miles east of Koksia, on the Lochas River (Svingen & Dumroese, 1997).

Population: National Breeding Bird Survey trend data are not available. There are about 165,000 breeding harlequins in western North America (Kear, 2005). Its status in Wyoming is marginal, with perhaps fewer than 60 breeding pairs estimated in 1998 (Faulkner, 2010).

Harlequin duck, adult male.
Drawing by author.

Surf Scoter *(Melanitta perspicillata)*

Status: A rare migrant or winter vagrant in the region, more frequent in northern areas.

Habitats and Ecology: Like the other scoters, this species is most likely to be encountered on large lakes or reservoirs while on migration, and full-plumaged males are almost never seen in this region.

Suggested Viewing Locations: This species occurs erratically on deeper marshes and lakes throughout the region during migration and in winter.

Population: National Breeding Bird Survey trend data are not available. Rose & Scott (1997) suggested a stable North American population of 765,000 birds.

White-winged Scoter *(Melanitta fusca)*

Habitats and Ecology: An uncommon to rare migrant. It is most likely to be seen on lakes, reservoirs and large rivers in the region, mainly during fall and winter in southern areas, but present through the summer in Alberta, especially around large prairie lakes where nesting occurs.

Comments: This is the most southerly nesting of the three scoters, and it often nests on islands of large lakes in forested as well as grassland habitats.

Suggested Viewing Locations: This species occurs erratically on deeper marshes and lakes throughout the region during migration. In Wyoming, both scoters are most likely to be seen at places such as Hutton Lake National Wildlife Refuge, Lake DeSmet (Johnson County) or Healey Reservoir (Scott 1993). Black scoters (*M. nigra*) are very rare throughout the general region.

Population: National Breeding Bird Survey trend data are not available. The North American population (*deglandi*) may consist of about 1,000,000 birds (Rose & Scott, 1997; Kear, 2005).

Long-tailed Duck (Oldsquaw) *(Clangula hyemalis)*

Status: A rare migrant or winter vagrant over much of the region; probably more common to the north.

Habitats and Ecology: Likely to be observed on reservoirs, lakes, or large rivers, usually far from shore, while on migration. During the breeding season the birds occur on arctic tundra in the vicinity of lakes, ponds, coastlines, or islands. Most wintering occurs in coastal areas, although wintering birds often use deep and large inland lakes.

Suggested Viewing Locations: This species occurs erratically on deeper marshes and lakes throughout the region during migration and in winter. In Wyoming, long-tailed ducks are likely to be seen at places such as Lake DeSmet (Johnson County) or Yellowstone Lake (Scott 1993).

Population: National Breeding Bird Survey trend data are not available. The world population of long-tailed ducks may include about 2.7 million birds in North America (Rose & Scott, 1997),

Bufflehead *(Bucephala albeola)*

Status: A local breeding summer resident from the Tetons northward, mainly in wooded wetlands where tree cavities (especially woodpecker holes) offer nesting sites.

Habitats and Ecology: This species is so small that females can use the old nest holes of flickers (which are also used by bluebirds, starlings, and similar-sized hole-nesters) for nesting. In the interior Northwest, buffleheads begin breeding in mixed coniferous forests during the mature forest stage of succession (Sanderson, Bull and Edgerton, 1980). At other times the birds are generally found on large and deep waters.

Suggested Viewing Locations: This species occurs commonly on deeper marshes and lakes throughout the region during migration and during winter. The densest regional breeding concentrations are probably in the forested wetlands of the Greater Yellowstone Ecosystem (including Park and Teton counties of Wyoming), and in similar wetlands of northern Montana (Glacier National Park) and adjacent Alberta.

Population: National Breeding Bird Survey trend data indicate a significant annual population increase of 3.1%. There is a population estimate of one million birds for all of North America (Wetlands International, 2002).

Common Goldeneye *(Bucephala clangula)*

Status: A local summer resident from western Montana northward, becoming common in Alberta, and a widespread migrant and wintering species throughout.

Habitats and Ecology: During the breeding season goldeneyes of both species are usually found in forested wetland habitats, where cavities in large trees offer nesting sites. At other times they occur on deeper and larger bodies of water such as lakes. Because of confusion with Barrow's goldeneye, breeding reports from Grand Teton and Yellowstone parks need confirmation, as do reports from the Wapati and Dubois areas of Wyoming. Faulkner (2010) does not mention any proven Wyoming breeding records.

Suggested Viewing Locations: This species occurs commonly on deeper marshes and lakes throughout the region during migration and during winter. It is Wyoming's most common wintering duck, but is not yet proven to breed there. The densest regional breeding concentrations are probably in the forested wetlands of northern Idaho (*e.g.,* Kootenai National Wildlife Refuge), northern Montana (Glacier National Park), and adjacent Canada.

Population: National Breeding Bird Survey trend data indicate a significant annual population increase of 3.7%. Eastern North American breeding grounds surveys in 2009 indicated a total population of 369,000 goldeneyes, nearly all of which would be commons (U.S.F.W.S., 2009a). The entire North American population might total about 1.5 million birds (Kear, 2005).

Barrow's Goldeneye *(Bucephala islandica)*

Status: A common breeder through the montane wetlands of the region, from central Wyoming northward. Elsewhere a common migrant or resident in most locations:

Habitats and Ecology: Breeding birds are associated with forested montane lakes, beaver ponds, and slowly flowing rivers in this region; elsewhere nesting in cliff or rock crevices also is frequent.

Suggested Viewing Locations: This species occurs commonly on deeper marshes and lakes throughout the region during migration and during winter. Nesting is very common in the Jackson Hole area of Wyoming, and occurs throughout the Greater Yellowstone ecosystem. It also breeds locally in northwestern Colorado (Flat Tops Wilderness area). The densest regional breeding concentrations are probably in the forested wetlands of the Greater Yellowstone region and similar wetlands in northern Montana (*e.g.,* Swan River National Wildlife Refuge, Glacier National Park) and adjacent Canada.

Population: National Breeding Bird Survey trend data are not available. Minimum estimated population totals exist for Alaska (45,000), British Columbia (70,000–126,000) and the Pacific Coast states (under 8,000) (Kear, 2005); smaller but unknown numbers also occur along the Rocky Mountains

Hooded Merganser (*Lophodytes cucullatus*)

Status: A local summer resident in wooded areas from western Montana northward; southern breeding limits uncertain. Elsewhere a generally rare migrant, although breeding has been reported for the Powell vicinity of Colorado, and there is a single Wyoming breeding record.

Habitats and Ecology: Generally found in river areas bounded by woods and supporting good fish populations associated with clear water.

Suggested Viewing Locations: This species occurs commonly on deeper marshes and lakes throughout the region during migration and more rarely during winter. The densest regional breeding concentrations are probably in the forested wetlands of northern Montana (*e.g.,* Lee Metcalf and Swan River national wildlife refuges; Glacier National Park).

Population: National Breeding Bird Survey trend data are not available, nor are any continental population estimates. Rangewide population estimates are not available, but hunter harvest data suggest a population during the 1990's of at least 270,000–385,000 (Dugger, Dugger and Fredrickson, 1994).

Red-breasted Merganser (*Mergus serrator*)

Status: A rather uncommon to rare migrant over most of the region, with a few scattered breeding records.

Habitats and Ecology: Generally found in similar habitats as the common merganser, but with a more northerly breeding distribution and a more coastal wintering distribution.

Suggested Viewing Locations: This species occurs commonly on deeper marshes, rivers and lakes throughout the region during migration. In Wyoming, Ocean Lake (Fremont County) is a favorite location for migrants (Scott,

Hooded merganser, adult male,
Drawing by author.

1993). There is a reported breeding from the Dubois area (Luce *et al.*, 1997), but Faulkner (2010) considers Wyoming breeding as hypothetical.

Population: National Breeding Bird Survey trend data are not available. World population estimates of this Holarctic species include 237,000 birds in North America (Rose & Scott, 1997; Kear, 2005).

Common Merganser *(Mergus merganser)*

Status: A common resident in most montane rivers during the breeding season, and extending out into non-forested rivers, lakes and reservoirs at other times.

Habitats and Ecology: This fish-eating species occurs in areas of clear water supporting large fish populations, and is much the commonest merganser of the region. Nesting occurs in tree cavities, rock crevices, and sometimes under boulders or dense shrubbery. Breeding occurs from the mountains of western Colorado northward.

Suggested Viewing Locations: This species occurs commonly on most deeper marshes, rivers and lakes throughout the region during migration and winter. Many regional rivers and lakes are also used for nesting. The densest regional breeding concentrations are probably in the forested wetlands of the Greater Yellowstone region, and similar wetlands in northern Montana (*e.g.*, Swan River National Wildlife Refuge, Glacier National Park) and southern Alberta.

Population: National Breeding Bird Survey trend data indicate a significant annual population increase of 2.8%. World population estimates of this Holarctic species include 640,000 birds in North America (Kear, 2005).

Ruddy Duck *(Oxyura jamaicensis)*

Status: An occasional to rare summer resident over much of the region, mainly in grassland marshy habitats, becoming rarer in montane areas, and a migrant more or less throughout. Breeding in the montane parks is known only for Grand Teton National Park.

Habitats and Ecology: Non-breeding birds are found on larger and generally deeper waters that have silty or muddy bottoms; breeding is on overgrown shallow marshes with abundant emergent vegetation and some open water.

Suggested Viewing Locations: This species occurs commonly on deeper marshes and lakes throughout the region during migration. In Wyoming, nesting occurs almost statewide on lower altitude (below 8,000 feet) wetlands having abundant emergent vegetation. The densest regional breeding concentrations are probably in the marshes east of the Rockies in Idaho (*e.g.*, Camas Lake National Wildlife Refuge), Montana (Medicine Lake National Wildlife Refuge) and Alberta.

Population: National Breeding Bird Survey trend data indicate a nonsignificant annual population increase of 0.5%. Trend estimates for the Rocky Mountains & Plains States region (U.S.F.W.S. Region 6) indicate a significant increase of 5,1%. The North American population has been estimated at about 500,000 birds (Kear, 2005).

New World Quails

Northern Bobwhite (*Colinus virginianus*)

Status: Highly local in the region, as a result of introductions, and absent from the montane parks.

Habitats and Ecology: Generally associated with brushy edge areas, where a combination of grassy, shrubby, and woody cover all occur in close proximity, and where water can usually be found nearby. Eastern Colorado and southeastern Wyoming (North Platte and Laramie valleys) represent the limits of the natural range of this species.

Suggested Viewing Locations. There are few places in Wyoming where bobwhites can be reliably found, but the North Platte and Laramie valleys near the Nebraska border provide the best possibilities. Idaho and Montana are both beyond the bobwhite's native range. The densest regional breeding populations probably occur in the mixed-grass prairies of northeastern and southeastern Colorado. Colorado's Prewett Reservoir near Sterling offers good viewing possibilities (Kingery, 2007).

Population: National Breeding Bird Survey trend data indicate a nonsignificant annual population decline of 3.0%. Trend estimates for the Rocky Mountains & Plains States region (U.S.F.W.S. Region 6) indicate a significant annual decline of 1.3%. The North American population (north of Mexico) has been estimated at 7.5 million birds (Rich *et al.*, 2004).

Partridges, Pheasants & Grouse

Gray Partridge (*Perdix perdix*)

Status: Widespread in the prairies and agricultural lands of the region, but rarely entering the montane areas. Only a vagrant in the montane parks except for the Jackson Hole area, where possibly a rare resident.

Habitats and Ecology: Generally associated with grainfields and nearby edge habitats, such as shelterbelts, but also sometimes extending into sagebrush areas, especially where local water supplies are present. Nesting usually occurs in grainfields or hayfields, under grassy or herbaceous cover.

Suggested Viewing Locations: In Wyoming, agricultural fields and grasslands around Sheridan, Lovell and Tensleep attract this species (Dorn and Dorn, 1990), which is most common in Sheridan and Fremont counties (Faulkner, 2010). Irrigated areas such as around Wyoming's Bighorn Mountains, support good populations (Scott, 1993). The densest regional breeding concentrations are probably in the grasslands and meadows of Idaho (*e.g.*, Minidoka National Wildlife Refuge), Montana (C. R. Russell National Wildlife Refuge, National Bison Range) and eastern Alberta.

Population: National Breeding Bird Survey trend data indicate a nonsignificant annual population decline of 0.1%.

Chukar (*Alectoris chukar*)

Status: A local permanent resident in arid lands of southern Idaho and Wyoming; present in the montane parks only as a vagrant. Once casual in Rocky Mountain National Park, but apparently not reported since 1957. It is possibly a rare resident in the Jackson Hole area.

Habitats and Ecology: Primarily associated with sagebrush habitats during the breeding season, but extending into grassland and sometimes also riparian habitats at other times. Nearly always found in hilly, rocky topography; unlike bobwhites the birds tend to flee rapidly on foot upslope rather than fly when frightened.

Suggested Viewing Locations: In Wyoming, this introduced species may be found in many arid, rocky areas, such as around Copper Mountain (Fremont County), the Bighorn Basin, and in Hot Spring State Park in Hot Springs County (Dorn and Dorn, 1990; Scott, 1993). Dense regional breeding concentrations also occur in the rocky scrublands of southwestern Wyoming (*e.g.,* Hagerman Wildlife Management Area and in the Green River valley) and south-central Idaho (City of Rocks National Reserve).

Population: National Breeding Bird Survey trend data indicate a significant annual population increase of 7.4%. Trend estimates for the Rocky Mountains & Plains States region (U.S.F.W.S. Region 6) indicate a significant increase of 7.1%.

Ring-necked Pheasant (*Phasianus colchicus*)

Status: A widespread resident throughout the general region, but rare or accidental in the montane parks, and mostly limited to low-altitude grasslands, croplands, and similar non-wooded environments.

Habitats and Ecology: Breeding occurs mainly in native grasslands and grain croplands or their edges, but sometimes also in marsh edges, hayfields, or shelterbelts, as well as roadside ditches.

Suggested Viewing Locations: In Wyoming, good birding locations include irrigated croplands the vicinities of Torrington, Hawk Springs, Sheridan and Lovell (Dorn and Dorn, 1990), and also areas around Riverton, Ocean Lake (Fremont County) and the Bighorn Basin (Scott, 1993). Southern Idaho supports good pheasant populations, in locations such as at Tex Creek Wildlife Management Area (16 miles east of Idaho Falls) and Sterling Wildlife Management Area (four miles east of Aberdeen (Svingen & Dumroese, 1997). The densest regional breeding concentrations are probably in the native grasslands and grainfields of Montana (*e.g.,* Bowdoin, Lee Metcalf and Ninepipe national wildlife refuges).

Population: National Breeding Bird Survey trend data indicate a nonsignificant annual population decline of 0.6%.

Spruce Grouse (*Falcipennis canadensis*)

Status: A permanent resident from southwestern Montana northward, mainly in montane forests. Although reported earlier for Yellowstone National Park, there have been no records for several decades, and any sightings are likely to be of vagrants.

Habitats and Ecology: Largely limited to coniferous forests throughout the year. In the interior Northwest, spruce grouse begin breeding in mixed coniferous forests during the pole-sapling stage of succession (Sanderson, Bull and Edgerton, 1980).Conifer needles are their primary food source, supplemented by insects and berries in late summer.

Suggested Viewing Locations: In Montana, spruce-fir forests in Glacier National Park (such as the Pike Creek logging road near Summit Siding) offer likely places for seeing spruce grouse which, unlike they are displaying, are hard to locate in shady woods. However, displaying males are often fearless, and might be approached at times almost to arm's length. In Idaho the western Selkirk Mountains (north of Priest River), and around Upper Payette Lake in the Payette National Forest provide good viewing possibilities (Svingen & Dumroese, 1997).

Population: National Breeding Bird Survey trend data are not available. The North American population has been estimated at 1.2 million birds (Rich *et al.,* 2004).

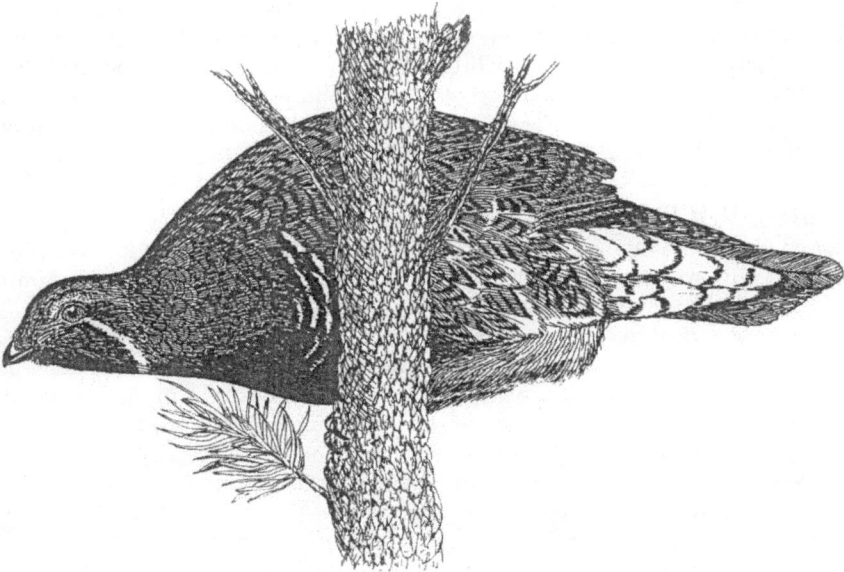

Spruce grouse, adult male,
Drawing by author.

Dusky (Blue) Grouse (*Dendragapus obscurus*)

Status: Relatively widespread in coniferous forests throughout the region, from Colorado to Alberta, breeding in all the montane parks. In Wyoming, present in the major ranges except the Black Hills (where extirpated). Recently separated specifically from a more western population, now called the sooty grouse, because of many plumage and behavioral differences.

Habitats and Ecology: Closely associated with coniferous forests, but also reaching alpine timberline during the breeding season, and found as low as the ponderosa pine zone. In the interior Northwest, dusky grouse begin breeding in mixed coniferous forests during the shrub-seedling stage of succession (Sanderson, Bull and Edgerton, 1980). Nonbreeders move in late spring and summer into the sagebrush zone.

Suggested Viewing Locations: In Wyoming, birding locations include coniferous woods throughout the state except the northern corner. Rocky Mountain, Grand Teton, Yellowstone and Glacier National Parks all support good populations of dusky grouse, and from spring to early summer "hooting" periods the males' low pitched display calls can often be heard and the birds thus located. They are hard to find when perched in trees, but actively displaying males are often quite fearless and can be easily approached. The densest regional breeding concentrations are probably in the montane forests of western Montana (*e.g.*, National Bison Range, Red Rock Lakes National Wildlife Refuge) and eastern and northern Idaho (Grays Lake National Wildlife Refuge, Western Selkirk Mountains north of Priest Lake).

Population: Classified as a Partners-in-Flight Watch List species of continental conservation importance (Rich et al., 2004). National Breeding Bird Survey trend data indicate a nonsignificant declining annual population trend of 2.1%. The North American population has been estimated at 2.6 million birds (Rich *et al.*, 2004).

White-tailed Ptarmigan (*Lagopus leucurus*)

Status: Occurs in alpine areas from Alberta southward more or less continuously to western Montana, and on scattered alpine areas farther south throughout the region.

Habitats and Ecology: Limited to the alpine and timberline zones, moving slightly lower during winter, especially where willows remain exposed above the snow. In Alberta the birds sometimes descend to mountain valleys in severe winters, where they may feed on seeds and waste grain.

Suggested Viewing Locations: The rocky tundra along Trail Ridge Road in Rocky Mountain National Park, and Summit County in western Colorado are good places to look for this very hard-to-see species (Kingery, 2007). Other favorable Colorado sites for finding these birds during summer include Guanella Pass, near Georgetown, the alpine trails to nearby Grays and Torrey's Peaks, the Mount Evans Highway, Independence Pass near Twin Lakes, and Twin Lakes Reservoir (Young, 2000). In Idaho, alpine trails in the Selkirk Range are among the few areas where this species might be encountered (Svingen and Dumroese, 1997). In Wyoming, Quadrant Mountain in Yellowstone Park and Ishawooa Mesa in Park County have had reported sightings (Dorn and Dorn, 1990), as has Medicine Bow peak, but documented sightings from the Snowy Range are fairly old (Faulkner, 2010). The most extensive regional breeding concentrations are probably in the alpine zones of Colorado and the Canadian montane parks.

Population: National Breeding Bird Survey trend data are not available. The North American population has been estimated at two million birds (Rich *et al.,* 2004).

White-tailed ptarmigan, adult male, winter plumage.
Drawing by author.

Ruffed Grouse (*Bonasa umbellus*)

Status: Widespread and relatively common in wooded areas from southwestern Wyoming (Uinta County) northward. Absent from Colorado, but fairly common over most of Idaho, except for the arid southwestern corner. In Wyoming, present in the Black Hills, Uintas, and all the major ranges except the Sierra Madre, Medicine Bow and Laramie ranges, and most common west of the continental divide.

Habitats and Ecology: Especially associated with aspen woodlands, the buds and catkins of which provide a major food source. However, it also occurs up to the spruce-fir zone of coniferous forest. Nesting is often in or near aspen clumps.

Suggested Viewing Locations: Deciduous woods, especially aspen groves, are excellent places to find ruffed grouse, In Wyoming, the Pacific Creek area of Grand Teton National Park is recommended (Dorn and Dorn, 1990), as are the Valley Trail and Two Ocean Lake (Scott, 1993). The densest regional breeding concentrations are probably in aspen–poplar groves or mixed aspen–conifer forests from the Greater Yellowstone region north to Jasper National Park.

Population: National Breeding Bird Survey trend data indicate a nonsignificant annual population decline of 0.6%. The North American population has been estimated at 8.3 million birds (Rich *et al.*, 2004).

Greater Sage-grouse (*Centrocercus urophasianus*)

Status: Relatively common on sage habitats in the plains and foothills; rare or absent in the montane parks except for the Jackson Hole area of Grand Teton National Park. In Wyoming, breeding has been reported from all counties.

Habitats and Ecology: Closely associated with sagebrush, which is the primary food and also is used for nesting cover. Occurs locally in sage to 9,000 feet elevation.

Suggested Viewing Locations: The range of this species is inexorably decreasing, as sagebrush areas are being cleared and converted to irrigated cultivation. However, Wyoming still supports the nation's largest sage-grouse population. Sage-grouse display socially in spring, with as many as 50 or more males "strutting" in local display grounds, or leks. The densest regional breeding concentrations are probably in the sagebrush scrublands of south-central and southwestern Wyoming. Grand Teton National Park has long had a display ground near the Jackson Airport, but numbers have been low in recent years. Hat Six Road near Casper is an easily reached site from which birds can be seen from a parked vehicle. The Farson area, along the road to Big Sandy Reservoir, is perhaps the best location in Wyoming for watching display. From Casper I-25 exit 182, go south on Wyo. Hwy. 253 for 8.2 miles to Natrona Co. Rd. 605, then go left to lek a few hundred yards on the south side (Scott, 1993). A major threat to Wyoming sage-grouse is the proposed 98,000-acre wind-turbine farm "Chokecherry and Sierra Madre Wind Energy Project" south of Rawlins in the Red Desert, which would result in the building of about 470 miles of roads. Sites having significant greater sage-grouse populations in western Colorado include Brown's Park National Wildlife Refuge. About 75% of Colorado's greater sage-grouse population occurs n Moffat County. In Montana one of the state's bigger leks may be seen near Big Sheep Creek (Fischer and Fischer, 1995). In Idaho, Sand Creek Wildlife Management Area (208/624-7065), nine miles north of St. Anthony, is one of the state's best areas for watching leks of both sage-grouse and sharp-tailed grouse (Svingen & Dumroese, 1997).

Population: Classified as a Partners-in-Flight Watch List species of continental conservation importance (Rich et al., 2004). National Breeding Bird Survey trend data indicate a nonsignificant annual population decline of 1.6%. However, trend estimates for the Rocky Mountains & Plains States region (U.S.F.W.S. Region 6) indicate a significant increase of 3.1%. The North American population has been estimated at 150,000 birds, of which 80% breed in the Intermountain West Avifaunal Biome (Rich *et al.*, 2004).

Greater sage-grouse, adult male in strutting display.
Drawing by author.

Sharp-tailed Grouse (*Tympanuchus phasianellus*)

Status: Widespread on plains and foothills, but rare or absent in the montane parks except at Watertown Lakes National Park, where local on grassland areas.

Habitats and Ecology: Associated with grasslands and grassy sagebrush areas, and sometimes also mountain meadows during the breeding season, and extending into cultivated fields during fall and winter. Brushy foothills and similar edge habitats are often used in Alberta.

Suggested Viewing Locations: In spring, male sharp-tails "dance" on traditional lek display areas in groups of from a few to 20 or more males, during which dominance is determined and the relative access of males to females for fertilization is established. These activities begin in late winter, and may continue until May. In Wyoming, grasslands around Sheridan, the Four Corners area of Weston County and the rim of Goshen Hole in Laramie and Goshen counties are good locations for locating leks (Dorn and Dorn, 1990). Other good viewing locations include Wagon Box Road south of Story, Bird Farm Road south of Big Horn, and along State Road 335 west of Big Horn (Scott, 1993). Pine Butte Swamp Preserve, a Nature Conservancy property near Choteau, Montana, and Blackleaf Wildlife Management Area, near Bynum, are among the many sites in central Montana where this grouse may be seen displaying in spring (Fischer & Fischer, 1995). In southeastern Idaho, Sand Creek Wildlife Management Area (see previous account), Daniels Reservoir in Bannock County, and Tex Creek Wildlife Management Area (208/525-7290) in Bonneville County provide viewing opportunities for sharp-tailed grouse, but the best location may be in Curlew National Grassland, in Oneida County. Information on Idaho's lek locations may be obtained from the Idaho Fish & Game Dept. (208/232-4703, or the Forest Service office (208/766-4743) (Svingen & Dumroese, 1997). The densest regional breeding concentrations are probably in northeastern Montana.

Population: National Breeding Bird Survey trend data indicate a nonsignificant annual population increase of 1.0%. The North American population has been estimated at 1.2 million birds (Rich *et al.*, 2004). The westernmost Columbian race (*P. p. columbianus*) of this species is extremely rare if not wholly extirpated in Montana, and its Natural Heritage status category in Wyoming is critically imperiled, and is probably now limited to western Lincoln and Teton counties, and Yellowstone National Park (Faulkner, 2010).

Wild Turkey (*Meleagris gallopavo*)

Status: Locally present in the region as a result of introduction efforts, mainly in open forests of ponderosa pines or mixed woods, especially those with oaks or other mast-bearing trees. Virtually absent from the montane parks. In Wyoming, present in the northern half of the state, from Park southeast to Goshen, and probably locally elsewhere, as a result of introductions.

Suggested Viewing Locations. In Wyoming, the western Black Hills area of Crook County is a good birding location. There are also good populations in the

Bighorns, in the ranges near Casper and Laramie, and across all of north-eastern Wyoming. The densest regional breeding concentrations are proba-bly in northeastern Wyoming and adjacent southeastern Montana.

Population: National Breeding Bird Survey trend data indicate a significant an-nual population increase of 12.6%. Trend estimates for the Rocky Mountains & Plains States region (U.S.F.W.S. Region 6) indicate a significant annual in-crease of 15%, one of the highest rates of increase for any native species. The North American population has been estimated at 1.3 million birds (Rich *et al.*, 2004), but the U.S. population probably now exceeds seven million birds. Matings with domesticated turkeys have blurred range limits.

Loons

Red-throated Loon (*Gavia stellata*)

Status: A vagrant throughout the region, varying from rare to accidental.

Habitats and Ecology: Like the other loons, this species is found on larger rivers and lakes that support good fish populations.

Suggested Viewing Locations: During migration larger Wyoming lakes such as Yellowstone Lake, Lake DeSmet (Johnson County), and Keyhole Reservoir (Crook County) sometimes attract this species (Dorn and Dorn, 1990). Rare and erratic occurrences have also been noted in other parts of the Rocky Mountain region.

Population: National Breeding Bird Survey trend data are not available.

Pacific Loon (*Gavia arctica*)

Status: This is a vagrant species throughout the region, and is rare to accidental in the region.

Habitats and Ecology: To be expected on large rivers and lakes supporting good fish populations.

Suggested Viewing Locations: What has been written for the red-throated loon applies equally to this species.

Population: National Breeding Bird Survey trend data are not available.

Common Loon (*Gavia immer*)

Status: A local breeder on large montane lakes.

Habitats and Ecology: Breeding typically occurs on clear and sometimes deep mountain lakes, where fish are abundant, human disturbance is at a min-imum, and where small islands provide nest sites. In some areas muskrat houses or similar artificial islands may also be used.

Suggested Viewing Locations McEneaney (1988) judged the Yellowstone Park population to be less than 15 pairs during the late 1980's. Yellowstone Lake is probably the best location for finding breeding loons in that park, espe-cially its southwestern arm (McEneaney, 1988; Scott, 1993). All but seven of Wyoming's 27 known breeding sites are from Yellowstone Park (Faulkner,

2010). Migrating loons may be seen at the sites mentioned above for the other loons, especially Lake DeSmet (Johnson County) and some of the larger and deeper reservoirs. Nesting loons may also be seen on Montana's Seeley Valley's chain-of-lakes driving route, or along the Clearwater Canoe Trail, both near Seeley Lake, Montana (Fischer and Fischer, 1995). In Idaho, loons nest along the Coeur d'Alene River's chain of lakes in Kootenai County, between Rose Lake and St. Maries (Svingen and Dumroese, 1997). The densest regional breeding concentrations are probably in Jasper Park.

Population: National Breeding Bird Survey trend data indicate a significant annual population increase of 2.1%. In Wyoming, this is a Priority I conservation species, and has a Natural Heritage Program status of rare.

Grebes

Pied-billed Grebe (*Podilymbus podiceps*)

Status: A summer resident on overgrown marshes nearly throughout the region, except in high mountain and semi-desert areas. Uncommon to rare in the mountain parks; not yet reported from Yoho and Kootenay parks.

Habitats and Ecology: Primarily found on heavily overgrown ponds or marshes, especially during the breeding season.

Suggested Viewing Locations: During migration, many marshes and lakes throughout the region attract this species, such as Wyoming's Hutton Lake National Wildlife Refuge, Lovell Lakes in Big Horn County, Goldeneye Wildlife Area in Natrona County, and Cliff Graham Reservoir in Uinta County (Dorn and Dorn, 1990). The densest regional breeding concentrations are perhaps in eastern Montana (Bowdoin and Medicine Lake national wildlife refuges), but apparent breeding densities of this inconspicuous, non-gregarious and territorial species are relatively low everywhere.

Population: National Breeding Bird Survey trend data indicate a significant annual population increase of 1.35%. Trend estimates for the Rocky Mountains & Plains States region (U.S.F.W.S. Region 6) indicate a significant annual increase of 4.9%.

Horned Grebe (*Podiceps auritus*)

Status: Breeds in varying abundance in the northern parts of the region, mainly on marshes, ponds and shallow lakes. Has bred on Swiftcurrent Lake, Glacier National Park. In Wyoming it may once have bred at Yellowstone National Park, and in 1978 a pair was found at nearby Beck Lake (Luce *et al.*, 1997).

Habitats and Ecology: Breeding typically occurs on fairly small ponds (under 2.5 acres), with the pairs scattered and nesting in clumps of rather sparse emergent vegetation.

Suggested Viewing Locations: During migration, marshes and lakes throughout the region attract this species, such as Wyoming's Hutton Lake National

Wildlife Refuge, Grayrocks Reservoir in Platte County and Lake DeSmet in Johnson County (Dorn and Dorn, 1990). The densest regional breeding concentrations are probably in northern Montana (Swan Lake and Medicine Lake national wildlife refuges) and southern Alberta.

Population: National Breeding Bird Survey trend data indicate a significant annual population decline of 3.3%. Trend estimates for the Rocky Mountains & Plains States region (U.S.F.W.S. Region 6) indicate a stable population, with a trend of 0.0%.

Red-necked Grebe (*Podiceps grisegena*)

Status: Limited as a breeder to the northwestern portion of our region, with park breeding records so far available only for Banff, where it is also regular on the Bow River in spring. It has also been found nesting near Jasper and Watertown lakes, south into western Montana and southeastern Idaho. On the lakes of central Alberta it is probably the commonest diving bird, even in areas of rather heavy human use.

Habitats and Ecology: Associated with larger ponds and lakes during the breeding season, and nesting either as solitary pairs or in loose colonies.

Suggested Viewing Locations: During migration, marshes and lakes throughout the region attract this species. A few might be found at Wyoming sites such as Yellowstone Lake, Lake DeSmet (Johnson County) and Keyhole Reservoir (Crook County) (Dorn and Dorn, 1990). The densest regional breeding concentrations are probably in Idaho (*e.g.,* Kootenai and Deer Flat national wildlife refuges), northwestern Montana (Ninepipe National Wildlife Refuge) and southern Alberta.

Population: National Breeding Bird Survey trend data indicate a nonsignificant annual population increase of 0,4%.

Eared Grebe (*Podiceps nigricollis*)

Status: A generally widespread breeder in the region, except in the montane lakes. The only park for which there are breeding records is Yellowstone, here breeding has occurred on many lakes and ponds. Occasional breeding in other parks is probable, at least in the foothill areas.

Habitats and Ecology: Associated in the breeding season with rather shallow marshes and lakes having extensive reedbeds and submerged aquatic plants. Generally found in larger and more open ponds than either pied-billed grebes or horned grebes and, unlike these species, typically nesting in large colonies.

Suggested Viewing Locations: During migration, marshes and lakes throughout the region attract this species, such as Wyoming's Hutton Lake National Wildlife Refuge, Grayrocks Reservoir in Platte County, Goldeneye Wildlife Area in Natrona County, and Cliff Graham Reservoir in Uinta County (Dorn and Dorn, 1990). Nesting in Wyoming occurs locally at sites such as Loch Katrine (Park County) (near Cody)(Scott, 1993). The densest regional breeding concentrations are probably in the prairie marshes of northern Montana (Benton Lake and Medicine Lake national wildlife refuges) and southern Alberta.

Population: National Breeding Bird Survey trend data indicate a significant annual population increase of 3.9%. Trend estimates for the Rocky Mountains & Plains States region (U.S.F.W.S. Region 6) indicate a significant annual increase of 6.5%.

Western Grebe (*Aechmophorus occidentalis*) and Clark's Grebe (*Aechmophorus clarkii*)

Status: Both species are local summer breeders on lowland marshes over much of the entire region, with breeding records in the parks apparently confined to Yellowstone (no current evidence of breeding) and Glacier (historic record for North Fork of Flathead River).

Habitats and Ecology: Breeding typically occurs on permanent ponds and shallow lakes that are often slightly brackish and have large areas of open water as well as semi-open growths of emergent vegetation.

Suggested Viewing Locations: During migration, marshes and lakes throughout the region attract both of these two closely related species, such as Wyoming's Hutton Lake National Wildlife Refuge, Grayrocks Reservoir in Platte County, Flaming Gorge Reservoir in Sweetwater County, and Lovell Lakes in Big Horn County (Dorn and Dorn, 1990). In Wyoming, Clark's grebes have been found nesting in Sweetwater, Bighorn and Fremont counties, and westerns nest in these counties and several others, the largest colonies being in Big Horn, Carbon, Fremont, Park, and Uinta (Faulkner, 2010). Cascade Reservoir near Donelly, Colorado, is a major nesting area for both species. Both also nest at Deer Flat National Wildlife Refuge near Nampa, and Market Lake Wildlife Management Area near Roberts, Colorado. Both species also occur at Lake Cheraw, and at Nee Noshe and Neeso Pah reservoirs in southeastern Colorado. The densest regional breeding concentrations are probably in southeastern and southwestern Idaho (*e.g.*, Minidoka and Bear Lake national wildlife refuges) and Montana (Medicine Lake National Wildlife Refuge).

Population: National Breeding Bird Survey trend data (combined species) indicate a significant annual population increase of 1.1%.

Pelicans

American White Pelican (*Pelecanus erythrorhynchus*)

Status: Yellowstone Park is the only regional national park to support a breeding colony of this species, but it is regularly seen in the Teton area as well, especially on Jackson Lake and the adjoining Snake River.

Habitats and Ecology: Associated with lakes and rivers having large fish populations that can be reached by surface-feeding. Gregarious, typically foraging and nesting in groups, and sometimes foraging well away from the nesting grounds, which are typically low islands. The Molly Islands on the southern part of Yellowstone Lake are small and low islands, with an extremely limited nesting area that is often subject to high wave effects.

Suggested Viewing Locations: During migration, marshes, reservoirs and lakes throughout the region attract this species, especially those with good fish populations. During summer In Wyoming, non-breeders may appear at almost any large reservoir (Dorn and Dorn, 1990), such as Ocean Lake, Wheatland Reservoir, Keyhole Reservoir and Jackson Lake (Scott, 1993). Nesting occurs on Yellowstone Lake, Pathfinder Reservoir (Natrona County) and Bamforth National Wildlife Reservation (Albany County). The densest regional summer concentrations are in central Montana (Medicine Lake National Wildlife Refuge has the region's largest nesting colony), and eastern Idaho (*e.g.,* Minidoka. Camas and Bear Lake national wildlife refuges).

Population: National Breeding Bird Survey trend data indicate a significant annual population increase of 2.2%.

Cormorants

Double-crested Cormorant *(Phalacrocorax auritus)*

Status: Generally a migrant only in the mountain parks; only Yellowstone National Park has a few breeding birds, on the Molly Islands of Yellowstone Lake.

Habitats and & Ecology: Associated with lakes and rivers with good fish populations, often nesting on islands or on cliffs, and sometimes also in trees.

Suggested Viewing Locations: During migration, marshes, reservoirs and lakes throughout the region attract this species, especially those with good fish populations. Nesting in Wyoming occurs at Pathfinder Reservoir, Ocean Lake, and elsewhere across the state except in the arid southwest. The densest regional breeding concentrations occur in eastern Idaho (*e.g.,* Minidoka National Wildlife Refuge and other impoundments) and Montana (Benton Lake and Medicine Lake national wildlife refuges).

Population: National Breeding Bird Survey trend data indicate a significant annual population increase of 4.8%. Trend estimates for the Rocky Mountains & Plains States region (U.S.F.W.S. Region 6) indicate a significant annual increase of 10.2%.

Bitterns and Herons

American Bittern *(Botaurus lentiginosus)*

Status: A widespread but rather inconspicuous breeder, with a few montane nesting records for Banff, Yellowstone and Grand Teton parks. Breeds locally elsewhere, especially along overgrown edges of beaver ponds or marshes.

Habitats and Ecology: Associated with reedbeds and other emergent marsh vegetation, and rarely observed feeding in open water in the manner of other herons. Foods include frogs, snakes, and other animal life in addition to fish, and thus the species is not limited to areas where fish occur.

Suggested Viewing Locations: There are scattered nesting records for Wyoming, such as on the Laramie Plains, Goshen Hole, and Cokeville Meadows National wildlife refuge (Faulkner, 2010). I (Johnsgard, 1982) found a nest at Christian Pond, Grand Teton National Park, and nesting also may occur at Hutton Lake National Wildlife Refuge and Table Mountain Wildlife Unit (Scott, 1993). The densest regional breeding concentrations are probably in Montana (*e.g.*, Bowdoin and Swan River national wildlife refuges) in prairie or montane marshes.

Population: National Breeding Bird Survey trend data indicate a significant annual population decline of 1.9%. In Wyoming, this is a Priority II conservation species, and it is considered rare in Colorado.

Great Blue Heron (*Ardea herodias*)

Status: Occurs throughout the region, nesting locally wherever conditions permit, but absent from high montane lakes. It nests regularly in Grand Teton, Yellowstone, and Glacier parks.

Habitats and Ecology: This species occurs in a variety of habitats supporting fish life, but usually breeds where there are trees. However, it rarely nests on the ground, on rock ledges, or among bulrushes. Large cottonwoods are a favored location for nesting colonies in the Tetons, such as in the Oxbow area.

Suggested Viewing Locations: In Wyoming, Hutton Lake National Wildlife Refuge (Albany County), Table Mountain Wildlife Unit (Goshen County), Oxbow Lake (Grand Teton National Park) and Seedskadee National Wildlife Refuge (Sweetwater County) are good places to look for this species (Dorn and Dorn, 1990). The densest regional breeding concentrations are probably in Wyoming along the Bighorn, Green, North Platte, Powder and Snake rivers (Faulkner, 2010) and in Montana (all major refuges).

Population: National Breeding Bird Survey trend data indicate a significant annual population decline of 1.3%.

Snowy Egret (*Egretta thula*)

Status: A regular but rare migrant over much of the region, but more common to the south. There are no national park breeding records, although summering birds have been seen in Yellowstone and Grand Teton parks. Breeding has occurred in Montana near Choteau and is regular during summer at Bowdoin National Wildlife Refuge. Breeding has also occurred in several Wyoming counties (Albany, Fremont, Hot Springs, Lincoln and Natrona).

Habitats and Ecology: These birds occur in a wide range of aquatic habitats, but seem to prefer somewhat sheltered locations for breeding, and often occur in company with other larger heron species. When foraging the birds are fairly active, and sometimes rush about in shallow water in an apparent attempt to flush out their prey.

Suggested Viewing Locations: In Wyoming, Hutton Lake National Wildlife Refuge (Albany County) and Table Mountain Wildlife Unit (Goshen County)
are good places to look for this species (Dorn and Dorn, 1990). Nesting occurred at Hutton Lake National Wildlife Refuge in 2005. The greatest regional breeding concentrations are probably in southeastern Idaho (*e.g.,*
Bear Lake and Minidoka national wildlife refuges).

Population: National Breeding Bird Survey trend data indicate a significant annual population increase of 4.4%.

Cattle Egret *(Bubulcus ibis)*

Status: A rare vagrant throughout the region, with no definite nesting records
for the national parks. There are non-park breeding records for at least
Idaho and Colorado. The species is gradually expanding its range in North
America. Nesting occurred at Wyoming's Hutton Lake National Wildlife
Refuge in 1996, the state's first breeding record.

Habitats and Ecology: Besides various aquatic habitats, this species also is regularly observed on agricultural lands, especially where there are cattle present. The birds forage on grasshoppers and other insects that are stirred up
by the movements of the livestock. The species is highly social, and often
nests among colonies of other herons.

Suggested Viewing Locations: This species is still too rare in Wyoming to suggest any reliable birding sites, but any pasture with grazing cattle may attract cattle egrets. Regular breeding occurs in south-central and southeastern Idaho, at locations such as at Camas National Wildlife Refuge.

Population: National Breeding Bird Survey trend data indicate a nonsignificant
annual population decline of 0.2%.

Green Heron *(Butorides striatus)*

Status: Generally a rare migrant or vagrant throughout the region, but becoming commoner to the southeast, and breeding locally in Colorado east of the
Front Range.

Habitats and Ecology: A wide variety of habitats are used by this adaptable species, which is usually found near trees but also sometimes breeds well
away from tree cover. The birds are not very gregarious, and generally are
seen as single individuals or territorial pairs. Foraging is done in shallow
water; actual baiting of the water to attract prey has been observed in this
species.

Suggested Viewing Locations: This species is too rare in the region to suggest any
reliable birding sites.

Population: National Breeding Bird Survey trend data indicate a significant annual population decline of 1.3%.

Black-crowned Night Heron *(Nycticorax nycticorax)*

Status: Generally rare to accidental in the montane parks, but much more common and widespread in prairie and semiarid areas of the region, wherever

suitable habitat occurs. The only national park where breeding has been reported is Yellowstone, but the species is rare even there.

Habitats and Ecology: This is a highly adaptable species that can use a wide variety of habitats, but in our region it is likely to be associated with shallow bulrush or cattail marshes, often well away from woodlands. The species has very large eyes and, as its name implies, often forages in dim light when it is too dark for most herons to see their prey.

Suggested Viewing Locations: In Wyoming, Hutton Lake National Wildlife Refuge (Albany County) and Table Mountain Wildlife Unit (Goshen County) are good places to look for this species (Dorn and Dorn, 1990). Both are known nesting sites (Scott, 1993); the species now nests in Wyoming only below 7,000 feet and in the southern half of the state (Faulkner, 2010). The densest regional breeding concentrations are probably in central and eastern Montana (*e.g.,* Freezeout Lake, Benton Lake and Bowdoin Lake national wildlife refuges) and southern Alberta.

Population: National Breeding Bird Survey trend data indicate a nonsignificant annual population increase of 1.2%.

Ibises

White-faced Ibis (*Plegadis chihi*)

Status: A relatively uncommon migrant in southern portions of the region, becoming rarer
northwardly. Rarely seen north of Yellowstone Park. A common local breeder at lower altitudes, and has bred at Colorado's Brown's Park National Wildlife Refuge.

Habitats and Ecology: Generally associated with freshwater or brackish marshes having an abundance of cattails, bulrushes or phragmites.

Suggested Viewing Locations: In Wyoming, Hutton Lake National Wildlife Refuge (Albany County), Hawk Springs Reservoir near La Grange and Table Mountain Wildlife Unit (Goshen County) are good places to look for this species during migration (Dorn and Dorn, 1990). Nesting occurs along the Bear River drainage, and at Hutton Lake National Wildlife Refuge (the state's largest colony), as well as a few other sites on the Laramie plains and probably as far north as Ocean Lake (Faulkner, 210). The densest regional breeding concentrations are probably in eastern and southeastern Idaho (e.g., Camas, Minidoka and Bear Lake national wildlife refuges).

Population: National Breeding Bird Survey trend data indicate a significant annual population increase of 9.4%. In Wyoming, this has been considered a Priority I conservation species and critically imperiled (Finch, 1992), but has been increasing in recent years.

American Vultures

Turkey Vulture (*Cathartes aura*)

Status: A summer resident nearly throughout the region, but less common northwardly, and rare in Canada. Generally rare in the montane parks, but common on the arid plains.

Habitats and Ecology; A scavenger species that consumes only dead remains of large animals such as livestock, which it finds visually or by using its fine olfaction. Generally found below 8,000 feet, and at lower altitudes farther north.

Suggested Viewing Locations: Generally, hilly areas near reservoirs are good places to look for this species, or in range country where road-kill carcasses are likely to be found. Nests in Wyoming are usually located on cliffs, such as Casper Mountain or the Mendicino Hills near Guernsey (Scott, 1993). The densest regional breeding concentrations are probably in Wyoming and Montana, but breeding densities are hard to measure, as the nests are well hidden.

Population: National Breeding Bird Survey trend data indicate a significant annual population increase of 1.8%. Trend estimates for the Rocky Mountains & Plains States region (U.S.F.W.S. Region 6) indicate a significant annual increase of 4.8%. Long-term migration studies in North America also indicate increasing populations (Bildstein *et al.*, 2008).The North American population (north of Mexico) has been estimated at 1.3 million birds (Rich *et al.*, 2004).

Kites, Hawks & Eagles

Osprey (*Pandion haliaetus*)

Status: A common summer resident in montane areas near lakes or streams, probably breeding in all the montane parks. Mostly a migrant on the plains, except around reservoirs, rivers and lakes

Habitats and Ecology: Commonly seen along clear rivers and lakes, these birds sometimes nest on rock pinnacles (as in Yellowstone Canyon), but more often nest in tall trees near water, in large nests resembling those of eagles. In the interior Northwest, ospreys begin breeding in mixed coniferous forests during the mature forest stage of succession (Sanderson, Bull and Edgerton, 1980). The erection of artificial nesting platforms in areas lacking good natural sites has helped expand the breeding range.

Suggested Viewing Locations: Artist and Lookout Points are excellent locations for observing nesting by these birds in Yellowstone Park (McEneaney, 1988). McEneaney judged the Yellowstone National Park population at 50–60 pairs during the late 1980's, but Wyoming populations have increased considerably sincee 1978 (Faulkner, 2010). Torrey Lake near Dubois, Wyoming is another good location (Dorn and Dorn, 1990). Ospreys nest commonly along the Green River from near Pinedale northwest, and north through the national parks. In Idaho, ospreys nest along the Coeur d'Alene River's chain of lakes in Kootenai County, between Rose Lake and St. Maries (Svingen and Dumroese, 1997). The densest regional breeding concentrations are probably in northwestern Montana (such as at Lee Metcalf and C. M. Russell national wildlife refuges), northern and central Idaho (deltas of Pack River and Clark Fork River, Coeur d'Alene chain of lakes, upper Salmon and Lemhi rivers) and southern British Columbia.

Population: National Breeding Bird Survey trend data indicate a significant annual population increase of 5.4%. Trend estimates for the Rocky Mountains & Plains States region (U.S.F.W.S. Region 6) indicate a significant annual increase of 9.0%. Long-term migration studies in North America also indicate increasing populations (Bildstein *et al.,* 2008).The North American population (north of Mexico) has been estimated at 212,000 birds (Rich *et al.,* 2004).

Bald Eagle *(Haliaeetus leucocephalus)*

Status: A resident throughout the forested montane areas, breeding in nearly all the montane parks. Otherwise a migrant, and in late fall (especially early November) up to several hundred birds gather in Glacier Park to forage on salmon, producing a unique spectacle.

Habitats and Ecology: This species feeds almost exclusively on nongame fish, such as squawfish, during the breeding season, and also on dying salmon following spawning. In the interior Northwest, bald eagles begin breeding in mixed coniferous forests during the mature forest stage of succession (Sanderson, Bull and Edgerton, 1980). In Wyoming, the presence of carrion provided by wolves has influenced the distribution and abundance of eagles.

Suggested Viewing Locations: McEneaney (1988) judged the Yellowstone National Park population at about 15 pairs during the late 1980's, but eagle populations have increased since then. Yellowstone River and Lake are excellent locations for finding bald eagles in Yellowstone National Park during summer, while in winter Mammoth and Gardiner are good locations (McEneaney, 1988). Commissary Ridge, north of Kemmerer, is a major migration corridor for eagles and other raptors. Winter roosts are in areas of open water, such as on the Bear and North Platte Rivers. Nesting also occurs along Wyoming's North Platte and Bighorn Rivers (Scott, 1993). Nesting bald eagles can be readily seen at many Montana locations, such as Swan River National Wildlife Refuge, near Swan Lake, and along the Kootenai River near Libby (Fischer and Fischer, 1995). Idaho nestings can be seen at Kootenai National Wildlife Refuge, and around Lake Pend Oreille and Lake Coeur d'Alene. Wolf Lodge Bay of Lake Coeur d'Alene attracts 40-80 eagles each fall during the kokanee salmon spawning period (information can be obtained at the B.L.M. office, (208/765-1511). Regional breeding concentrations are apparently variable throughout the northern Rocky Mountains, and may largely depend on local fish availability.

Population: National Breeding Bird Survey trend data indicate a significant annual population increase of 5.4%. The North American population (north of Mexico) has been estimated at 330,000 birds (Rich *et al.,* 2004). Delisted in 2007 from the list of nationally threatened species, bald eagle breedings

Bald eagle, adult,
Drawing by author.

in Wyoming have increased substantially, from 35 nests in 1980 to more than 185 pairs (Faulkner, 2010). Long-term migration studies in North America also indicate increasing populations (Bildstein *et al.*, 2008).

Northern Harrier (*Circus cyaneus*)

Status: Occurs throughout the region, nesting locally, especially in non-forested habitats such as grasslands, croplands, and meadows.

Habitats and Ecology: Grassy areas, especially those near water, are favored by these birds, which nest on the ground rather than in trees, as with most hawks.

Suggested Viewing Locations: In Wyoming, Hutton Lake National Wildlife Refuge (Albany County) and Table Mountain Wildlife Unit (Goshen County) are good places to look for this species (Dorn and Dorn, 1990), which breeds almost statewide on later elevation grasslands and marshes. The densest regional breeding concentrations are probably in northern Montana (*e.g.*, Bowdoin and Medicine Lakes national wildlife refuges) and prairie marshes in southern Alberta.

Population: National Breeding Bird Survey trend data indicate a significant annual population decline of 1.1%. Trend estimates for the Rocky Mountains & Plains States region (U.S.F.W.S. Region 6) indicate a significant annual decline of 1.5%. Long-term migration studies in North America also indicate decreasing populations (Bildstein *et al.*, 2008). The North American population (north of Mexico) has been estimated at 209,000 birds (Rich *et al.*, 2004).

Sharp-shinned Hawk (*Accipiter striatus*)

Status: A common to rare summer resident or year-round resident of montane woodlands of the region; breeding in most and perhaps all the montane parks. Rarely found far from woodland, even on migration.

Habitats and Ecology: Fairly dense forests, either mixed or coniferous, are the preferred habitats of this species, which is swift and elusive, and usually nests in dense groves of trees. Aspens, riparian woodlands, and coniferous forests are all used for breeding. In the interior Northwest, sharp-shinned hawks begin breeding in mixed coniferous forests during the pole-sapling stage of succession (Sanderson, Bull and Edgerton, 1980). In Wyoming they breed in aspen and mid-elevation coniferous forests, probably favoring young, dense and even-aged stands (Faulkner, 2010).

Suggested Viewing Locations: This widespread species may occur anywhere in wooded areas. In Wyoming, a migration corridor occurs from the Bighorns south to Pine Mountain, then to Casper Mountain, the Laramie ranges, and on south into Colorado (Scott, 1993). During winter it regularly visits bird feeders in search of prey. Regional breeding concentrations are similar through the northern Rocky Mountain range.

Population: National Breeding Bird Survey trend data indicate a nonsignificant annual population increase of 2.1%. Trend estimates for the Rocky Mountains & Plains States region (U.S.F.W.S. Region 6) indicate a significant annual increase of 6.6%. However, long-term migration studies in North

America indicate decreasing populations (Bildstein *et al.*, 2008). The North American population (north of Mexico) has been estimated at 583,000 birds (Rich *et al.*, 2004).

Cooper's Hawk (*Accipiter velox*)

Habitats and Ecology: Associated with mature forests, especially deciduous or mixed, and less often in pure coniferous stands. Aspen groves are favored breeding locations; nonbreeders use riparian woodlands, scrub oaks, and mountain meadows.

Suggested Viewing Locations: This is a widely distributed species, and can occur anywhere in wooded areas, especially in montane forests. In the interior Northwest, Cooper's hawks begin breeding in mixed coniferous forests during the pole-sapling stage of succession (Sanderson, Bull and Edgerton, 1980). It also visits suburban bird feeders in winters, like sharp-shined hawks. Regional breeding concentrations are similar through the northern Rocky Mountain range. In Wyoming they are most common in mid-elevation aspen and coniferous forests, especially in riparian habitats (Faulkner, 2010).

Population: National Breeding Bird Survey trend data indicate a significant annual population increase of 6.0%. Trend estimates for the Rocky Mountains & Plains States region (U.S.F.W.S. Region 6) indicate a significant annual increase of 10.2%. Long-term migration studies in North America also indicate increasing populations (Bildstein *et al.*, 2008). The North American population (north of Mexico) has been estimated at 530,000 birds (Rich *et al.*, 2004).

Cooper's hawk, adult,
Drawing by author.

Northern Goshawk (*Accipiter gentilis*)

Status: An uncommon to rare permanent resident in woodland and montane forests of the region.

Habitats and Ecology: This species is found in many habitats from aspen groves to timberline, but favors dense conifers or aspens near water for breeding, and ranges into low woodlands, riparian woods, and sage areas at other times. In the interior Northwest, goshawks begin breeding in mixed coniferous forests during the mature forest stage of succession (Sanderson, Bull and Edgerton, 1980). In south-central Wyoming the birds favor trees of large diameter, in mixed lodgepole–quaking aspen forests (Faulkner, 2010).

Suggested Viewing Locations: An elusive species, mature montane coniferous forests are the best locations for finding this hawk, with most Wyoming records from the northwestern part of the state. Regional breeding concentrations are similar through the northern Rocky Mountain range.

Population: National Breeding Bird Survey trend data indicate a nonsignificant annual population increase of 1.2%. The North American population (north of Mexico) has been estimated at 240,000 birds (Rich *et al.*, 2004).

Broad-winged Hawk (*Buteo platypterus*)

Status: A rare vagrant or migrant in most of the region, but breeding in the Cypress Hills of Alberta, as well as north-central Alberta. A rare migrant east of the Front Range in Colorado, but has bred in the vicinity of Fort Collins.

Habitats and Ecology: Associated with deciduous woodlands, including riparian woods and aspen grovelands.

Suggested Viewing Locations: This species is too rare in the Rocky Mountain region to suggest any reliable birding sites. Woodlands along plains streams east of the Rockies offer the best chances of finding it.

Population: National Breeding Bird Survey trend data indicate a significant annual population increase of 1.4%. Long-term migration studies in North America also indicate increasing populations (Bildstein *et al.*, 2008). The North American population (north of Mexico) has been estimated at 1.7 million birds (Rich *et al.*, 2004).

Swainson's Hawk (*Buteo swainsoni*)

Status: A common to rare summer resident over most of the region; least common in the heavily forested montane parks, especially to the north.

Habitats and Ecology: Associated with open grasslands, sagebrush, agricultural lands, and rarely with riparian areas, typically nesting in isolated trees, but sometimes in bushes, on man-made structures, or on cliffs.

Suggested Viewing Locations: Grasslands east of the Rockies offer the best chances of finding it. In Wyoming, such areas in Albany and Laramie counties provide such possibilities (Dorn and Dorn, 1990). In Wyoming nesting occurs statewide at elevations under 9,000 feet (Faulkner, 2010). The densest regional breeding concentrations are probably in the plains of northern Montana and southern Alberta.

Northern goshawk, adult,
Drawing by author.

Population: Classified as a Partners-in-Flight Watch List species of continental conservation importance. The North American population (north of Mexico) has been estimated at 460,000 birds (Rich *et al.*, 2004). National Breeding Bird Survey trend data indicate a nonsignificant annual population decline of 0.2%. However, long-term migration studies in North America indicate increasing populations (Bildstein *et al.*, 2008).

Red-tailed Hawk *(Buteo jamaicensis)*

Status: A permanent resident nearly throughout the region, although somewhat migratory to the north, and becoming rare in more open plains, where replaced by the Swainson's hawk and, in nearly treeless areas, the ferruginous hawk.

Habitats and Ecology: This typically tree-nesting buteo also extends to open woodlands and even treeless areas, where nesting may occur on cliffs. However, trees, especially large cottonwoods and pines, are favored nest sites.

Suggested Viewing Locations: This nearly ubiquitous buteo is likely to be found anywhere in open country from sage or greasewood scrub to subalpine areas. Telephone poles or scattered trees offer attractive perching and lookout sites. The densest regional breeding concentrations are probably in western Montana, southern Alberta, and northern Idaho.

Population: National Breeding Bird Survey trend data indicate a significant annual population increase of 2.1%. Trend estimates for the Rocky Mountains & Plains States region (U.S.F.W.S. Region 6) indicate a significant annual increase of 2.2%. The North American population (north of Mexico) has been estimated at 1.96 million birds (Rich *et al.*, 2004).

Ferruginous Hawk *(Buteo regalis)*

Status: A relatively uncommon to rare summer or year-round resident in the region; primarily in open-country habitats, and rare in the montane parks, with only Yellowstone reporting breeding.

Habitats and Ecology: Found during the breeding season in grasslands, sagebrush, and sometimes also mountain meadows, and nesting in pygmy conifers, cliff ledges, rock outcrops, and sometimes on man-made structures such as windmills.

Suggested Viewing Locations: In Wyoming, favorable viewing possibilities exist in open country around Laramie, Pine Tree Junction, east of McFadden, and north of Baggs (Dorn and Dorn, 1990). Wyoming's breeding population about 800 pairs is considered the continent's second largest, but may be declining owing to habitat loss and alteration (Faulkner, 2010). The densest regional breeding concentrations are probably in northern and central Montana (*e.g.*, C. M. Russell National Wildlife Refuge), southeastern Idaho (Grays Lake National Wildlife Refuge), and southern Alberta.

Population: National Breeding Bird Survey trend data indicate a significant annual population increase of 2.5%. Trend estimates for the Rocky Mountains

& Plains States region (U.S.F.W.S. Region 6) indicate a significant annual increase of 2.7%. Long-term migration studies in North America indicate decreasing populations (Bildstein *et al.*, 2008).The North American population has been estimated at 23,000 birds (Rich *et al.*, 2004). In Wyoming, this species is considered a Priority III conservation species.

Rough-legged Hawk (*Buteo lagopus*)

Status: A regular winter visitor throughout the region, especially in open habitats.

Habitats and Ecology: Usually found hunting in grasslands, sagebrush, or sometimes over marshes or mountain meadows.

Suggested Viewing Locations: During winter this species might be found in open country across the region. In Wyoming, favorable locations include Goshen Hole (5–10 miles southwest of LaGrange), Campstool Road east of Cheyenne, Sweetwater Valley below Sweetwater Station, the Bridger Valley (Uinta County), south of Buffalo, and northeast of Sundance (Dorn and Dorn, 1990). The plains of eastern Montana attract large numbers in winter, and some wintering occurs in grasslands as far south as southern Colorado, such as at Monte Vista and Alamosa national wildlife refuges (Kingery, 2007).

Population: National Breeding Bird Survey trend data are not available. The North American population has been estimated at 285,000 birds (Rich *et al.*, 2004).

Golden Eagle (*Aquila chrysaetos*)

Status: A permanent resident throughout the region, but most common in montane or rimrock country that is relatively open.

Habitats and Ecology: This is a mountain- and plains-adapted species, that often occurs in grasslands, semi-desert areas, pinyon–juniper woodlands, the ponderosa pine zone of coniferous forests, and sometimes forages above mountain meadows or alpine tundra. It nests over a broad altitudinal range, usually on cliffs or in trees, rarely on the ground. In the interior Northwest, golden eagles begin breeding in mixed coniferous forests during the mature forest stage of succession (Sanderson, Bull and Edgerton, 1980).

Suggested Viewing Locations: In Wyoming, cliff and canyon country across the state provides viewing possibilities year-around, and the Sarasota and Monet areas are favored in winter (Dorn and Dorn, 1990). Nesting occurs virtually throughout Wyoming, which may have the largest population of golden eagles in the U.S., and estimated in the 1970's to total about 10,000 birds. The densest regional breeding concentration of golden eagles is in the Great Divide Basin of south-central Wyoming. A famous winter eagle roost (goldens and balds) is located in Jackson Canyon (about five miles southwest of Casper, on Wyoming Highway 220). In Montana, the region between Clyde Park and Ringling (U.S. Highway 89) is famous for its eagles; from mid-March to mid-May large numbers of golden and bald eagles

Golden eagle, adult,
Drawing by author.

migrate through it, as do other raptors. A 27-mile stretch along Highway 200, from Lincoln over Rogers Pass is a major spring migration route for up to a thousand or more golden and bald eagles from March to early May. Golden eagles are widespread in Colorado year-around, and in Idaho they are most abundant in the panhandle (Svingen & Dumroese, 1997). The largest known raptor migration in Idaho occurs near Boise, where the Idaho Bird Observatory tallies and bands raptors each autumn at Lucky Peak. The Hawkwatch Migration Association of North America has three observation sites in the Rocky Mountain region, including Dinosaur Ridge, west of Denver, Colorado, the Bridger Mountains, N.N.E. of Bozeman, Montana, and Commissary Ridge, near Cokeville, Wyoming.

Population: National Breeding Bird Survey trend data indicate a significant annual population increase of 1.5%. The North American population (north of Mexico) has been estimated at 80,000 birds (Rich *et al.*, 2004).

Falcons

American Kestrel *(Falco sparverius)*

Status: A relatively common summer or permanent resident over nearly all the region, becoming more seasonal in occurrence farther north, and probably breeding in all the montane parks.

Habitats and Ecology: This is an open-country falcon, occurring in agricultural areas, grasslands, sagebrush, desert scrub, and nesting in tree cavities, rock or building crevices or cavities, and rarely in earthen holes. It avoids forests, but sometimes forages as high as mountain meadows. In Wyoming nesting occurs up to about 8,500 feet.

Suggested Viewing Locations: For most of the year this species can be found in any fairly open-country habitat, especially where there are some trees with woodpecker holes for nesting, and perching sites such as overhead lines. In Wyoming, the Torrington and Lovell areas attract good numbers of kestrels (Dorn and Dorn, 1990). The densest regional breeding concentrations are probably in southeastern Montana and northeastern Wyoming (*e.g.*, Black Hills National Forest, Bighorn Mountains).

Population: National Breeding Bird Survey trend data indicate a significant annual population decline of 0.7%. Long-term migration studies in North America also indicate significantly decreasing populations (Bildstein *et al.*, 2008). The North American population (north of Mexico) has been estimated at 4.35 million birds (Rich *et al.*, 2004).

Merlin *(Falco columbarius)*

Status: A relatively rare migrant or summer resident in the region, mainly in montane wooded areas, and breeding locally.

Habitats and Ecology: This is a forest and woodland-adapted falcon, usually breeding in tree clumps of open woodlands, often in bottomlands or valleys. In the interior Northwest, merlins begin breeding in mixed coniferous forests during the mature forest stage of succession (Sanderson, Bull and Edgerton, 1980). In one Wyoming study, merlins nested preferentially in old magpie nests near major river systems, and areas of mixed-grass and ponderosa pine savannas (Faulkner, 2010). During the non-breeding season they also appear over grasslands, agricultural lands, desert scrub, and marshes or shorelines.

Suggested Viewing Locations: In Wyoming, this species breeds in woodlands but moves to open grasslands during winter, especially where horned larks are abundant (Dorn and Dorn, 1990). Merlins are notably common in Wyoming's Green River valley above the town of Green River (Scott, 1993). The densest regional breeding concentrations are probably in southern Alberta.

Population: National Breeding Bird Survey trend data indicate a significant annual population increase of 8.2%. Long-term migration studies in North America also indicate increasing populations (Bildstein *et al.*, 2008). The North American population (north of Mexico) has been estimated at 650,000 birds (Rich *et al.*, 2004).

Peregrine Falcon *(Falco peregrinus)*

Status: Once a nester throughout the montane areas and nesting in at least the Rocky Mountain and Yellowstone parks. Efforts have been underway since the 1980's to re-establish the species in some of these areas by releasing hand-reared birds, which has proven very successful in the Rocky Mountain region.

Habitats and Ecology: This species is largely a cliff-nesting species, typically in woodland habitats. Non-breeders occur over a wide habitat range, from mountain meadows to grasslands, marshes, and riparian habitats. McEneaney (1988) judged the Yellowstone National Park population to be about 12 birds during the late 1980's, but Rocky Mountain populations have greatly increased since then. In Wyoming, restoration efforts resulted in the first documented nestling in 1984, increasing to more than 60 nesting pairs by 2004 (Faulkner, 2010).

Suggested Viewing Locations: Hayden Valley is an excellent location for finding these birds in Yellowstone National Park (McEneaney, 1988), and Clarks Fork Valley in Park County is also a possibility (Dorn and Dorn, 1990). In Montana the Gates of the Mountains region south of Helena and Red Rock Lakes National Wildlife Refuge are very good locations for seeing peregrines (Fischer and Fischer, 1995).

Population: National Breeding Bird Survey trend data indicate a significant annual population increase of 8.8%. Long-term migration studies in North America also indicate increasing populations (Bildstein *et al.*, 2008). The North American population (north of Mexico) has been estimated at 276,000 birds (Rich *et al.*, 2004). In Wyoming, and Colorado this species has been considered as critically imperiled by the Natural Heritage program (Fitch, 1992), but its population is increasing in both states.

Prairie Falcon (*Falco mexicanus*)

Status: A widespread summer resident or year-round resident throughout the region, mainly in mountain or rimrock areas offering open country for hunting.

Habitats and Ecology: Breeding birds are largely associated with plains, sagebrush, or desert scrub habitats with steep cliffs nearby for nesting; sometimes tundra areas also support breeders, and foraging may be done on mountain meadows or similar alpine habitats.

Suggested Viewing Locations: In Wyoming, good locations during the breeding season include Flaming Gorge Reservoir and buttes around Rock Springs and Green River (Sweetwater County), Beaver Rim (Fremont County), Wind River Canyon (Washakie County), and the western slope of the Bighorn Mountains (Dorn and Dorn, 1990). In Colorado, Castlewood Canyon State Park, the Garden of the Gods park near Colorado Springs, and Sheep Mountain in the North Park region are likely places for finding prairie falcons (Kingery, 2007). In Montana the Terry Badlands and Makoshika State Park are very good locations for seeing prairie falcons (Fischer and Fischer, 1995). Outside the breeding season they might be seen anywhere. Regional breeding concentrations are probably densest in open country canyon-and-rimrock habitats.

Population: National Breeding Bird Survey trend data indicate a significant annual population increase of 3.2%. The North American population (north of Mexico) has been estimated at 34,500 birds (Rich *et al.*, 2004).

Rails & Coots

Virginia Rail (*Rallus limicola*)

Status: A local summer resident and breeder in the general region, especially at lower elevations; generally rare or lacking in the montane parks except for Rocky Mountain National Park, where a rare breeder in the Colorado River valley.

Habitats and Ecology: Inhabits marshes with dense stands of emergent vegetation, nesting on wet ground or over shallow water in such stands.

Suggested Viewing Locations: Marshy areas at lower altitudes throughout the region provide viewing (or at least hearing) opportunities. In Wyoming, nesting occurs in marshes such as Table Mountain, Sarasota, Cokeville Meadows National Wildlife Refuge, and Flat Creek marsh in the National Elk Refuge (Scott, 1993). They are known to nest generally in the southern third of the state, along the western border, and around Sheridan (Faulkner, 2010). The densest regional breeding concentrations are probably in southeastern Idaho (*e.g.*, Camas, Grays Lake and Bear Lake national wildlife refuges).

Population: National Breeding Bird Survey trend data indicate a significant annual population increase of 2.1%.

Sora (*Porzana carolina*)

Status: An uncommon to occasional summer resident in the region; infrequent in the montane parks, but known to nest in several of them.

Habitats and Ecology: Found in essentially the same marshy habitats as the Virginia rail, and apparently having very similar niche adaptations, but perhaps somewhat more vegetarian in its diet. More surface-feeding and less probing-for food is also done by soras than Virginia rails.

Suggested Viewing Locations: Marshy areas at lower altitudes throughout the region provide viewing (or at least hearing) opportunities. In Wyoming, these birds nest in many lower elevation marshy areas across the state, such as those mentioned for the Virginia rail, and others such as Loch Katrine (Park County) and Seedskadee National Wildlife Refuge (Scott, 1993). The densest regional breeding concentrations are probably in southern Alberta and adjacent northwestern Montana (*e.g.*, Medicine Lake and Bowdoin national wildlife refuges).

Population: National Breeding Bird Survey trend data indicate a nonsignificant annual population increase of 0.1%.

American Coot (*Fulica americana*)

Status: Widespread and a summer resident on wetlands throughout the region, especially at lower elevations. Present and probably breeding in all the montane parks.

Habitats and Ecology: Associated with ponds and marshes having a combination of open water and emergent reedbeds, in which nesting occurs. Besides foraging on aquatic plants, the birds sometimes also graze on nearby shorelines and meadows.

Suggested Viewing Locations: Wetlands at lower altitudes throughout the region provide viewing opportunities. Breeding probably occurs in all larger marshes of Wyoming (Scott, 1993), except for high montane wetlands. Dense regional breeding concentrations are widespread, especially in prairie marshes east of the Rocky Mountains.

Population: National Breeding Bird Survey trend data indicate a nonsignificant annual population decline of 0.5%.

Cranes

Sandhill Crane (*Grus canadensis*)

Status: A local summer resident in the more remote wetlands of the region, north almost to Glacier National Park. The range is probably now expanding, and may soon include the Glacier area, where the last reported breeding occurred in 1899.

Habitats and Ecology: In the Rockies, sandhill cranes are especially associated with beaver impoundments, where the birds nest along shorelines or sometimes on beaver lodges, often in dense willow thickets. The birds are highly territorial and nests usually are well scattered. Their loud calls serve to advertise territories and to communicate over long distances.

Suggested Viewing Locations: In Grand Teton National Park favored crane habitats include Willow Flats near Jackson Lake dam, sedge meadows behind Christian Pond, and beaver ponds below Teton Point and along the Buffalo Fork River. In Yellowstone National Park, Lamar and Hayden valleys are highly favored habitats, and other good crane habitats exist at Willow Park, Swan Lake Flats, Blacktail Ponds, Antelope Creek and near Fishing Bridge. Outside of these areas, breeding occurs along the Snake River and Green River drainages, and locally east to the Bighorn Range (Faulkner, 2010). Grays Lake and Bear Lake national wildlife refuges in Idaho support large breeding crane populations. In southwestern Montana sandhill cranes nest at sites such as Red Rock Lakes National Wildlife Refuge, Blackfoot Waterfowl Production Area near Ovando, and at Mount Haggin Wildlife Management Area, near Anaconda (Fischer & Fischer, 1995). Morgan Bottoms, on the Yampa River in northwestern Colorado, is an important fall stopover area for migrating sandhill cranes (Young, 2000), as is Teton Valley in Teton County, Idaho (Svingen & Dumroese, 1997). Many other excellent crane-watching sites exist in the Rocky Mountain region (Johnsgard, 2011).

Population: National Breeding Bird Survey trend data indicate a significant annual population increase of 6.2%. Trend estimates for the Rocky Mountains & Plains States region (U.S.F.W.S. Region 6) indicate a significant annual increase of 4.5%. Counting all races, the total population of sandhill cranes may exceed 700,000 birds. The Rocky Mountain population has been estimated at about 20,000 birds and appear to be stable (Johnsgard, 2011).

Whooping Crane (*Grus americana*)

Status: A very rare and local migrant east of the Rocky Mountains.

Habitats and Ecology: Once widespread in central North America, the whooping crane currently breeds only in a limited area of northwestern Canada. It nests in remote wetlands there, in a muskeg-rich wilderness area (Wood Buffalo National Park). On migration the birds utilize a variety of habitat types, but they are usually include shallow and broad roosting sites, and nearby wetland foraging opportunities.

Suggested Viewing Locations: This species is too rare in the Rocky Mountain region to suggest any good birding sites, especially now that the Grays Lake–Bosque del Apache population has died out. Shallow wetlands along the eastern boundary of the region in eastern Montana provide the best chances of seeing whooping cranes, especially during late April. The historic migratory population in the Great Plains was approaching 300 birds as of 2010, its largest recorded size (Johnsgard, 2011).

Plovers

Black-bellied Plover (*Pluvialis squatarola*)

Status: A rare migrant or vagrant throughout the region, more prevalent on the plains than in montane regions.

Habitats and Ecology: Migrant birds are likely to be found along lakes or reservoirs, but sometimes also are seen on plowed fields or forage on short meadows or pasturelands.

Suggested Viewing Locations: This species is too rare in the Rocky Mountain region to suggest any reliable birding sites. In Wyoming, Keyhole Reservoir is among the best locations for finding these birds (Scott, 1993). Elsewhere, the species is probably most common on the eastern plains of Montana and Colorado.

Population: National Breeding Bird Survey trend data are not available. The North American population has been estimated at 200,000 birds (Morrison *et al.*, 2001).

American Golden Plover (*Pluvialis dominica*)

Status: A local migrant in the region, especially on the plains and prairie areas, but only a vagrant in the montane parks.

Habitats and Ecology: Migrant birds are usually found along lakes or reservoirs, or on agricultural lands, during migration. They often occur on plowed or recently burned fields on migration, where surface-foraging opportunities are available.

Suggested Viewing Locations: This species is too rare in the Rocky Mountain region to suggest any reliable birding locations, but the sites mentioned for the previous species are perhaps among the best (Scott, 1993).

Population: National Breeding Bird Survey trend data are not available. The North American population has been estimated at 150,000 birds (Morrison *et al.*, 2001).

Semipalmated Plover (*Charadrius semipalmatus*)

Status: A local migrant throughout the region, mainly on the plains, and a rare migrant or vagrant in the montane parks.

Habitats and Ecology: Migrating birds are usually observed on open sandy or gravelly habitats along rivers or beaches, where they feed by running about and picking up morsels from the surface, rather than probing for foods.

Suggested Viewing Locations: In Wyoming, Table Rock Wildlife Unit (Goshen County) is a regular migration stopover point (Dorn and Dorn, 1990). Lake DeSmet, Table Mountain and Keyhole Reservoir are among the best Wyoming locations for finding this plover (Scott, 1993).

Population: National Breeding Bird Survey trend data are not available. The North American population has been estimated at 150,000 birds (Morrison *et al.*, 2001).

Killdeer (*Charadrius vociferus*)

Status: A common summer or permanent resident throughout the region, both on the plains and montane areas, but more common at lower elevations and not reaching alpine areas.

Habitats and Ecology: Widely distributed in open-land habitats, including pastures, roadsides, gravel pits, golf courses, airports, and sometimes suburban lawns. Gravelly areas are favored, and graveled rooftops are sometimes used for nesting in urban areas. Migrating and wintering birds are more closely associated with water, but also use mud flats and open fields.

Suggested Viewing Locations: This abundant plover can be seen on grassy meadows or shorelines almost anywhere in the region at lower altitudes. Dense breeding populations occur in all such habitats of non-montane areas.

Population: National Breeding Bird Survey trend data indicate a nonsignificant annual population decline of 0.6%. Trend estimates for the Rocky Mountains & Plains States region (U.S.F.W.S. Region 6) indicate a significant annual decline of 1.1%. Morrison *et al.* (2001b) estimated the species' total population at one million birds.

Mountain Plover (*Charadrius montanus*)

Status: A local summer resident on the plains east of the mountains; a rare migrant or vagrant in the montane parks.

Habitats and Ecology: This species breeds exclusively in early spring on arid grasslands, where grasses are usually no more than three inches in height, and sometimes in semi-desert areas with cacti and scattered shrubs far from water. During the nonbreeding seasons the birds are also found in relatively dry habitats.

Suggested Viewing Locations: In Wyoming, breeding birds are most likely to be seen on the Laramie Plains, especially near Bamforth Lake, and along the Carbon-Albany county line (Dorn and Dorn, 1990). Shirley Rim between Wyoming Hwys. 77 and 487 is a good location (Scott, 1993). Most Wyoming nesting occurs in the Bighorn, Great Divide, Laramie, Shirley, and Washakie Basins (Faulkner, 2010. In Montana the Charles M. Russell National Wildlife Refuge (especially around Old Manning Corrals) offers viewing opportunities (McEneaney, 1993). The densest regional breeding concentrations are probably in eastern Colorado, especially in shortgrass prairie sites such as the Pawnee National Grassland (Kingery, 2007).

Population: National Breeding Bird Survey trend data indicate a significant annual population decline of 2.8%. The North American population has been estimated at 9,000 birds (Morrison *et al.*, 2001). It is classified as imperiled in Colorado by the Natural Heritage Program, but is seemingly secure in Wyoming (Finch, 1992).

Stilts & Avocets

Black-necked Stilt (*Himantopus mexicanus*)

Status: A local migrant or vagrant through much of the region, except for southeastern Idaho, where breeding regularly occurs. Absent or accidental in the montane parks. Breeding occurs locally in central Montana, and in Wyoming breeding has occurred in at least seven scattered locations.

Habitats and Ecology: Breeding in this species usually occurs in the grassy shoreline areas of shallow freshwater or brackish pools of wetlands having extensive mudflats, or sometimes along the shorelines of salt lakes where vegetation is essentially lacking. Often found in company with American avocets, which use similar habitats.

Suggested Viewing Locations: In Wyoming, important migration stopping points include Lovell Lakes (Big Horn County), Hutton Lake National Wildlife Refuge (Albany County), Goldeneye Wildlife Area (Natrona County), and the Bridger Power Plant near Point of Rocks (Dorn and Dorn, 1990). Breedings have occurred periodically in Sweetwater County (Bridger Power Plant, Old Eden Reservoir), and at Lovell Lakes, Loch Katrine (Park County), Ocean Lake (Fremont County), and Table Mountain Wildlife Unit (Goshen County) (Faulkner 2010). In Montana, Freezeout Lake Wildlife Management Area and Benton Lake National Wildlife Refuge are good birding choices. In Colorado stilts nest in the San Luis Valley and a few other locations (Kingery, 1998). The densest regional breeding concentrations are probably in southeastern Idaho, in shallow freshwater or saline wetlands such as at Camas and Bear Lake national wildlife refuges (Svingen & Dumroese, 1997).

Population: National Breeding Bird Survey trend data indicate a significant annual population increase of 2.8%. The North American population has been estimated at 150,000 birds (Morrison *et al.*, 2001).

American Avocet (*Recurvirostra americana*)

Status: A local summer resident over much of the region, mainly on shallow marshes of the plains. Rare in the montane parks, with breeding reported only for Yellowstone (no recent records).

Habitats and Ecology: During breeding this species favors ponds or shallow lakes with exposed and sparsely vegetated shorelines, and somewhat saline waters that have large populations of aquatic invertebrates, which are gathered by making scythelike movements of the curved bill through the water.

Suggested Viewing Locations: In Wyoming, the sites listed for the black-necked stilt also attract avocets, as does Dave Johnson Power Plant east of Glenrock, and Carmody Lake (about three miles northwest of Sweetwater Station in Fremont County)(Dorn and Dorn, 1990). The alkaline lakes in the Red Desert region have high numbers, and breeding has most frequently occurred at Yant's Puddle near Casper (Scott, 1993; Faulkner 2010). In Montana avocets breed commonly on many prairie marshes, but Freezeout Lake and Benton Lake National Wildlife Refuge are good birding choices. Many marshes in eastern Colorado also support breeding avocets, especially in the San Luis Valley, but also in eastern Colorado from the Arkansas River north to the South Platte (Kingery, 1998). The densest regional breeding concentrations are probably in southeastern Idaho, in shallow wetland sites such as at Bear Lake National Wildlife Refuge, and along the Montana-Alberta border.

Population: National Breeding Bird Survey trend data indicate a nonsignificant annual population decline of 0.4%. Trend estimates for the Rocky Mountains & Plains States region (U.S.F.W.S. Region 6) indicate a significant annual decline of 3.4%. The North American population has been estimated at 450,000 birds (Morrison *et al.*, 2001).

Sandpipers, Snipes & Phalaropes

Spotted Sandpiper (*Tringa macularius*)

Status: A common summer resident throughout the region, breeding in all the montane parks.

Habitats and Ecology: Associated with forest streams, pools, and rivers, usually at lower elevations, but extending locally to alpine timberline and utilizing a wide array of open terrains with water present, and rarely even in the absence of nearby water. Shaded watercourses are favored, and sometimes the birds are found along rapidly flowing mountain torrents.

Suggested Viewing Locations: In Wyoming, good observation areas in summer include Lovell Lakes (Big Horn County), Hutton Lake National Wildlife Refuge (Albany County), Goldeneye Wildlife Area (Natrona County), and Yellowstone Lake (Dorn and Dorn, 1990). The densest regional breeding concentrations are probably in central and northern Idaho, and in the Greater Yellowstone region, but regional breeding locations are widespread.

Population: National Breeding Bird Survey trend data indicate a nonsignificant annual population decline of 0.8%. Morrison *et al.* (2001b) estimated the North American population at 150,000 birds.

Solitary Sandpiper (*Tringa solitaria*)

Status: A migrant throughout the region, breeding along the northern edge of this book's boundaries, in central and northern Alberta (south to about the North Saskatchewan River and, in the montane parks, to Kootenay National Park. In Jasper National Park known breeding areas include Willow, Blue, and Isaac Creeks, Topaz and Southesk Lakes, and Rocky Forks.

Habitats and Ecology: Breeding is done around muskeg ponds, along woodland lakes, and near forest ponds, where the old nests of tree-nesting birds such as American robins are utilized.

Suggested Viewing Locations: This species migrates across the plains mostly to the east of the Rockies, stopping in Wyoming at sites such as Hutton Lake National Wildlife Refuge (Albany County) and Table Mountain Wildlife Unit (Goshen County). Beaver ponds and farm or woodland ponds with fairly dense surrounding vegetation in Wyoming are attractive to migrants (Scott, 1993), as are many other similar wetlands from Colorado to Alberta.

Population: National Breeding Bird Survey trend data indicate a nonsignificant annual population decline of 4.5%. The North American population has been estimated at 25,000 birds (Morrison *et al.,* 2001).

Greater Yellowlegs (*Tringa melanoleuca*)

Status: A migrant over most of the region, but a summer resident in the mountains of Alberta, occasionally breeding south to Banff Park. Breeds rather commonly in Jasper Park (Miette and Athabasca valleys, Willow and Blue Creeks, and others).

Habitats and Ecology: In migration these birds occupy the edges of marshes and slow-moving rivers, foraging along the shorelines and sometimes wading out belly-deep to probe in the mud or skim the surface for invertebrates. On the breeding grounds the birds favor muskeg areas, with a mix of ponds, trees, and clearings, and sometimes extend into subalpine scrub near timberline. In Alberta a favored nesting habitat consists of muskeg with spruce and tamarack.

Suggested Viewing Locations: This species migrates across the plains mostly to the east of the Rockies, stopping in Wyoming at sites such as Hutton Lake National Wildlife Refuge (Albany County), and many other similar wetlands from Colorado to Alberta are used by migrants.

Population: National Breeding Bird Survey trend data indicate a nonsignificant annual population increase of 1.7%. Morrison *et al.* (2001) estimated the North American population at 100,000 birds.

Willet (*Tringa semipalmata*)

Status: A migrant or summer resident over much of the region, breeding mainly in grassland marshes; generally rare or lacking in the montane parks. Although once reported as nesting in Yellowstone National Park, current evidence indicates that it is only a spring and fall migrant.

Habitats and Ecology: Breeding habitats of this species consist of prairie marshes, usually brackish to semi-alkaline, seasonal ponds, and sometimes also intermittent streams in grassland areas. The birds are effective probers, and spend much of their time feeding in this way, but also peck at objects on the water surface.

Suggested Viewing Locations: In Wyoming, This species can be seen at sites such as Hutton Lake National Wildlife Refuge (Albany County), and Table Mountain Wildlife Unit (Goshen County). Breeding has occurred locally at places such as Yant's Puddle (north of Casper), and ponds at the eastern edge of the Red Desert (Scott, 1993). Wyoming breedings are most frequent in the southwestern counties, but regularly extend east to Fremont and Natrona counties (Faulkner, 2010). In Colorado, Arapaho National Wildlife Refuge, and Walden Reservoir in the North Park region, are the only regular breeding sites for willets (Kingery, 1998). The densest regional breeding concentrations are probably in southern Alberta and adjacent Montana, as well as more locally in southeastern Idaho and southwestern Montana.

Population: National Breeding Bird Survey trend data indicate a nonsignificant annual population decline of 0.9%. The North American population has been estimated at 250,000 birds (Morrison *et al.*, 2001).

Lesser Yellowlegs (*Tringa flavipes*)

Status: A migrant nearly throughout the entire region, breeding locally in central Alberta, but not known to breed in either Jasper or Banff parks, where it is a rare migrant or vagrant.

Habitats and Ecology: Breeding typically occurs in habitats that have a combination of rather open and tall woodlands, with low and sparse brushy undergrowth, and fairly close to grassy or marshy ponds. Broken hills, covered with burned or fallen timber, and low poplar second growth, are a favored Alberta nesting habitat. Outside the breeding season the birds occur along mud flats and shallow ponds, often with vegetated shorelines, and sometimes visit flooded fields.

Suggested Viewing Locations: This species migrates across the plains mostly to the east of the Rockies, stopping in Wyoming at sites such as Hutton Lake National Wildlife Refuge (Albany County) (Dorn and Dorn, 1990). Wyoming's Table Mountain Wildlife Unit (Goshen County), Loch Katrine (Park County) and the Bridger Power Plant ponds (Sweetwater County) also attract migrating yellowlegs (Scott, 1993), as do many similar freshwater wetlands from Colorado to Alberta.

Population: National Breeding Bird Survey trend data indicate a significant annual population decline of 9.4%. The North American population has been estimated at 500,000 birds (Morrison *et al.*, 2001).

Upland Sandpiper (*Bartramia longicauda*)

Status: A summer resident in native grassland areas east of the Rockies; a rare migrant or vagrant in the montane parks.

Habitats and Ecology: Generally associated with wet meadows, hayfields, mowed prairies, or mid-length prairies, avoiding both shortgrass steppe areas and extremely tall grasses. Often found far from water, and rarely if ever wading for its foods.

Suggested Viewing Locations: Most Wyoming breeding occurs east of a line extending from Sheridan to Laramie, and locally farther west in Carbon, Natrona and Big Horn counties (Faulkner, 2010). Favored nesting sites include the North Platte River valley northwest of Douglas, areas south of Lusk along Highway 85, and meadows just west of LaGrange (Dorn and Dorn, 1990). Breeding also occurs on the high plains north of Shirley Basin, near the headwaters of Bate's Creek (Scott, 1993). Very few breedings have been reported for Idaho, and in Colorado the best chances of finding these birds nesting are in native grasslands in the northeastern corner of the state, such as in the sandhills north of the Arikaree River in Yuma and Washing-

ton counties (Kingery, 1998). In Montana, places such as Benton Lake National Wildlife Refuge provide good breeding-season viewing (McEneaney,
1993). The densest regional breeding concentrations are probably in native
grasslands of central Montana and the easternmost parts of Montana and
Wyoming.

Population: National Breeding Bird Survey trend data indicate a nonsignificant annual population increase of 0.4%. Trend estimates for the Rocky
Mountains & Plains States region (U.S.F.W.S. Region 6) indicate a significant annual increase of 0.8%. The North American population has been
estimated at 350,000 birds (Morrison *et al.*, 2001). It is considered rare in
Colorado's Natural Heritage Program, and a Priority II species in Wyoming
(Finch, 1992).

Long-billed Curlew (*Numenius americanus*)

Status: A summer resident in grassland areas over much of the region, but
mostly absent in the montane parks except Grand Teton and Yellowstone
national parks, where the birds are rare.

Habitats and Ecology: On the breeding grounds this species occurs in shortgrass
areas, grazed taller grasslands, and overgrazed grasslands with scattered
shrubs or cacti. Hilly or rolling areas seem favored over flatlands, and the
birds often nest rather far from standing water. However, migrating birds
usually are found on beaches or other shoreline habitats.

Suggested Viewing Locations: In Wyoming, breeding occurs in the meadows
west of Pinedale and Daniel Junction, and in the Bear River marshes of
Cokeville Meadows National Wildlife Refuge (Scott, 1993). Curlews also
breed near Lusk (Niobrara County), at Chapman Bench (Park County) and
in northern Sublette County (Faulkner, 2010). They may also be seen at
the National Elk Refuge (Dorn and Dorn, 1990). In Colorado, the densest
breeding occurs from Baca County west to the Purgatoire River in Las Animas County (Kingery, 1998). The East Unit of Comanche National Grassland is a prime summer birding location (Kingery, 2007). In Idaho the birds
are fairly common in southern Idaho, such as at Henrys Lake State Park
(Fremont County), Harriman State Park, and Camas and Minidoka national
wildlife refuges (Svingen and Dumroese, 1997). In Montana, Centennial
Valley is a prime area for seeing these birds in summer (McEneaney, 1997),
as are Makoshika State Park and Benton Lake National Wildlife Refuge. The
densest regional breeding concentrations are probably in native grasslands
of southern Alberta, southeastern Idaho and western and central Montana.

Population: National Breeding Bird Survey trend data indicate a nonsignificant
annual population decline of 0.8%. Trend estimates for the Rocky Mountains & Plains States region (U.S.F.W.S. Region 6) indicate a significant annual decline of 2.3%. The North American population has been estimated
at 20,000 birds (Morrison *et al.*, 2001), but more recent estimates are closer
to 150,000. It is considered a Priority III species in Wyoming (Finch, 1992).

Hudsonian Godwit (*Limosa haemastica*)

Status: A rare to occasional local migrant in the region, mostly in the Great Plains region, and rare or accidental in montane areas.

Habitats and Ecology: On migration this species is likely to be found along shorelines of prairie marshes, singly or in small numbers, and usually probing dowitcher-like for food. They breed in subarctic areas where woods and scrub tundra intermix, and where wet meadows or ponds are nearby.

Suggested Viewing Locations: This species is too rare in the Rocky Mountain region to suggest any reliable birding sites, but grassy wetlands in the plains to the east of the mountains offer the most likely viewing possibilities.

Population: National Breeding Bird Survey trend data are not available. The North American population has been estimated at 50,000 birds (Morrison *et al.*, 2001).

Marbled Godwit (*Limosa fedoa*)

Status: A local migrant and summer resident in northern parts of the region, including eastern Alberta and northern Montana. In Alberta it breeds from the foothills north to St. Paul and Athabasca, but is absent from the montane parks.

Habitats and Ecology: On the breeding grounds this species occupies wetlands associated with prairies, including intermittent streams, ponds, and shallow lakes ranging from fresh to strongly alkaline. Semipermanent ponds and lakes are especially preferred, with nesting occurring in grassy flats nearby.

Suggested Viewing Locations: In Wyoming, Hutton Lake National Wildlife Refuge (Albany County) and Table Mountain Wildlife Unit (Goshen County) attract migrating birds (Dorn and Dorn, 1990), but the only known state breeding record occurred in Yellowstone National Park. The densest regional breeding concentrations are probably in native grasslands of southern Alberta and adjacent Montana. In Montana, key summer viewing locations include Benton Lake and Bowdoin national wildlife refuges, and the Centennial Valley region west of Yellowstone National Park (McEneaney, 1997).

Population: National Breeding Bird Survey trend data indicate a nonsignificant annual population decline of 0.8%. The North American population has been estimated at 171,500 birds (Morrison *et al.*, 2001).

Ruddy Turnstone (*Arenaria interpres*)

Status: A local migrant east of the mountains; an accidental vagrant in the montane parks.

Habitats and Ecology: On migration, these birds are likely to be found foraging on stubble fields where they sometimes forage with other species of shorebirds such as plovers, or on sandy shorelines of lakes or reservoirs.

Suggested Viewing Locations: This species is too rare in Wyoming to suggest reliable birding sites, but rocky or gravelly shorelines are favored wherever it occurs.

Population: National Breeding Bird Survey trend data are not available. The North American population has been estimated at 235,000 birds (Morrison *et al.*, 2001).

Sanderling (*Calidris alba*)

Status: A rare migrant or vagrant in the region, more common on the plains, and rare or accidental in the montane parks.

Habitats and Ecology: Migrating birds are usually seen around the larger lakes, especially those with wave-swept sandy beaches.

Suggested Viewing Locations: This species is too rare in Wyoming to suggest reliable birding sites, but sandy shorelines of larger bodies of water are favored. Lake DeSmet (Johnson County) and Table Mountain Wildlife Unit (Goshen County) are often used, as is Loch Katrine (Park County) (Scott, 1993).

Population: National Breeding Bird Survey trend data are not available. The North American population has been estimated at 300,000 birds (Morrison *et al.*, 2001).

Semipalmated Sandpiper (*Calidris pusilla*)

Status: A migrant throughout the region, fairly common in the plains, but only a vagrant or rare migrant in the montane parks.

Habitats and Ecology: A very close relative of the western sandpiper, this species is more prone to occur on wet and dry mud, where it often picks up surface organisms, while the western sandpiper is more often found standing in water or in wet mud, where it probes for food. The semipalmated sandpiper is also less prone to move out into grassy flats to forage than is the Baird's sandpiper.

Suggested Viewing Locations: In Wyoming, Hutton Lake National Wildlife Refuge (Albany County), the Bridger Power Plant ponds (Sweetwater County), Loch Katrine (Park County) and Table Mountain Wildlife Unit (Goshen County) attract migrating birds (Dorn and Dorn, 1990; Scott, 1993).

Population: National Breeding Bird Survey trend data are not available. The North American population has been estimated at 3.5 million birds (Morrison *et al.*, 2001).

Western Sandpiper (*Calidris mauri*)

Status: A migrant nearly throughout the region, mainly in the plains, and rare or accidental in the montane parks.

Habitats and Ecology: Migrants are likely to be seen in the same areas as semipalmated and least sandpipers, and frequently mingle with both these species, allowing for each comparison. Their breeding areas are farther to the west than those of these other two species, and thus they are more likely to be seen west of the Rockies than to the east.

Suggested Viewing Locations: This is a common migrant on shallow wetlands across Wyoming and the other Rocky Mountain states.

Population: National Breeding Bird Survey trend data are not available. The North American population has been estimated at 3.5 million birds (Morrison *et al.*, 2001).

Least Sandpiper *(Calidris minutilla)*

Status: A migrant throughout the region; more common on the plains, where it is generally among the commonest of the "peeps," but rare to accidental in the montane parks.

Habitats and Ecology: While on migration these sandpipers are found on a variety of moist habitats, often in company with semipalmated, Baird's, or western sandpipers, and probably feeding on much the same invertebrate foods as these species.

Suggested Viewing Locations: In Wyoming, Hutton Lake National Wildlife Refuge (Albany County), Table Mountain Wildlife Unit (Goshen County) and the south end of Boysen Reservoir (Fremont County) attract migrating birds (Dorn and Dorn, 1990).

Population: National Breeding Bird Survey trend data are not available. Morrison *et al.* (2001) estimated the North American population at 600,000 birds.

White-rumped Sandpiper *(Calidris /fuscicollis)*

Status: A local migrant through the eastern portions of the region, east of the Rockies; absent from the montane parks.

Habitats and Ecology: Migrants utilize the same kinds of prairie ponds as do the other "peeps," but on the breeding grounds the birds seek out wet tundra around the edges of ponds or lakes.

Suggested Viewing Locations: This is a local migrant on shallow, grassy wetlands across Wyoming and the other Rocky Mountain states.

Population: National Breeding Bird Survey trend data are not available. The North American population has been estimated at 400,000 birds (Morrison *et al.*, 2001).

Baird's Sandpiper *(Calidris bairdii)*

Status: A migrant throughout the region; more common on the plains and rare in the montane parks.

Habitats and Ecology; Migrants are associated with wet meadows and shallow ponds, often feeding in grassy areas somewhat away from water, but also along muddy shorelines, where they tend to peck at food sources rather than to probe for them.

Suggested Viewing Locations: In Wyoming, Hutton Lake National Wildlife Refuge (Albany County), Table Mountain Wildlife Unit (Goshen County), the south end of Boysen Reservoir (Fremont County) and Yellowstone Lake area attract migrating birds (Dorn and Dorn, 1990). Bigger lakes, such as Keyhole Reservoir, and Loch Katrine (Park County) are favored by migrants in Wyoming (Scott, 1993). Migrants throughout the region use similar shallow wetland habitats.

Population: National Breeding Bird Survey trend data are not available. The North American population has been estimated at 300,000 birds (Morrison *et al.*, 2001).

Pectoral Sandpiper (*Calidris melanotos*)

Status: A migrant throughout the region, more common on the plains; rare or accidental in the montane parks.

Habitats and Ecology; Migrants are commonly seen along prairie marshes or potholes, where they wade in shallow water and probe or peck for food. Often found near grassy cover rather than on open mud flats.

Suggested Viewing Locations: In Wyoming, Hutton Lake National Wildlife Refuge (Albany County), Table Mountain Wildlife Unit (Goshen County) and Hawk Springs Reservoir (near La Grange) attract migrating birds (Dorn and Dorn, 1990), as do many wetlands with small grassy areas (Scott, 1993).

Population: National Breeding Bird Survey trend data are not available. The North American population has been estimated at 400,000 birds (Morrison *et al.*, 2001).

Dunlin (*Calidris alpinis*)

Status: A rare migrant in the region, with few records, and none for the montane parks.

Habitats and Ecology: Migrant birds are likely to be seen with other small sandpipers such as the "peeps," and usually occur on mud flats or sandy beaches, where they probe for food.

Suggested Viewing Locations: This species is too rare in Wyoming to suggest reliable birding sites, but sandy or muddy shorelines are favored. It has been reported at places such as the Bridger Power Plant ponds (Sweetwater County) and Lake DeSmet (Johnson County) (Scott, 1993).

Population: National Breeding Bird Survey trend data are not available. The North American population has been estimated at 1.5 million birds (Morrison *et al.*, 2001).

Stilt Sandpiper (*Calidris himantopus*)

Status: A local migrant, mainly east of the mountains; very rare to absent from the montane parks.

Habitats and Ecology: Migrants are usually found in company with the typical "peeps," but usually are wading in belly-deep water, and thrusting their bills at organisms or probing the bottom with their rather long bills. They are fairly gregarious on migration, and often occur in moderately large flocks.

Suggested Viewing Locations: In Wyoming, Table Mountain Wildlife Unit (Goshen County) attracts migrating birds (Dorn and Dorn, 1990). Hutton Lake National Wildlife Refuge is a favored Wyoming site in spring (Scott, 1993).

Population: National Breeding Bird Survey trend data are not available. The North American population has been estimated at 200,000 birds (Morrison *et al.*, 2001).

Short-billed Dowitcher (*Limnodromus griseus*)

Status: Usually less common on migration than the long-billed, although in Alberta it is likely that the reverse is true. Breeding occurs in northern Alberta, south to about Edmonton.

Habitats and Ecology: On migrations, dowitchers are found in grassy marshes, where they feed by probing their long bills in belly-deep water.

Suggested Viewing Locations: In Wyoming, this species is too rare to suggest good birding areas, but during migration it forages in the same shallow-water habitats as those used by long-billed dowitchers. See that species' account for suggested locations.

Population: National Breeding Bird Survey trend data are not available. The North American population has been estimated at 320,000 birds (Morrison *et al.*, 2001).

Long-billed Dowitcher (*Limnodromus scolopaceus*)

Status: A migrant throughout the region, rarer in montane areas and virtually absent from the montane parks.

Habitats and Ecology: Migrating birds use the same habitats as do short-billed dowitchers.

Suggested Viewing Locations: In Wyoming, Hutton Lake National Wildlife Refuge (Albany County), Table Mountain Wildlife Unit (Goshen County) and Lowell Lakes (Bighorn County) attract migrating birds (Dorn and Dorn, 1990), as do Keyhole Reservoir (Crook County), the Bridger Power Plant ponds (Sweetwater County), and Loch Katrine (Park County) (Scott, 1993). Prairie marshes east of the mountains are the best regional birding sites, such as Monte Vista and Alamosa national wildlife refuges in Colorado, or Bowdoin and Medicine Lake refuges in Montana.

Population: National Breeding Bird Survey trend data are not available. The North American population has been estimated at 500,000 birds (Morrison *et al.*, 2001).

Wilson's Snipe (*Gallinago gallinago*)

Status: A summer or year-round resident nearly throughout the region, both in mountains and plains wetlands, and probably breeding in all of the montane parks.

Habitats and Ecology: Wilson's (previously known as "Common") snipes nest in marshy areas, often using beaver ponds in the Rocky Mountains, and muskeg ponds or other heavily vegetated marshes elsewhere in their extensive range. Peatland habitats are especially favored, but the birds may also occur along slow-moving rivers, marshy shorelines of lakes, or sometimes even wet hayfields.

Suggested Viewing Locations: This species is a fairly common migrant on lower-altitude wetlands of Wyoming and a local but inconspicuous breeder. It is common in montane meadows, such as in Grand Teton and Yellowstone National Parks. Elsewhere in Wyoming it is widespread, especially along the floodplains of larger rivers and in intermountain valleys (Faulkner, 2010). In Colorado the birds are most abundant in wet mountain parks and meadows at altitudes of 7,500 to 10,000 feet, especially northwardly (Kingery, 1998). In Idaho they are abundant nesters at Grays Lake National Wildlife Refuge, and common at Camas and Bear Lake refuges. In Montana they are abundant nesters at Red Rock Lakes National Wildlife Refuge, and common at Bowdoin, C. R. Russell, Lee Metcalf, and Swan River refuges. The densest regional breeding concentrations are probably in wetlands of southern Alberta, and throughout much of Montana and Idaho.

Population: National Breeding Bird Survey trend data indicate a nonsignificant annual population decline of 0.3%. Morrison *et al.* (2001b) estimated the North American population at about two million birds.

Wilson's Phalarope *(Phalaropus tricolor)*

Status: A summer resident over most of the region's lowlands, becoming rarer in the mountains, and a rare migrant in most of the montane parks. Breeding there has been reported only from Yellowstone National Park.

Habitats and Ecology: Breeding habitats are typically wet meadows adjoining shallow marshes, which range from fresh to highly saline. Ditches, river edges, and shallow lakes are sometimes also used for breeding. Migrating birds use similar areas.

Suggested Viewing Locations: This species is a fairly common migrant on lower-altitude wetlands of Wyoming (Ocean Lake in Fremont County, and most shallow lakes and marshes), and a local breeder on freshwater and especially saline wetlands wherever crustaceans such as brine shrimp are often abundant. In Colorado, the highest breeding concentrations are probably in North Park, the San Luis valley, the Gunnison valley, and the Yampa watershed (Kingery, 1998). The densest regional breeding concentrations are probably in wetlands of southern Alberta, northern and central Montana (abundant at Medicine Lake National Wildlife Refuge, and common at Bowdoin), and southeastern Idaho (common at Bear Lake, Camas, and Grays Lake refuges).

Population: National Breeding Bird Survey trend data indicate a nonsignificant annual population decline of 0.3%. The North American population has been estimated at 1.5 million birds (Morrison *et al.,* 2001).

Red-necked Phalarope *(Phalaropus lobatus)*

Status: An uncommon to rare migrant throughout the region, mainly in the plains, with most montane records in the Alberta parks. The nearest breeding region is in northern Saskatchewan or possibly extreme northern Alberta.

Habitats and Ecology: Breeding habitats of this species are subarctic ponds, marshes, and lagoons having adjacent grassy or sedge vegetation, where nesting occurs. Proximity to lakes or other fairly permanent bodies of water may also be a part of the habitat characteristics. On migration the birds are found in the same areas as Wilson's phalaropes, and often are seen in company with them.

Suggested Viewing Locations: This species is too rare in Wyoming to suggest reliable birding sites, but migrants favor saline wetlands that also used by the Wilson's phalarope. Migrants are often seen around larger lakes such as Table Mountain Wildlife Unit (Goshen County) , Ocean Lake, Hutton Lake, and wetlands such as the Bridger Power Plant ponds (Sweetwater County) (Scott, 1993).

Population: National Breeding Bird Survey trend data are not available. The North American population has been estimated at 2.5 million birds (Morrison et al., 2001).

Gulls and Terns

Sabine's Gull *(Xema sabini)*
Status: A rare migrant or vagrant in the region; accidental in the montane parks. A high-arctic breeder, the nearest nesting areas are in extreme northern Canada. It is most likely to be encountered in northern portions of the region.

Habitats and Ecology: This arctic nester is most likely to be observed in flocks of migrating Franklin's or Bonaparte's gulls, but it is extremely rare south of Canada.

Suggested Viewing Locations: This species is too rare regionally to suggest reliable birding sites.

Population: National Breeding Bird Survey trend data are not available.

Bonaparte's Gull *(Chroicocephalus philadelphia)*
Status: An uncommon to rare migrant throughout the region, mainly in the plains regions, and rare or accidental in the montane parks. The nearest breeding areas are in central Alberta (from Battle Lake and Edmonton northward), in muskeg forests.

Habitats and Ecology: This gull is unique in its tree-nesting adaptations; it typically nests in small coniferous trees well above ground level, but at times also nests in reedbeds of marshes. Jack pines, spruces, and other conifers are the usual nesting site; typically the mate stands watch in a nearby tree as the other bird incubates. The birds often nest in loose colonies, and outside the breeding season they are highly gregarious, often forming flocks numbering in the hundreds.

Suggested Viewing Locations: In Wyoming, migrants often may be found at Hutton Lake National Wildlife Refuge (Albany County), Grayrocks Reservoir (Platte County), and Goldeneye Wildlife Area (Natrona County) (Dorn and Dorn, 1990). In Colorado possible viewing sites include Lake Pueblo State Park (near Pueblo), Prewitt Reservoir (near Sterling) and several Larimer County reservoirs near Fort Collins (Kingery, 2007). In Montana Freezeout Lake Wildlife Management Area is a potential viewing location (McEneaney, 1993).

Population: National Breeding Bird Survey trend data are not available. An estimate of its total population in the 1990's was 85,000–175,000 pairs (Burger and Gochfeld, 2002).

Franklin's Gull (*Larus pipixcan*)

Status: A migrant and local summer resident in the region, primarily on the plains; relatively rare in the montane parks, but there is an undocumented breeding record from Yellowstone National Park, where it is rare in summer.

Habitats and Ecology: Breeding occurs in large, relatively permanent prairie marshes having extensive stands of emergent vegetation, where the birds nest in colonies. Unlike other gulls of the region, the nest is constructed over water, in dense vegetation. On migration they typically feed on dry land, often in fields that are being cultivated prior to planting.

Suggested Viewing Locations: In Wyoming, migrants often may be found at Hutton Lake National Wildlife Refuge (Albany County) and Goldeneye Wildlife Area (Natrona County). During summer the Cokeville Meadows/Bear River area (Lincoln County) should have breeding Franklin's gulls, but this remains unproven because of difficulty of access (Dorn and Dorn, 1990; Faulkner. 2010). The densest regional breeding concentrations are probably in wetlands of southern Alberta, northern and central Montana (abundant at Benton, Bowdoin and Medicine Lake refuges), and eastern Idaho, where they are abundant at Bear Lake (up to 13,000 nesting) and Grays Lake national wildlife refuges.

Population: National Breeding Bird Survey trend data indicate a nonsignificant annual population increase of 3.3%.

Ring-billed Gull (*Larus delawarensis*)

Status: A summer resident and local colonial breeder over most of the region, primarily on the plains; reported breeding in the montane parks is limited to Yellowstone, where the last known breeding was in 1949. The range is gradually expanding in western North America, and there are now numerous colonies in the region.

Habitats and Ecology: This is a highly adaptable gull, exploiting new habitats in the form of reservoirs. Breeding usually occurs on isolated and sparsely vegetated islands of lakes and reservoir impoundments, sometimes in colonies of a thousand pairs or more.

Suggested Viewing Locations: This widespread species may be seen during migration at many Wyoming lakes and reservoirs, such as at Yellowstone Lake, Hutton Lake National Wildlife Refuge (Albany County), Ocean Lake (Fremont County), and Grayrocks Reservoir (Platte County) (Dorn and Dorn, 1990). Breeding records in Wyoming have been limited to birds periodically nesting among California gulls at Yant's Puddle near Casper, and two years of nesting at Ocean Lake (Faulkner, 2010). The birds are likely to be seen in any wetland in the entire region, or even around city parks or landfills. The densest regional breeding concentrations are probably in wetlands of southern Alberta, northern and central Montana (*e.g.*, Bowdoin, C. R. Russell, Medicine Lake and Ninepipe national wildlife refuges), and southern Idaho (Camas and Deer Flat national wildlife refuges).

Population: National Breeding Bird Survey trend data indicate a significant annual population increase of 1.3%. Trend estimates for the Rocky Mountains & Plains States region (U.S.F.W.S. Region 6) indicate a significant annual increase of 4.7%. Its total world population has been estimated at 3–4 million birds, most of which nest in Canada (Olsen and Larson, 2004; Alderfer, 2006).

California Gull (*Larus califomicus*)

Status: A summer resident and local breeder over much of the region, mainly on the plains; the only breeding in the montane parks is in Yellowstone National Park, where 200–300 pairs breed yearly on the Molly Islands of Yellowstone Lake. The breeding range of the species in the general region is increasing.

Habitats and Ecology: Like the ring-billed gull, this species usually nests on gravelly islands of large lakes or reservoirs or along their shorelines, and in many areas the two species nest in close proximity. In Alberta the California gulls tend to nest on more elevated and boulder-strewn sites, while ring-bills occupy more level terrain. Ring-billed gulls also tend to cluster their nests, while California gulls space their nests more randomly.

Suggested Viewing Locations: Wyoming locations for seeing this species include the lakes on the Laramie Plains, Ocean Lake (Fremont County), Yellowstone Lake, Flaming Gorge Reservoir, and wetlands in the Casper area (Dorn and Dorn, 1990). A colony of up to 2,000 nests has been reported at Yant's Puddle near Casper, and colonies have been regular at Bamforth and Ocean lakes, Pathfinder Reservoir, and Yellowstone Lake (Scott, 1993; Faulkner, 2010). For a few years nesting occurred at Hutton Lake National Wildlife Refuge. The densest regional breeding concentrations are probably in wetlands of southern Idaho (especially Camas, Deer Flat and Minidoka refuges), central and southeastern Wyoming, southern Alberta, and north-central Montana (abundant at Bowdoin, C. R. Russell, Medicine Lake and Red Rock Lakes refuges. In Colorado large nesting colonies (up to 1,500 birds) have been reported at Riverside and Antero reservoirs, and smaller ones at Elevenmile, Adobe Creek, North Park and John Martin (Kingery, 1998). Like the ring-billed gull, this is an adaptable and easily seen species.

Population: National Breeding Bird Survey trend data indicate a nonsignificant annual population decline of 1.2%. Trend estimates for the Rocky Mountains & Plains States region (U.S.F.W.S. Region 6) indicate a significant annual increase of 7.9%. Its total North American population has been estimated at 50,000–100,000 birds (Olsen and Larson, 2004; Alderfer, 2006).

Herring Gull (*Larus argentatus*)

Status: An uncommon to rare migrant or vagrant in much of the region, becoming more common northwardly in the montane parks. The nearest regular breeding areas are in northern Alberta, south to Namur Lake and Lower

Therien Lake, but some local breeding has occurred in central Alberta, and an extra-limital breeding in southeastern Wyoming has been reported.

Habitats and Ecology: This is primarily a coastal gull, but it also breeds in small colonies across northern Canada on the islands of larger lakes, sometimes among colonies of ring-billed or California gulls where they often nest as single pairs. They usually winter coastally, but sometimes spend the winter on ice-free lakes or impoundments in the more southerly states.

Suggested Viewing Locations: In Wyoming, and elsewhere, diverse lakes, reservoirs and rivers attract these birds as migrants and non-breeding residents, such as Ocean Lake (Fremont County) and the lakes of the Laramie Plains (Dorn and Dorn, 1990).

Population: National Breeding Bird Survey trend data indicate a nonsignificant annual population decline of 4.7%. Its Atlantic Coast population was estimated at 500,000 pairs in the 1980's (Olsen and Larson, 2004; Alderfer, 2006).

Caspian Tern *(Hydroprogne caspia)*

Status: An uncommon to rare migrant and local summer resident in the region. It is a regular breeder in Yellowstone National Park (on the Molly Islands), Pathfinder Reservoir, Yant's Puddle, and Ocean Lake, and has periodically nested elsewhere, such as at Bamforth Lake (Faulkner, 2010). It has bred at least once in Montana, and nests regularly in southern Idaho.

Habitats and Ecology: This species usually nests near coastlines, but has also nested interiorly on shorelines or islands of large lakes or reservoirs, usually on sandy or stony beaches. Often in these locations only one or two pairs nest among other terns or gulls, but normally nesting is done in colonies.

Suggested Viewing Locations: In Wyoming, non-breeding birds gather on various lakes, such those on the Laramie Plains, in the Casper area, and around regular nesting areas such as Yellowstone Lake and Pathfinder Reservoir. In Idaho, breeding is regular at Bear Lake and Deer Flat national wildlife refuges.

Population: National Breeding Bird Survey trend data indicate a nonsignificant annual population increase of 0.8%.

Black Tern *(Chlidonias niger)*

Status: A summer resident over much of the region, mainly in plains marshlands; rarer in the montane parks but breeding occasionally or regularly in several.

Habitats and Ecology: Typical nesting habitat consists of small to large marshes with extensive stands of emergent vegetation and some areas of open water. Fish populations are not necessary, as the birds feed mostly on insects while on the nesting grounds. Nests are more often placed among emergent vegetation than on muskrat houses, although the latter are sometimes used.

Suggested Viewing Locations: In Wyoming, migrants or summering birds may be found at the lakes on the Laramie Plains, the Table Mountain Wildlife Unit

(Goshen County) and the Cokeville/Bear River area (Dorn and Dorn, 1990). They breed regularly at sites on the Laramie Plains, such as at Hutton Lake National Wildlife Refuge and Cokeville National Wildlife Refuge, and periodically probably nest elsewhere (Faulkner, 2010). The densest regional breeding concentrations are probably in wetlands of southern Alberta, and locally in Montana (common at Bowdoin, Medicine Lake, Ninepipe, Swan Lake and Red Rock Lakes refuges), southeastern British Columbia, and Idaho (common at Bear Lake, Camas, Deer Flat, Grays Lake, Kootenai, and Minidoka refuges). Black terns now nest sparingly in Colorado, such as at Alamosa and Arapaho national wildlife refuges, and at San Luis Lake State Wildlife Area (Kingery, 1998).

Population: National Breeding Bird Survey trend data indicate a significant annual population increase of 2.0%. Trend estimates for the Rocky Mountains & Plains States region (U.S.F.W.S. Region 6) indicate a significant annual increase of 2.7%. It is considered a Priority II species in Wyoming, and classified as imperiled by the Wyoming Natural Heritage Program (Finch, 1992).

Common Tern (*Sterna hirundo*)

Status: A regional migrant and local summer resident from central Montana northward, mainly on plains lakes; rare in the montane parks. Although breeding has reportedly occurred in Yellowstone National Park this was not documented, and there is no current evidence of this.

Habitats and Ecology: Islands in large lakes are favored breeding grounds in this region; sparsely vegetated areas are used for colonial nesting. Occasionally a pair or two will also build nests in reedy vegetation over water, but this behavior is much more typical of Forster's terns. Sometimes the two species will nest in close proximity, but normally they are well isolated ecologically.

Suggested Viewing Locations: In Wyoming, this species might be seen during migration on many larger lakes.). There is some localized breeding in southern Alberta, and in Montana such as Benton, Bowdoin and Medicine Lake national wildlife refuges. They also breed at Freezeout Lake Wildlife Management Area and Nelson Reservoir (McEneaney, 1993).

Population: National Breeding Bird Survey trend data indicate a nonsignificant annual population decline of 6.1%.

Forster's Tern (*Sterna forsteri*)

Status: A summer resident locally in the region, mainly on plains marshes; rare or accidental in the montane parks, with no breeding records.

Habitats and Ecology: Large marshes having extensive reedbeds or muskrat houses for nest sites are the typical breeding habitats of this species, which breeds colonially in such locations, with as many as five nests sometimes situated on a single muskrat house. Such sites that are close to open water areas for foraging are especially favored nesting locations.

Suggested Viewing Locations: In Wyoming, migrants and summering birds may be found at the lakes on the Laramie Plains, Ocean Lake (Fremont County),

Lovell Lakes (Big Horn County), Cliff Graham Reservoir (Uinta County), and Goldeneye Wildlife Area (Natrona County) (Dorn and Dorn, 1990). Breeding has been documented in only a few Wyoming sites, such as on the Laramie Plains, at Ocean Lake, and at Cokeville National Wildlife Refuge (Faulkner, 2010). The densest regional breeding concentrations are probably in wetlands of southern Idaho (common at Minidoka National Wildlife Refuge), but extensive breeding also occurs in southern Alberta and Montana (common at C. R. Russell, Ninepipe and Red Rock Lakes refuges).

Population: National Breeding Bird Survey trend data indicate a nonsignificant annual population increase of 0.6%. Trend estimates for the Rocky Mountains & Plains States region (U.S.F.W.S. Region 6) indicate a significant annual decline of 1.0%.

Pigeons and Doves

Rock Pigeon (*Columba livia*)

Identification: This is the familiar barnyard pigeon, well known to everyone, living in the wild state. Wild-type rock doves rather resemble band-tailed pigeons, but lack yellowish bills and a white band, although the plumage patterns sometimes vary greatly. Rarely found far from humans, but at times living on cliffs or other natural sites.

Status: Present virtually throughout the region, although rare or lacking in high montane areas, and declining northwardly.

Habitats and Ecology: Largely associated with cities and farms in North America, and infrequent in forested areas. Buildings that provide narrow nesting ledges are preferred for nesting, but cliff ledges or crevices are sometimes also used.

Suggested Viewing Locations: Found throughout the region in all cities and towns.

Population: National Breeding Bird Survey trend data indicate a nonsignificant annual population decline of 0.2%.

Band-tailed Pigeon (*Columba fasciata*)

Status: Limited to the southernmost part of the region, breeding in Rocky Mountain National Park and its vicinity, north to the Wyoming border. It is a vagrant farther north, rarely to Montana and southern Alberta.

Habitats and Ecology: This species is generally associated with western oak woodlands or mixed oak and pine woodlands, and extending into the ponderosa pine zone locally, especially where Gambel's oaks are also present. Available foods in the form of acorns are an important determinant of local distributions. During July and August it is found up to 10,000 feet in Rocky Mountain National Park, and gradually moves to lower altitudes in late summer.

Suggested Viewing Locations: In Wyoming, this species is most often seen along the southern edge of the state, in arid-adapted pines and scrub oak woodlands. The densest regional breeding concentrations are probably in the

woodlands of southern Colorado; they are generally most abundant near
the New Mexico border. Junction Creek and Hermosa Creek, near Durango,
and Glenwood Canyon, near Glenwood Springs, offer good viewing opportunities (Kingery, 1998).

Population: Classified as a Partners-in-Flight Watch List species of continental
conservation importance (Rich *et al.*, 2004). National Breeding Bird Survey trend data indicate a nonsignificant annual population decline of 1.4%.
The North American population (north of Mexico) has been estimated at
975,000 birds (Rich *et al.*, 2004).

Mourning Dove (*Zenaida macroura*)

Status: A widespread and common breeder in the region, occupying nearly all
vegetational zones up to the lower coniferous forest zone. Present and probably breeding in all of the montane parks.

Habitats and Ecology: Breeds from riparian woodlands and cultivated areas
through grasslands and sagebrush to woodlands, aspen, and open coniferous forest habitats, as well as in cities and farmsteads. Nests either on the
ground or, preferentially, in shrubs or trees.

Suggested Viewing Locations: This is an extremely widespread species that is
easily found throughout the region, both in towns and countryside. The
densest regional breeding concentrations are probably in the riparian
woodlands of eastern Montana, eastern Wyoming and eastern Colorado.

Population: National Breeding Bird Survey trend data indicate a nonsignificant
annual population decline of 0.01%. The North American population (north
of Mexico) has been estimated at 110.5 million birds (Rich *et al.*, 2004).

Eurasian Collared-Dove (*Streptopelia decaocto*)

Status: A recently invasive and rapidly expanding species that now covers the
entire region at lower altitudes. During the 2009 Audubon Christmas Bird
Count more than 1,000 were seen around Pueblo, Colorado, several hundred were found at Casper and Lander, Wyoming, over 200 were counted
at Bozeman, Montana, and 168 at Salmon, Idaho. The Jackson Hole, Wyoming, count reported 38 birds; the species was first seen there in 2007.
There is also at least one Yellowstone National Park record (Faulkner, 2010).

Habitats and Ecology: Usually found in smaller towns and villages, especially
around feed lots or granaries where waste grain is abundant. It is a year-
round resident in Colorado and Wyoming, and a probable seasonal resident
in Montana, northern Idaho and southern Canada.

Suggested Viewing Locations: This species is most likely to be seen near grain el-
evators in an small towns of the plains, or in other locations where waste
grain is available.

Population: National Breeding Bird Survey trend data indicate a significant annual population increase of 5.4%. Trend estimates for the Rocky Mountains & Plains States region (U.S.F.W.S. Region 6) indicate a significant (and
amazing) annual increase of 65.1%, which must reflect major range expan-
sions as well as high reproductive rates.

Cuckoos

Yellow-billed Cuckoo (*Coccyzus americanus*)

Status: A summer resident in the southern and eastern parts of the region, mainly east of the mountains, but breeding occurs west locally at least to southern Idaho, and north to southeastern Montana. Rare at higher elevations, and absent from the montane parks. Declining nationally at a substantial rate.

Habitats and Ecology: Associated with thickety areas near water, second-growth woodlands, deserted farmlands, and brushy orchards. Dense woodlands are avoided.

Suggested Viewing Locations: In Wyoming, the North Platte riverbottoms of Goshen County, the Powder River bottoms in Sheridan and Campbell counties, and Ash Creek north of Sheridan provide good birding sites for this species (Dorn and Dorn, 1990). The densest (but increasingly rare) regional breeding concentrations are apparently in the North Platte woodlands of southeastern Wyoming and eastern Colorado, the latter in the riparian woodlands of the South Platte and Arkansas river valleys.

Population: National Breeding Bird Survey trend data indicate a significant annual population decline of 1.7%. Trend estimates for the Rocky Mountains & Plains States region (U.S.F.W.S. Region 6) indicate a significant annual decline of 0.9%. The North American population (north of Mexico) has been estimated at 8.46 million birds (Rich *et al.*, 2004). It is classified as a Priority II species in Wyoming.

Black-billed Cuckoo (*Coccyzus erythropthalmus*)

Status: An uncommon summer resident east of the mountains in plains woodlands; a rare vagrant in the montane parks. Breeds most commonly on the plains of southern Alberta and northern Montana. Declining nationally at a substantial rate.

Habitats and Ecology: Associated during the breeding season with somewhat dense woodland cover, such as upland woods with a variety of trees, shrubs, and vines, offering shady hiding places and nest sites.

Suggested Viewing Locations: In Wyoming, the North Platte River bottoms of Goshen County, the Bighorn River bottoms in the Big Horn Basin, and Ash Creek north of Sheridan provide good birding sites for this species (Dorn and Dorn, 1990). The greatest number of Wyoming's Breeding Bird Survey observations have come from Big Horn and Natrona counties (Faulkner, 2010). In Montana the Bighorn National Recreation Area is a good birding location for this species; it is an uncommon nester at Medicine Lake and C. R. Russell national wildlife refuges. The densest regional breeding concentrations are probably in the riparian woodlands of eastern Montana and northeastern Wyoming.

Population: National Breeding Bird Survey trend data indicate a nonsignificant annual population decline of 1.8%. Trend estimates for the Rocky Mountains & Plains States region (U.S.F.W.S. Region 6) indicate a significant annual decline of 4.5%. The North American population has been estimated at 1.1 million birds (Rich *et al.*, 2004).

Barn Owls

Barn Owl (*Tyto alba*)

Status: Barn owls are unreported from the region's national parks. They are mostly limited to lower altitudes in the eastern halves of Wyoming and Colorado, but also occur in the valleys of western Colorado. They are resident or summer migrants in southwestern Idaho, north almost to the Washington State line, and are local in western Montana.

Habitats and Ecology: Open country, which provide ready-made nesting sites such as abandoned buildings, or road cuts in fairly soft earth that offer the possibility of nesting burrows may attract barn owls. They also need a reliable supply of small rodents.

Suggested Viewing Locations: In Wyoming, barn owls are rare, but a local population exists in the North Platte valley from Casper eastward (Scott, 1993). Most of the state's breeding records have come from Goshen County, (Faulkner, 2010). In Colorado the eastern reservoirs such as the Queens and Two Buttes State Wildlife Areas near Lamar offer some of the best chances of encountering these shy birds (Kingery, 2007). In Idaho, barn owls are most widespread and likely to be encountered along the Snake River plain, and are common at Deer Flat National Wildlife Refuge.

Population: National Breeding Bird Survey trend data indicate a nonsignificant annual population decline of 1.9%. The North American population (north of Mexico) has been estimated at 343,000 birds (Rich *et al.*, 2004).

Typical Owls

Flammulated Owl (*Otus flammeolus*)

Status: Probably more common and widespread than currently known. Breeds in western Idaho, and present during the breeding season in the River of No Return Wilderness area, and in Sawtooth and Caribou national forests. There is at least one breeding record for Rocky Mountain National Park (Kingery, 1998), and the birds are widespread in Colorado's pine forests. They also occur along the Idaho–Montana border, and in the mountains of southeastern Wyoming.

Habitats and Ecology: Associated with aspen and ponderosa pine forests in both breeding and non-breeding periods, particularly ponderosa pines. In the interior Northwest, this species begins breeding in mixed coniferous forests during the young forest stage of succession (Sanderson, Bull and Edgerton, 1980). The only summer reports for Wyoming are from Carbon county, with the first actual nesting record not obtained until 2005

(Faulkner, 2010). In Colorado nesting from 6,000–10,000 feet has been noted, usually in pines or aspens with woodpecker holes about ten to twenty feet above ground. Old-growth pinyon–juniper woodlands are sometimes also used.

Suggested Viewing Locations: This inconspicuous and highly nocturnal owl is notoriously hard to find, but Maclay Flat Recreation Area near Missoula, Montana, is one place where it might profitably be sought out (McEneaney, 1997). In Colorado, where the species has best been studied, most breeding records come from the southwestern corner of the state (Kingery, 1998).

Population: National Breeding Bird Survey trend data are not available. The North American population (north of Mexico) has been estimated at 28,500 birds (Rich *et al.*, 2004).

Western Screech-Owl (*Otus kennicottii*) and Eastern Screech-Owl (*Otus asio*)

Status: Range limits of these two owls are still uncertain, but at least in most Western Slope montane areas of the region the western form is present. From the eastern slope of Montana's Continental Divide and the eastern slopes of Wyoming's Wind River Range eastward the eastern species is apparently the resident form. Screech-owls probably breed in all of the montane parks south of Canada, but become rarer northwardly, and are apparently absent from the Alberta parks.

Habitats and Ecology: Associated with a variety of wooded habitats, including farmyards, cities, orchards, *etc.*, and from riparian edges through pinyon–juniper and oak-mahogany woodlands to aspens and ponderosa pine forests. In Wyoming the eastern species is found exclusively in large riparian cottonwood trees, where average annual temperatures are above 44° F., while the western species has been reliably reported from Park, Teton and Sublette counties (Faulkner, 2010).

Suggested Viewing Locations: Screech-owls are often seen or heard in city parks, woodlots, or other areas that are fairly close to humans. They are most easily seen by imitating their calls, or using playbacks of recordings to lure them closer. Breeding density data are not available for this species. In Wyoming, eastern screech-owls have been reported around Casper, Wheatland, in Sybille Canyon (between Wheatland and Bosler), and in Sheridan (Scott, 1993). In Montana the eastern species probably extends west across the plains and montane forests to the continental divide, and in Colorado they also occur on the eastern slope of the Front Range at altitudes up to at least 9,000 feet. In all three states riparian woodlands or city parks well grown to large trees such as cottonwoods often provide nesting sites for this owl. Western screech-owls use essentially the same habitats, such as developed urban areas and woodlots to riparian woods with large cottonwoods. Some range contact or overlap may occur in southeastern Colorado, where both species reportedly occur in Bent County (Kingery, 1998).

Population: National Breeding Bird Survey trend data indicate a significant annual population decline of 6.3% for the western species, and a nonsignificant annual population increase of 1.2% for the eastern species. The North American population (north of Mexico) has been estimated at 511,000 birds for the western species, and 739,000 for the eastern (Rich *et al.*, 2004).

Great Horned Owl *(Bubo virginianus)*
Status: A common resident in wooded habitats throughout the region; probably breeds in all the montane parks.
Habitats and Ecology: A powerful and adaptable owl, this species occurs everywhere from riparian woodlands through the coniferous forest zones, and extends into city parks, farm woodlots, and rocky canyons well away from trees. Nesting is thus highly variable, but often occurs in abandoned bird or squirrel nests, or on tree crotches, rock ledges, or even on the ground.
Suggested Viewing Locations: Like screech-owls. great horned owls sometimes live quite close to humans, such as in well-wooded city parks, and can be detected by using play-backs or imitations of their calls. Breeding densities are quite uniform throughout the entire region.
Population: National Breeding Bird Survey trend data indicate a nonsignificant annual population decline of 0.3%. The North American population (north of Mexico) has been estimated at 2.28 million birds (Rich *et al.*, 2004).

Snowy Owl *(Bubo scandiaca)*
Status: A wintering migrant over most of the region, becoming rarer farther south. Rarely reported from the montane parks, it is usually found on open plains.
Habitats and Ecology: An arctic breeding species that periodically is forced south in winter when food supplies on the breeding grounds are limiting.
Suggested Viewing Locations: Open fields attract this arctic visitor, but its abundance and choice of locations are unpredictable from year to year.
Population: National Breeding Bird Survey trend data are not available. The North American population has been estimated at 145,000 birds (Rich *et al.*, 2004).

Northern Pygmy-Owl *(Glaucidium gnoma)*
Status: An inconspicuous species that probably occurs through the northern montane forests south in Wyoming at least to Jackson Hole. It also breeds in the mountains of western Colorado, although apparently is rare in Rocky Mountain National Park.
Habitats and Ecology: Found in similar habitats to those of saw-whet and flammulated owls, but apparently ranging higher, and more active during daylight than are these species. Nesting is done in woodpecker holes or similar tree cavities. In the interior Northwest, this species begins breeding in mixed coniferous forests during the young forest stage of succession (Sanderson, Bull and Edgerton, 1980). No breeding records yet exist for Wyo-

ming, but it has been reported from the Bighorn Mountains and from along the Colorado border in Carbon and Laramie counties (Faulkner, 2010).

Suggested Viewing Locations: This tiny owl is not easily seen, but in Wyoming Grand Teton and Yellowstone National Parks are perhaps the best places to search. In Colorado, where the birds range from at least 5,400–10,100 feet, sites such as Mesa Trail, near Boulder, and East Brush Creek in Sylvan Lake State Park (Kingery, 2007, 2008) might be visited. In Montana, Kirk Hill, a site near the Museum of the Northern Rockies, offers sighting possibilities (Fischer and Fischer, 1995). In Idaho this owl is an occasional breeder at Kootenai National Wildlife Refuge, and is rare at Deer Flat National Wildlife Refuge.

Population: National Breeding Bird Survey trend data indicate a nonsignificant annual population increase of 1.9%. The North American population (north of Mexico) has been estimated at 84,000 birds (Rich *et al.,* 2004).

Burrowing Owl (*Athene cunicularia*)

Status: A summer resident on the plains over much of the region; rare in most of the montane parks, and not proven to breed in any.

Habitats and Ecology: This is the only North American owl closely associated with plains rodents such as prairie dogs, and as the range and abundance of these mammals have decreased, so too has the status of the burrowing owl. It is largely an insectivorous species, often eating large beetles, but also takes many small mice.

Suggested Viewing Locations: In Wyoming, prairie dog towns near Cheyenne, north of Pine Mountain in Sweetwater County, and east of Continental Peak along the Sweetwater–Fremont county line are good places to look for this relatively tame owl (Dorn and Dorn, 1990). They are most abundant in the eastern third of Wyoming (Faulkner, 2010). The C. R. Russell National Wildlife Refuge in Montana has a good population, as does Deer Flat National Wildlife Refuge in Idaho. The densest regional breeding concentrations are probably in the shortgrass prairies of eastern and southeastern Colorado, in locations such as Arapaho National Wildlife Refuge, and Pawnee and Comanche national grasslands.

Population: National Breeding Bird Survey trend data indicate a nonsignificant annual population decline of 1.1%. The North American population (north of Mexico) has been estimated at 620,000 birds (Rich *et al.,* 2004). It is classified as a Priority II species in Wyoming.

Barred Owl (*Strix varia*)

Status: A summer resident on the plains over much of the region; rare in most of the montane parks, and not proven to breed in any.

Habitats and Ecology: Dense bottomland woods are this owl's favorite haunts, and it occurs in both coniferous and deciduous woods, possibly preferring the former. It often nests in tree cavities, but at times also uses old hawk or crow nests. In the interior Northwest, this species begins breeding in mixed coniferous forests during the mature forest stage of succession (Sanderson, Bull and Edgerton, 1980). No Wyoming breeding records are yet known, but their discovery along the North Platte river, or in the Black Hills region, would not be surprising.

Suggested Viewing Locations: Barred owls are rarely seen but their distinctive calls are often heard, and they can be readily stimulated to respond to an imitated call. The densest regional breeding concentrations are probably in the montane forests of northeastern Idaho and (probably) the Canadian montane parks.

Population: National Breeding Bird Survey trend data indicate a significant annual population increase of 1.8%. The North American population has been estimated at 560,000 birds (Rich *et al.*, 2004).

Great Gray Owl (*Strix nebulosa*)

Status: A rare resident in the northern montane forest areas, south at least to Grand Teton National Park and the Wind River Range of Wyoming. It is known to breed in several of the montane parks, but is uncommon to rare in all.

Habitats and Ecology: In Alberta these birds usually nest in poplar woodlands, often near muskeg areas. Nests are usually in old hawk nests of various large species, from 10–80 feet above ground, in conifers or hardwood trees. In the interior Northwest, this species begins breeding in mixed coniferous forests during the young forest stage of succession (Sanderson, Bull and Edgerton, 1980).

Suggested Viewing Locations: In Wyoming, Yellowstone National Park is probably the best place to search for this owl. McEneaney (1988) judged the Yellowstone population to be less than 100 birds during the late 1980's. He noted that Canyon Junction and the Tower–Roosevelt area are excellent locations for finding these birds in Yellowstone National Park. The Jackson Hole area also is known to support breeding birds, but elsewhere in the state there are only suggestive indications of breeding (Faulkner, 2010). In Montana, great gray owls are often seen at the Blackfoot-Clearwater Wildlife Management Area south of Seeley Lake (Fischer and Fischer, 1995). Other Montana locations known to support great gray owls include Kirk Hill (see northern pygmy-owl account above)(Fischer and Fischer, 1995), Glacier National Park (near the Summit Siding campground), around Georgtown Lake, and Kings Hill (south of Anaconda). In Idaho, some of the places where this owl has been found include Bear Basin (Valley County), Cascade Reservoir (Valley County), Henrys Lake State Park, between Aston and lower Mesa Falls along ID Highway 47 (Fremont County), and Teton Valley (Teton County)(Svingen & Dumroese, 1997).

Population: National Breeding Bird Survey trend data are not available. The North American population has been estimated at 31,500 birds (Rich *et al.*, 2004).

Long-eared Owl (*Asio otus*)

Status: Resident over much of the region, except in Alberta, where relatively rare in montane areas, but known to have bred at Banff National Park and a regular breeder in aspen parklands and prairie coulees.

Habitats and Ecology: A widespread species, often associated with coniferous or deciduous forests, but also found in woodlots, orchards, large wooded parks, and even sagebrush or pinyon–juniper woodlands during the breeding season. Trees surrounded by open country seem to be favored for nesting. In the interior Northwest, this species begins breeding in mixed coniferous forests during the young forest stage of succession (Sanderson, Bull and Edgerton, 1980). In Colorado they have been found at elevations as high as 10,000 feet.

Suggested Viewing Locations: In Wyoming, these elusive owls might be found almost anywhere in woodlands, ranging from riparian shrubs to mature and dense montane forests. They are sparsely distributed across the state, usually being found below 9,000 feet in areas of mixed woodlands and more open landscapes (Faulkner, 2010). They especially favor isolated stands of conifers during winter (Dorn and Dorn, 1990; Kingery, 1998). The same is generally true elsewhere in the entire region.

Population: National Breeding Bird Survey trend data are not available. The North American population (north of Mexico) has been estimated at 36,000 birds (Rich *et al.*, 2004).

Short-eared Owl (*Asio flammeus*)

Status: Associated with open meadows and marshes.

Habitats and Ecology: This is a prairie-adapted species, usually breeding in areas of grassland, marshes, arctic tundra, and low brushland. Nests are usually on the ground, but sometimes in burrows. More diurnal than most owls, these owls are often seen hunting during daylight.

Suggested Viewing Locations: In Wyoming, this species occurs widely across the state, mainly in meadows, grasslands and marshes. There it is usually found in a patchy and irregular distribution (Faulkner, 2010). It might be searched-for in various parts of Wyoming, such as in the grasslands south of Van Tassel, or in the National Elk Refuge (Scott, 1993). It similarly occurs at low but fairly uniform densities throughout the entire Rocky Mountain region. In Colorado, native grasslands attract these owls, such as the Pawnee National Grassland, North Park, and the San Luis valley. They occur widely in eastern Montana, and are common at Benton Lake, Bowdoin, C. R. Russell, Medicine Lake, and Red Rock Lakes national wildlife refuges, and at the National Bison Range. In Idaho they are common at Camas and uncommon at Grays Lake national wildlife refuges.

Population: National Breeding Bird Survey trend data indicate a significant annual population decline of 4.4%. The North American population (north of Mexico) has been estimated at 696,000 birds (Rich *et al.*, 2004).

Boreal Owl (*Aegolius funereus*)

Status: A rare to uncommon resident in Alberta's montane parks; local farther south and a rare breeder in Glacier National Park. Known to occur locally in eastern Idaho's River of No Return Wilderness area, in north-central Colorado (Cameron Pass area), and in the Bighorn Mountains of Wyoming. Apparently only a vagrant at Rocky Mountain National Park, but it has been reported from Estes Park village, and has been heard calling at Bear Lake, Colorado.

Habits and Ecology: Associated with old-growth forest, often of mixed coniferous species (often of Douglas-fir, ponderosa pine, and lodgepole pine) and subalpine spruce–fir forests, in places where red-backed voles (*Clethrionomys gapperi*), their major prey, are abundant. Nocturnal and much more often heard than seen.

Suggested Viewing Locations: In Wyoming, this species has been seen or heard at Teton Pass and Togwotee Pass, and at Lake Marie in the Snowy Range (Scott, 1993). Breeding is known to occur in the Bighorn Mountains, and probably also occurs in the Greater Yellowstone region and Sierra Madre–Snowy Range region (Faulkner, 2010). In Colorado, Rabbit Ears Pass near Steamboat Springs, is a good place to look for boreal owls, as are Cameron Pass above Poudre Canyon, Endovalley Picnic Area in Rocky Mountain National Park, and the high-altitude forests of Grand Mesa (Kingery, 2007). Trappers Lake, in Flat Tops Wilderness Area of Routt National Forest, is also known to support boreal owls (Young, 2000). In Montana, Lolo Pass (near Missoula, U.S. Hwy, 12) and Lost Trail Pass (south of Missoula, U.S. Hwy. 93) might be visited (McEneaney, 1993). In Idaho, Bear Basin (Valley County), the Ketchum–Sun Valley area (Blaine County) Henry's Lake State Park (Fremont County) and Teton Pass (Teton County) provide opportunities for finding this elusive owl (Svingen and Dumroese, 1997). Boreal owls might also be found in the Bitterroot Mountains, in the Kaniksu and Salmon National Forests (McClung, 1992).

Population: National Breeding Bird Survey trend data are not available. The North American population has been estimated at 600,000 birds (Rich *et al.*, 2004).

Northern Saw-whet Owl (*Aegolius acadicus*)

Status: Widespread permanent resident through the montane forests of the region, including the montane parks.

Habitats and Ecology: Occurs widely, from riparian woodlands through aspen groves to the coniferous forest zones, but not reaching timberline. The foothills and ponderosa pine zones are probably their favored habitats, where they nest in old woodpecker holes, but old-growth pinyon–juniper woodlands are also used. In the interior Northwest, this species begins breeding in mixed coniferous forests during the young forest stage of succession (Sanderson, Bull and Edgerton, 1980). Forests and woodlands with open understories are favored for foraging.

Suggested Viewing Locations: In Wyoming, it has been found in Grand Teton National Park, and has been seen at Teton Pass and below both sides of Togwotee Pass (Scott, 1993). There are few specific nesting areas known from Wyoming, but the species is apparently most abundant in the northwestern mountains ((Faulkner, 2010). This elusive owl might also be encountered along Deep Creek, in Colorado's White River National Forest (Young, 2000), or along the Fall River Road in Rocky Mountain National Park (Kingery, 2007). In Montana, Glacier National Park would be a promising location, while in Idaho the species is known to nest at Kootenai National Wildlife Refuge.

Population: National Breeding Bird Survey trend data are not available. The North American population (north of Mexico) has been estimated at 1.92 million birds (Rich *et al.*, 2004).

Goatsuckers

Common Nighthawk (*Chordeiles minor*)

Status: A summer resident throughout the region, mainly below the zone of coniferous forests, but breeding in most of the montane parks, especially the more southern ones.

Habitats and Ecology: This species forages entirely in the air, on flying insects, and is especially common over grassland and urban areas, sometimes extending to sagebrush and desert scrub. Nesting occurs on the ground, usually in grasslands, or at the edges of woods, and sometimes on the asphalt rooftops of buildings. In Wyoming the species is common statewide up to about 8,500 feet (Faulkner, 2010).

Suggested Viewing Locations: This is a widespread species at lower altitudes, and is usually seen or heard coursing low over towns and cities near sundown across the region. The densest regional breeding populations are in eastern Colorado and eastern Wyoming.

Population: National Breeding Bird Survey trend data indicate a significant annual population increase of 1.7%. The North American population (north of Mexico) has been estimated at 10.56 million birds (Rich *et al.*, 2004).

Common Poor-will (*Phalaenoptilus nuttallii*)

Status: A local summer resident in the region, mainly on drier habitats toward the south. The only montane park supporting the species as a breeder is Rocky Mountain National Park, where the birds occur around rocky outcrops at elevations up to about 8,000 feet.

Habitats and Ecology: Generally this species is associated with rocky habitats having a cover of arid-adapted shrubs or low trees, such as pinyon–juniper, saltbush, greasewood, sagebrush, and dry grasslands. The birds nest on the ground, often under scrub oaks, the leaves of which provide concealment for both adults and young.

Northern saw-whet owl, adults,
Drawing by author.

Suggested Viewing Locations: In Wyoming, the sage-dominated basins of Sweetwater, Carbon, Fremont, and Park counties support good populations of this
nocturnal species (Dorn and Dorn, 1990), which is fairly common statewide except in Yellowstone National Park ((Faulkner, 2010). Populations are
not unusually dense anywhere in the region, but are probably best developed in the southernmost and driest parts of it. In Colorado, poor-wills occur on both sides of the Continental Divide, most often in pinyon–juniper
and sagebrush–mountain mahogany habitats. They are uncommon breeders at Browns Park National Wildlife Refuge. In Montana they are abundant
at C. R. Russell National Wildlife Refuge, and in Idaho are occasional at
Grays Lake National Wildlife Refuge.

Population: National Breeding Bird Survey trend data indicate a nonsignificant
annual population increase of 1.2%. The North American population (north
of Mexico) has been estimated at 2.84 million birds (Rich *et al.*, 2004).

Swifts

Black Swift (*Cypseloides niger*)

Status: A local summer resident in montane areas from Alberta south to central
Montana, and again in western Colorado. Rocky Mountain National Park
and Banff National Park both support good populations.

Habitats and Ecology: Associated with mountains having steep, almost inaccessible cliffsides for nesting, often in narrow canyons, and almost always
close to waterfalls. In Rocky Mountain National Park nests have been found
at 10,500 feet at Loch Vale, and the birds have been seen flying over 14,000-
foot peaks. Nesting in caves has been reported in other areas.

Suggested Viewing Locations: In Wyoming, there were no definite state records
for this species as of 2010 (Faulkner, 2010). One of the few places to see
breeding black swifts in Colorado in Treasure Falls, about 15 miles north
of Pagosa Springs, where white-throated swifts also nest. Black swifts
might also be seen nesting at Rifle Falls State Park, near Rifle, Colorado
(Young, 2000). Colonies also may be seen at Box Cañon Falls (near Ouray),
Hanging Lake (near Glenwood Springs), and Zapata Falls (San Luis valley)
(Kingery, 1998). They also may be seen at Black Canyon, in Gunnison National Monument, and at Loch Vale, Rocky Mountain National Park (Zimmer, 2000). There are also a few good Montana viewing locations (McE-
neaney, 1993), including such sites in Glacier National Park as Avalanche
Creek picnic area and Libby Dam on the Kootenai River, The densest regional breeding populations are in northern Idaho and northwestern Montana. In Idaho they are uncommon during summer at Kootenai National
Wildlife Refuge and probably nest nearby. One of the few reliable Idaho
viewing areas there is along Smith Creek, accessible via Idaho Forest Road
281, from near the Canadian border to West Fork Lake (Svingen and Dumroese, 1997).

Population: National Breeding Bird Survey trend data are not available. The North American population (north of Mexico) has been estimated at 87,000 birds (Rich *et al.*, 2004).

Chimney Swift (*Chaetura pelagica*)

Status: Rare in the region, and limited regionally to eastern Montana, eastern Wyoming and northeastern Colorado. Absent from the montane parks except for Rocky Mountain National Park, where an accidental summer vagrant.

Habitats and Ecology: A familiar city bird over most of eastern North America, mainly in the vicinity of towns and cities where chimneys offer roosting and nesting sites. Caves and hollow trees were used before chimneys became available, and may occasionally still be used in some localities.

Suggested Viewing Locations: This species is likely to be seen only in towns and villages located in easternmost parts of the region. In Wyoming, a good place to see chimney swifts is around the main post office and St. Mary's Cathedral in Cheyenne, and in downtown Douglas (Dorn and Dorn, 1990). No state nesting records existed as of 2010 (Faulkner, 2010) The densest regional breeding populations are in eastern Colorado, where they nest in towns and villages west to the Front Range. In Montana they might be easily seen in towns such as Glendive and Sidney (McEneaney, 1993).

Population: National Breeding Bird Survey trend data indicate a nonsignificant annual population decline of 1.85%. The North American population has been estimated at 15 million birds (Rich *et al.*, 2004).

Vaux's Swift (*Chaetura vauxi*)

Status: A summer resident in the northwestern parts of the region, including Glacier National Park, where it breeds commonly. Unreported for Wyoming or Colorado.

Habitats and Ecology: Generally similar to the chimney swift, but this species seldom nests in chimneys, and instead uses hollow trees. It is often found in woodlands near rivers and lakes. The nests are often placed in western hemlocks with dead or broken-off tops, or sometimes on cliffs. In the interior Northwest, this species begins breeding in mixed coniferous forests during the mature forest stage of succession (Sanderson, Bull and Edgerton, 1980).

Suggested Viewing Locations: In Glacier National Park these birds are often seen near Avalanche Campground, as well as in the McDonald Valley, along the North Fork of the Flathead River, and near the Flathead Ranger Station. They are also occasional breeders at Montana's Lee Metcalf National Wildlife Refuge. The densest regional breeding concentrations are probably in the montane forests of northern Idaho, and are common breeders at Kootenai National Wildlife Refuge.

Population: National Breeding Bird Survey trend data are not available. The North American population (north of Mexico) has been estimated at 700,000 birds (Rich *et al.*, 2004).

White-throated Swift (*Aeronautes saxatilis*)

Status: A widespread summer resident in mountainous areas of the region, more common southwardly, and absent from the Canadian montane parks.

Habitats and Ecology: This swift is associated with steep cliffs, deep canyons, and generally mountainous terrain, sometimes observed as high as 13,000 feet. Nesting occurs in crevices of canyon walls, in completely inaccessible locations. In Colorado nesting mostly occurs between 5,500 and 8,200 feet, occasionally to 10,000 feet. Nesting occurs in Wyoming up to about 9,500 feet (Faulkner, 2010).

Suggested Viewing Locations: In Wyoming, these widespread swifts can be seen at the Yellowstone River canyon, and in canyons of the Bighorn Mountains and Wind River Canyon. (Washakie County). Hawk Springs Reservoir (Goshen County), Flaming Gorge Reservoir (Sweetwater County), along Beaver Rim (Fremont County), and the rim of Goshen Hole (Goshen County) also offer possibilities (Dorn and Dorn, 1990). In Colorado the birds may be seen throughout the Rocky Mountain chain, in western plateau and canyon country (*e.g.,* Black Canyon National Park), and east to Las Animas County (Kingery, 1998). The densest regional breeding populations are in northeastern Wyoming, adjacent southeastern Montana, and southwestern Idaho. There they are likely to be seen foraging widely, over towns as well as mountain and canyon habitats.

Population: National Breeding Bird Survey trend data indicate a nonsignificant annual population decline of 1.5%. Trend estimates for the Rocky Mountains & Plains States region (U.S.F.W.S. Region 6) indicate a significant annual decline of 2.8%. The North American population (north of Mexico) has been estimated at 283,000 birds (Rich *et al.,* 2004).

Hummingbirds

Ruby-throated Hummingbird (*Archilochus colubris*)

Status: A rare summer resident in eastern Montana; elsewhere in the region a rare migrant or vagrant; the only montane park reporting the species is Watertown Lakes N.P., where it is rare.

Habitats and Ecology: This is the familiar hummingbird of the eastern half of North America, where it is the only breeding species. It occupies many woodland habitats, but is mostly found in open hardwood forests or forest edges, and in similar habitats such as orchards or city parks where a variety of nectar-bearing flowers are to be found.

Suggested Viewing Locations: Migrants might appear along the eastern edge of the region, especially northeastern Montana, but Watertown Lakes National Park is probably the only national park where summer sightings are likely.

Population: National Breeding Bird Survey trend data indicate a significant annual population increase of 2.3%. The North American population (north of Mexico) has been estimated at 7.3 million birds (Rich *et al.,* 2004).

Black-chinned Hummingbird (*Archilochus alexandri*)

Status: A summer resident in the western and southwestern parts of the region, mainly at lower elevations. It is a rare migrant or vagrant in the montane parks.

Habitats and Ecology: Typically associated with riparian habitats in dry canyons, but also occurring in oak-juniper woodlands, edges of aspen groves, and other habitats that usually are near water and offer open areas with many flowering plants. In the interior Northwest, this species breeds in mixed coniferous forests during the shrub-seedling and pole-sapling stages of succession (Sanderson, Bull and Edgerton, 1980).

Suggested Viewing Locations: In Wyoming, a good place to look for this species is Richard's Gap (south of Rock Springs near the Utah border), but there were no proven nesting records for the state as of 2010 (Faulkner, 2010). The densest regional breeding populations are in western Colorado and western Idaho. In Colorado the birds are common nesters at Browns Park National Wildlife Refuge, and probably also breed at Dinosaur National Monument.

Population: National Breeding Bird Survey trend data indicate a significant annual population increase of 1.0%. The North American population (north of Mexico) has been estimated at two million birds (Rich *et al.*, 2004).

Calliope Hummingbird (*Stellula calliope*)

Status: A common summer resident over most of the region west of the plains; probably the commonest breeding species in most of the montane parks, but absent from Colorado.

Habitats and Ecology: Open meadow areas near coniferous forests, such as low willow or sage areas rich in plants such as Indian paintbrush or gilia, are favored areas for this species in the Jackson Hole area. In the interior Northwest, this species breeds in mixed coniferous forests during the shrub-seedling and pole-sapling stages of succession (Sanderson, Bull and Edgerton, 1980). Openings in woodlands, sometimes as high as timberline, are also frequented, and in late summer alpine meadows are commonly used by migrating birds.

Suggested Viewing Locations: In Wyoming, this species is easily found in meadows and willow flats of Grand Teton and Yellowstone National Parks, and they probably breed locally east to Sheridan County (Faulkner, 2010). The densest regional breeding populations are in the Greater Yellowstone region, and in west-central Idaho. In Idaho the birds are common nesters at Grays Lake and Kootenai refuges, and in mountain meadows statewide.

Population: Classified as a Partners-in-Flight Watch List species of continental conservation importance (Rich et al., 2004). National Breeding Bird Survey trend data indicate a nonsignificant annual population increase of 1.0%. The North American population has been estimated at one million birds, of which 95% breed in the Intermountain West Avifaunal Biome (Rich *et al.*, 2004).

Broad-tailed Hummingbird (*Selasphorus platycercus*)

Status: A summer resident in the southern parts of the region, mainly west of the plains and south of Montana. The common breeding hummingbird in Rocky Mountain National Park, but rare or absent from the more northerly ones.

Habitats and Ecology: Typically associated with ponderosa pine forests and aspen groves, but also extending into mountain meadows, pinyon–juniper woodland and riparian cottonwoods in this region. Willow-lined streams adjacent to meadows and parklands are favored foraging areas. In Colorado breeding birds are abundant in foothills or mountains with aspens, pines, or Douglas-fir at about 6,500 –7,500 feet elevation, but nests have been seen from 5,200–10,750 feet. During the summer the birds gradually move upwards, finally reaching alpine meadows in late summer.

Suggested Viewing Locations: In Wyoming, this hummingbird is present in the northwestern mountains, the Bighorns, and the Medicine Bow Mountains (Dorn and Dorn, 1990), but is absent from the Black Hills National Forest (Faulkner, 2010). It is rare to occasional in the Greater Yellowstone ecosystem, and uncommon in Seedskadee National Wildlife Refuge. The densest regional breeding populations are in central and western Colorado, including Rocky Mountain National Park. It is uncommon at Arapaho National Wildlife Refuge. In Montana it is essentially confined to the Beartooth Mountains, and to the Centennial Valley–Red Rock Lakes region. Among Idaho's national wildlife refuges, this species is most common at Grays Lake.

Population: National Breeding Bird Survey trend data indicate a nonsignificant annual population decline of 0.3%. The North American population (north of Mexico) has been estimated at 1.96 million birds (Rich *et al.*, 2004).

Rufous Hummingbird (*Selasphorus rufus*)

Status: A summer resident in the northern parts of the region, becoming less common southwardly and generally uncommon to rare in Wyoming. The common breeding hummingbird of the Alberta montane parks, but not known to breed south of the Jackson Hole area.

Habitats and Ecology: In general, coniferous forests are used for breeding, but the birds occupy a variety of forest-edge habitats including mountain meadows and burned over forest areas where flowers are abundant. In the interior Northwest, this species breeds in mixed coniferous forests during the shrub-seedling and pole-sapling stages of succession (Sanderson, Bull and Edgerton, 1980). Brushy areas in the foothills are also used on migration, as are urban gardens and alpine tundra.

Suggested Viewing Locations: In Wyoming, the Greater Yellowstone region and the Wind River range offer some possibilities for finding this generally uncommon species, which as of 2010 had not been a proven breeding species (Faulkner, 2010). The densest regional breeding populations are in western Idaho and western Montana. In Idaho it is uncommon to common at Koo-

tenai and Grays Lake national wildlife refuges, and in Montana it is uncommon at Grays Lake, Red Rock Lakes, and Swan River national wildlife
refuges.

Population: Classified as a Partners-in-Flight Watch List species of continental conservation importance (Rich et al., 2004). National Breeding Bird Survey trend data indicate a nonsignificant annual population decline of 2.3%.
The North American population has been estimated at 6.5 million birds
(Rich *et al.*, 2004).

Kingfishers

Belted Kingfisher (*Megaceryle alcyon*)

Status: A resident in southern parts of the region, but more migratory farther
north. It is common in all the montane parks and probably breeds in all of
them.

Habitats and Ecology: Found near water rich in small fish populations, usually
where road cuts, eroded banks, gravel pits, or other exposed earthen surfaces provide opportunities for nesting, and usually also where nearby trees
provide convenient perching and observation sites between flights.

Suggested Viewing Locations: This widespread species can be seen along most
Rocky Mountain rivers or lakes that have good fish populations, nearby
perching sites, and steep clay banks for nesting. The densest regional
breeding populations are in northern Idaho.

Population: National Breeding Bird Survey trend data indicate a nonsignificant
annual population decline of 1.8%. The North American population has
been estimated at 2.2 million birds (Rich *et al.*, 2004).

Woodpeckers

Lewis's Woodpecker (*Melanerpes lewis*)

Status: A local summer resident in forested montane areas of the region, especially at lower altitudes; probably breeding rarely in most of the montane
parks.

Habitats and Ecology: This unusual woodpecker is especially associated with
pine forests that are rather open, with burned over or otherwise dead-tree
areas having abundant snags or stumps, which means that its distribution
tends to be labile and adapted to local conditions. Streamside cottonwood
groves in the ponderosa pine or pinyon–juniper zones are also used, and
old cottonwoods are favorite nesting trees. In the interior Northwest, this
species begins breeding in mixed coniferous forests during the young forest
stage of succession (Sanderson, Bull and Edgerton, 1980).

They are more likely to be found at lowland and foothill sites than in montane
forests; pinyon–juniper and oak–mountain mahogany woodlands are often
used for foraging. The birds are mainly adapted to catching free-living in-

sects rather than excavating for insects in wood, and can often be observed flycatching.

Suggested Viewing Locations: The densest regional breeding populations of this species are in western Idaho, northwestern Montana and northeastern Wyoming, but local distributions depend greatly on local forest conditions. Most known breeding records are from the Bear Lodge, Bighorn, Laramie, and Medicine Bow Mountains (Faulkner, 2010). Potential Wyoming viewing sites include the Black Hills National Forest and the entire Greater Yellowstone ecosystem, both areas having been greatly affected by bark beetle damage. Some suggested local viewing sites include Stockade Beaver Creek Road (Weston County), Cottonwood Creek (Platte County), and Ash Creek (Sheridan County)(Dorn and Dorn, 1990), although specific viewing "hot spots" often last only 4–5 years before the birds move on to more newly disturbed sites. In Montana, this woodpecker commonly occurs in the National Bison Range and at Lee Metcalf National Wildlife Refuge. In Idaho, Deer Flat National Wildlife Refuge is one of the more reliable places for finding this species. Colorado concentrations occur along the eastern slope of the Front Range, in the San Juan Basin of southwestern Colorado, in pinyon–juniper woodlands of southeastern Colorado, and in cottonwood gallery forests along the Arkansas River valley (Kingery, 1998).

Population: The North American population has been estimated at 230,000 birds, of which 87% breed in the Intermountain West Avifaunal Biome. Classified as a Partners-in-Flight Watch List species of continental conservation importance (Rich et al., 2004). National Breeding Bird Survey trend data indicate a nonsignificant annual population decline of 1.0%.

Red-headed Woodpecker *(Melanerpes erythrocephala)*

Status: A local summer resident at lower elevations east of the mountains; rare or accidental in the montane parks except for Rocky Mountain National Park, where it is infrequent in summer.

Habitats and Ecology: Associated with open deciduous forests, woodlots, and riparian areas, sometimes extending into the ponderosa pine zone. Aspens and riparian cottonwood forests are the species' major habitats in this region. Like the Lewis' woodpecker, this species tends to nest in dead trees or the dead portions of living trees, and does less excavating for insects in wood than do most woodpeckers, but often forages in open habitats at ground level.

Suggested Viewing Locations: The densest regional breeding populations are in river valleys along its eastern boundaries. In Wyoming, riparian cottonwood woodlands of the eastern plains and foothills attract this species, such as the North Platte River (Goshen County), the Powder River and Clear Creek (Sheridan County), and Lance Creek (Niobrara County) (Dorn and Dorn, 1990). In Montana, the species is common at the C. R. Russell

National Wildlife Refuge, at Fort Peck, and in eastern towns such as Lewiston and Roundup (McEneaney, 1993). In eastern Colorado they occur in streams lined with cottonwoods up to about 5,500 feet (Kingery, 1998).

Population: The North American population has been estimated at 2.5 million birds.

Classified as a Partners-in-Flight Watch List species of continental conservation importance (Rich *et al.*, 2004). National Breeding Bird Survey trend data indicate a significant annual population decline of 2.6%. Trend estimates for the Rocky Mountains & Plains States region (U.S.F.W.S. Region 6) indicate a significant annual decline of 0.3%.

Williamson's Sapsucker *(Sphyrapicus thyroides)*

Status: A local and usually uncommon summer resident over much of the region; variably common in the U.S. montane parks, generally rare or absent in the Alberta parks.

Habitats and Ecology: Breeding in this region usually occurs in the aspen or coniferous zones, especially in ponderosa pine forests, with mixed aspens, mainly between about 7,000–8,500 feet in Colorado, but sometimes as high as 10,700 feet. High ridges in the Douglas-fir zone are also used for foraging, but nesting is usually done in aspens. In the interior Northwest, this species begins breeding in mixed coniferous forests during the mature forest stage of succession (Sanderson, Bull and Edgerton, 1980).

Suggested Viewing Locations: In Wyoming, good viewing areas include the Sierra Madre Mountains in Carbon County, the Brooks Lake area of Fremont County, and along the trail from Rendezvous Peak to Granite Canyon in Grand Teton National Park (Dorn and Dorn, 1990). McEneaney (1988) judged the Yellowstone Park population to be less than 400 pairs during the late 1980's, with the best observation opportunities being around Tower Fall and the Blacktail Plateau. Wyoming's densest populations are in the Greater Yellowstone region, with fewer records in the Laramie and Bighorn ranges, and none from the Black Hills National Forest (Faulkner, 2010). The densest regional breeding populations are in northern Idaho, western Montana and western Colorado. In Idaho, this hard-to-find species is most likely to be found in high-elevation coniferous forests near the Canadian border, but the best birding area may be Bear Basin (Valley County) (Scott, 1997). In Montana, the Centennial Valley offers a possible viewing area (McEneaney, 1993). In Colorado the eastern slope of the Rockies has the greatest density of birds, whereas the red-naped sapsucker is more broadly distributed and is more closely associated with aspens.

Population: National Breeding Bird Survey trend data indicate a significant annual population decline of 0.4%. The North American population has been estimated at 310,000 birds, of which 94% breed in the Intermountain West Avifaunal Biome (Rich *et al.*, 2004).

Red-naped Sapsucker (*Sphyrapicus nuchalis*) and Yellow-bellied Sapsucker (*Sphyrapicus varius*)

Status: The red-naped sapsucker is a breeding summer resident throughout the wooded portions of the region, common in montane areas and probably breeding in all the montane parks. The closely related (and previously considered conspecific) yellow-bellied sapsucker is a probable migrant at the extreme eastern edges of the region, and a likely breeder along the eastern slope of the Canadian montane parks.

Habitats and Ecology: Sapsuckers use coniferous forests, deciduous forests, and mixed woodlands, but aspens are a favorite habitat in this region. They excavate holes in these trees to drink the sap, and also nest in aspens, either in dead trees or living ones that have dead and rotting interiors. They breed as high in the mountains as the upper limit of aspens, providing ideal nesting cavities for at least six other species of hole-nesting birds.

Suggested Viewing Locations: In Wyoming, the red-naped sapsucker is widespread in aspen woodlands, especially westwardly (Dorn and Dorn, 1990). There are no verified nesting records for the yellow-bellied sapsucker in the state (Faulkner, 2010). The densest regional breeding populations of red-naped sapsuckers are in northwestern Montana and adjacent Idaho, northeastern Wyoming, and northern Colorado. In Colorado they are notably common in the mountains from Rocky Mountain National Park northward. In Wyoming, they are common across the Greater Yellowstone ecosystem, and also occur in the Medicine Bow, Laramie, and Bighorn ranges, plus the Black Hills (Faulkner, 2010). In Montana they are common in the National Bison Range, and in Idaho they are common at Kootenai National Wildlife Refuge, as well as in many of the forested parts of northern and central Idaho (Scott, 1997).

Population: National Breeding Bird Survey trend data indicate a nonsignificant annual population increase of 0.9% for the red-naped sapsucker, and a nonsignificant annual population increase of 0.5% for the yellow-bellied sapsucker. Trend estimates for the Rocky Mountains & Plains States region (U.S.F.W.S. Region 6) indicate a significant annual increase of 6.5% for the red-naped sapsucker. The North American population has been estimated at 9.2 million birds for the yellow-bellied, and 2.2 million for the red-naped, with 95% of the latter breeding in the Intermountain West Avifaunal Biome (Rich *et al.*, 2004).

Downy Woodpecker (*Picoides pubescens*)

Status: A resident throughout the region in wooded habitats; relatively common and probably breeding in all of the montane parks.

Habitats and Ecology: A wide variety of wooded habitats are used by downy woodpeckers, including farmlots, orchards, city parks, and natural habitats ranging from riparian forests to pinyon–juniper woodlands, oak–mountain mahogany scrub, and aspen or coniferous forests.

Suggested Viewing Locations: These woodpeckers are nearly ubiquitous throughout the region, and are easy to find around bird feeders in winter, or in riverbottom woodlands with cottonwood trees (Dorn and Dorn, 1990). In Colorado they are most common during summer among aspen stands in the mountains, and in mature cottonwoods along lower river valleys (Kingery, 199). In Montana they are widespread and common at C. R. Russell, Lee Metcalf, and Red Rock Lakes national wildlife refuges, as well as the National Bison Range. In Idaho they are common at Deer Flat and Kootenai national wildlife refuges.

Population: National Breeding Bird Survey trend data indicate a nonsignificant annual population decrease of 0.2%. The North American population has been estimated at 13 million birds (Rich *et al.*, 2004).

Hairy Woodpecker (*Picoides villosus*)

Status: A resident in wooded areas throughout the region, occurring in all the montane parks and probably breeding in all.

Habitats and Ecology: Optimum breeding habitat consists of fairly extensive areas of woodlands of conifers (especially lodgepole pines) or hardwoods, but nesting occurs in riparian forests, in aspen groves, and in various coniferous forests nearly to timberline. Generally, aspens and other hardwoods are preferred over conifers for breeding. In the interior Northwest, this species begins breeding in mixed coniferous forests during the young forest stage of succession (Sanderson, Bull and Edgerton, 1980).

Suggested Viewing Locations: This widespread species is often found in mature aspen stands in summer, and among cottonwoods in winter (Dorn and Dorn, 1990). Relatively dense regional breeding populations probably occur in southern Alberta, northwestern Montana, and northwestern Wyoming. The birds are found in the same areas as suggested for downies, but are more common than downies in more heavily wooded habitats, and with lower numbers than downies being typical in sparsely wooded areas.

Population: National Breeding Bird Survey trend data indicate a significant annual population increase of 1.3%. The North American population (north of Mexico) has been estimated at 7.52 million birds (Rich *et al.*, 2004).

American Three-toed Woodpecker (*Picoides dorsalis*)

Status: A local and variably common resident in montane areas of the region; present and breeding in all the montane parks.

Habitats and Ecology: Like the following species, this is a fire-adapted form, typically moving into a burned forest area immediately after the fire, breeding, and dispersing four or five years later. Bluebirds, nuthatches, and many other cavity nesters then use the nest cavities until the snags eventually topple. Partly open, high-elevation coniferous or aspen forests are typical habitats. Spruce–fir forests are favored habitats in the northern Rockies.

Suggested Viewing Locations: Relatively dense regional breeding populations probably occur in central Idaho, western Colorado, and in northern Wyoming and adjacent Montana. In Wyoming, recently burned areas and old-growth spruce-fir forests are likely sites for finding this species (Dorn and Dorn, 1990). It is widespread in mountain ranges across the state, but its status in the Black Hills National Forest is still uncertain (Faulkner, 2010). In Montana, the species is likewise most likely seen at burned-over forests (Fischer and Fischer, 1995). In Colorado, breeding season observations ranged in elevation from 7,000–12,000 feet, with most records about 9,000 feet, and with a preference for subalpine coniferous forests (Kingery, 1998). Rabbit Ears Pass near Steamboat Springs and Red Sandstone Road near Vail have been recently suggested as places to find it (Kingery, 2007). Other older recommended Colorado viewing sites have included Brainard Lake, Echo Lake Park, Grand Mesa and Rocky Mountain National Park (Zimmer, 2000).

Population: National Breeding Bird Survey trend data indicate a significant annual population increase of 9.9%. Trend estimates for the Rocky Mountains & Plains States region (U.S.F.W.S. Region 6) indicate a significant annual increase of 7.2%. Thee trends may reflect the great recent increase in forest fires over western North America. The North American population has been estimated at 830,000 birds (Rich *et al.*, 2004).

Black-backed Woodpecker *(Picoides arcticus)*

Status: A resident in wooded areas throughout the mountains south to about Jackson Hole, Wyoming, and breeding in most or all of the montane parks from Grand Teton National Park northward, but rare in the Alberta parks. Absent from Colorado, and of only limited occurrence in eastern Wyoming, in the western Black Hills.

Habitats and Ecology: This species is nearly identical to the preceding one as to its general adaptations to feeding and breeding in recently burned coniferous forest areas, especially those of lodgepole pine and subalpine spruce–fir. As such, it is highly local and eruptive. Apparently the birds feed exclusively on beetles in dead or dying conifers, but sometimes nest in nearby aspens. In the interior Northwest, this species begins breeding in mixed coniferous forests during the young forest stage of succession (Sanderson, Bull and Edgerton, 1980).

Suggested Viewing Locations: The same areas mentioned for the American three-toed woodpeckers are likely to support this species, although it is generally rarer and more habitat--limited. Relatively dense regional breeding populations probably occur in central Idaho and in northwestern Wyoming, at least where forest fires have occurred recently. Like the previous species, these birds should be searched-for in localities where forest fires or other forest disruptions such as logging or infestations by boring or bark beetle have occurred within the past four to five years.

Population: National Breeding Bird Survey trend data indicate a nonsignificant annual population decrease of 0.8%. The North American population has been estimated at 1.3 million birds (Rich *et al.*, 2004). It is classified as a Priority III species in Wyoming.

Northern Flicker (*Colaptes auritus*)

Status: A common resident or migrant throughout the region, more abundant in wooded areas, but also in open country where trees are scattered. Present in all the montane parks, and probably breeding in all of them.

Habitats and Ecology: Flickers are unusual among woodpeckers in that much of their food consists of insects such as ants that are obtained by probing in the ground rather than by excavating trees. However, they do excavate holes in trees for nesting, usually those that are already dead or have decaying interiors, especially in relatively soft-wood species such as cottonwoods and aspens. Open woodlands, such as orchards, parks, and similar areas offering foraging opportunities on grassy areas nearby are preferred over dense forests. In the interior Northwest, this species begins breeding in mixed coniferous forests during the young forest stage of succession (Sanderson, Bull and Edgerton, 1980). In Colorado nests have been found at elevations from 5,400–11,480 feet. There red-shafted forms were mostly found in coniferous and aspen habitats, while yellow-shafted forms were most common in cottonwood riverbottoms (Kingery, 1998).

Suggested Viewing Locations: This woodpecker is ubiquitous in wooded and parkland areas throughout the region, with the yellow-shafted form most abundant in riverbottom woods along its eastern boundary, and the red-shafted form present elsewhere from the foothills to subalpine forests.

Population: National Breeding Bird Survey trend data indicate a significant annual population decrease of 1.7%. The North American population (north of Mexico) has been estimated at 14.56 million birds (Rich *et al.*, 2004).

Pileated Woodpecker (*Dryocopus pileatus*)

Status: A local resident in the northern parts of the region, south to southwestern Montana and adjacent Idaho. Common in the montane parks of Alberta and also Glacier National Park, but only vagrants occur in the Yellowstone area.

Habitats and Ecology: This magnificent bird is associated with mature forests of various types, especially old-growth stands of Douglas-fir at middle altitudes. Preferred habitats are usually near water and include mature trees among which there are tall trees having dead stubs, where nesting occurs. Nests are usually in trees that are 15–20 inches in diameter at the place of excavation, and are from 15–70 feet above ground. In the interior Northwest, this species begins breeding in mixed coniferous forests during the mature forest stage of succession (Sanderson, Bull and Edgerton, 1980).

Suggested Viewing Locations: Very rare in Wyoming, with no nesting records. Relatively dense regional breeding populations probably occur in southern British Columbia and northern Idaho. Good Montana locations for finding this species are Vinal Creek in Kootenai National Forest, the Blackfoot–Clearwater Wildlife Management Area near Seeley Lake, and Greenbough Park in Missoula (Fischer and Fischer, 1995). In Idaho it is uncommon at Kootenai National Wildlife Refuge, and is present in several state parks, such as Farragut, McCrosky, Round Lake and Winchester (Scott, 1997).

Population: National Breeding Bird Survey trend data indicate a significant annual population increase of 1.7%. The North American population has been estimated at 930,000 birds (Rich *et al.*, 2004).

Tyrant Flycatchers

Olive-sided Flycatcher (*Contopus borealis*)

Status: Widespread in the region west of the plains, primarily in coniferous forests, but also in riparian woodlands. It occurs in all the montane parks, and probably breeds in all.

Habitats and Ecology: Associated with coniferous montane forests (especially Douglas-fir and lodgepole pines), burned over or logged forests with standing snags, and muskeg areas in the region. Typically tall conifers and open, often boggy or meadow-like areas are present in territories. In the interior Northwest, this species begins breeding in mixed coniferous forests during the pole-sapling stage of succession (Sanderson, Bull and Edgerton, 1980). In Colorado it has been found breeding from 7,000–11,000 feet elevation, most often in montane coniferous forests with closed canopies (Kingery, 1998).

Suggested Viewing Locations: In Wyoming, good viewing areas include the forested uplands of Fossil Butte National Monument (Lincoln County), Pine Mountain (Sweetwater County) and the Brooks Lake area (Fremont County) (Dorn and Dorn, 1990). Breeding occurs from the Greater Yellowstone region east at altitudes of 7,500–10,000 feet to the Bighorns and southeast to the Laramie and Medicine Bow ranges (Faulkner, 2010). The densest regional breeding populations probably occur in the montane forests of eastern Idaho and adjacent western Montana.

Population: The North American population (north of Mexico) has been estimated at 1.19 million birds. Classified as a Partners-in-Flight Watch List species of continental conservation importance (Rich *et al.*, 2004). National Breeding Bird Survey trend data indicate a significant annual population decrease of 3.3%.

Western Wood-Pewee (*Contopus sordidulus*)

Status: Widespread throughout nearly the entire region in summer; breeding occurs in most and probably all of the montane parks.

Habitats and Ecology: Breeds in most coniferous forest types, and also to a varying extent in aspens, riparian forests, and various open deciduous or mixed woodland habitats. Open forests are favored, especially those dominated by conifers. In the interior Northwest, this species begins breeding in mixed coniferous forests during the young forest stage of succession (Sanderson, Bull and Edgerton, 1980). In Colorado these birds nest from 3,800–10,000 feet elevation, in trees with dead tops or exposed branches (Kingery, 1998)

Suggested Viewing Locations: In Wyoming, this abundant species is ubiquitous statewide in wooded areas up to about 9,000 feet, especially in ponderosa pine or mixed conifers (Faulkner, 2010). The densest regional breeding populations probably occur in northern and west-central Idaho, such as at Minidoka and Kootenai national wildlife refuges.

Population: National Breeding Bird Survey trend data indicate a nonsignificant annual population decrease of 0.85%. However, trend estimates for the Rocky Mountains & Plains States region (U.S.F.W.S. Region 6) indicate a significant annual increase of 0.9%. The North American population (north of Mexico) has been estimated at 7.76 million birds (Rich *et al.*, 2004).

Alder Flycatcher (*Empidonax alnorum*)

Status: An uncommon summer resident in the northwestern portions of the region, breeding in the montane parks of Alberta and British Columbia south to at least Kootenay and Banff. Not reported elsewhere in the region as a breeder.

Habitats and Ecology: This species and the willow flycatcher are so similar in most regards that the literature on them is greatly confused. However, this species seems to breed over most of the same general area as the willow flycatcher, in birch or willow thickets, along the edges of muskegs, forest margins, streamside shrubbery, and wooded lakeshores. In the interior Northwest, this species begins breeds in mixed coniferous forests during the pole-sapling and young forest stages of succession (Sanderson, Bull and Edgerton, 1980).

Suggested Viewing Locations: The Canadian montane parks (and possibly Glacier National Park) offer the best chance of seeing this species except during migration, when it is very difficult to distinguish from the willow flycatcher.

Population: National Breeding Bird Survey trend data indicate a nonsignificant annual population change of 0.0%. The North American population has been estimated at 49 million birds (Rich *et al.*, 2004).

Willow Flycatcher (*Empidonax traillii*)

Status: A summer resident through the region except for some of the low plains areas in the east. Present and probably breeds in all the montane parks.

Habitats and Ecology: Especially associated with riparian or wetland habitats in this region, including willow thickets, low gallery forests along streams, prairie coulees, and, farther north, in woodland edge habitats such as muskegs and boggy openings. In Wyoming, it breeds generally west of a line from Sheridan to western Laramie County, and increases in abundance in the western mountains. There is also a small local population in the western Black Hills (Faulkner, 2010). In Colorado, it nests at altitudes of 6,000–10,000 feet, especially in riparian thickets along foothill streams. Most Colorado nesting occurs west of the Continental Divide, in diverse habitats (Kingery, 1998). In the interior Northwest, this species begins breeds in mixed coniferous forests during the pole-sapling and young forest stages of succession (Sanderson, Bull and Edgerton, 1980).

Suggested Viewing Locations: Widespread and abundant throughout the region in thickety areas and woodland edges. The densest regional breeding populations probably occur in western Montana (*e.g.*, Lee Metcalf National Wildlife Refuge) and northern Idaho (Kootenai National Wildlife Refuge).

Population: National Breeding Bird Survey trend data indicate a significant annual population decrease of 1.2% for the willow/alder species group. The North American population has been estimated at 3.3 million birds (Rich *et al.*, 2004). The Southwestern race (*E. t. extimus*) breeds in southern Colorado, and is a candidate for federal listing as a threatened or endangered subspecies.

Least Flycatcher (*Empidonax minimus*)

Status: A common summer resident in the northern and northeastern parts of the region, south through Montana to northeastern Wyoming. Most common in the montane parks of Alberta, where it regularly breeds. Possibly a rare breeder in Glacier National Park.

Habitats and Ecology: Associated with open and edge-dominated habitats such as mature deciduous floodplain forests with shrubby understories in prairie areas, scattered prairie grovelands, shelterbelts, woody lake margins, and urban parks or gardens.

Suggested Viewing Locations: Look for this species in the riparian forests of northeastern Wyoming, such as the upper canyon of the North Fork of Crazy Woman Creek (Johnston County), or in mature woods above LAK reservoir (Weston County) (Dorn and Dorn, 1990). The densest regional breeding populations probably occur in central Montana (*e. g.*, C. R. Russell National Wildlife Refuge, Missouri Headwaters State Park) and the southern parts of Alberta and British Columbia.

Population: National Breeding Bird Survey trend data indicate a nonsignificant annual population decrease of 1.3%. However, trend estimates for the Rocky Mountains & Plains States region (U.S.F.W.S. Region 6) indicate a significant annual increase of 2.1%. The North American population has been estimated at 14 million birds (Rich *et al.*, 2004).

Hammond's Flycatcher (*Empidonax hammondi*)

Status: A summer resident in coniferous forests over most of the region, occasional to common in all the montane parks and probably breeding in all of them.

Habitats and Ecology: Associated with tall, montane coniferous forests (especially Douglas-fir), from about 5,000 to 9,000 feet over much of this region, but from 7,000–10,000 feet in Colorado. Mature coniferous stands with limited understory vegetation is favored (Kingery, 1998). In the interior Northwest, this species begins breeding in mixed coniferous forests during the mature forest stage of succession (Sanderson, Bull and Edgerton, 1980). It is also found along willow- and alder-lined mountain streams, with males sometimes singing at elevations above timberline.

Suggested Viewing Locations: In Wyoming, and elsewhere in the region, moist montane coniferous forests provide the best places to hunt for this species. In Wyoming it occurs throughout the Greater Yellowstone region, as well as in the Bighorn and Medicine Bow ranges (Faulkner, 2010). The densest regional breeding populations probably occur in western Montana (*e.g.*, National Bison Range; Beartooth Mountains) and adjacent Idaho (Kootenai National Wildlife Refuge).

Population: National Breeding Bird Survey trend data indicate a significant annual population increase of 1.8%. The North American population has been estimated at 13 million birds (Rich *et al.*, 2004).

Dusky Flycatcher *(Empidonax oberholseri)*

Status: A summer resident in most areas west of the plains; present but variably common in all the montane parks, and breeding in most.

Habitats and Ecology: Associated with open woodland with dense shrubby understories, ranging from riparian edges through oak–mountain mahogany woodlands, to aspens and open ponderosa pine woods. In Montana, brushy, logged-over slopes seem to be favored habitats, while in Colorado areas dominated by shrubs or aspens are preferred nesting habitats (Kingery, 1998). In the interior Northwest, this species begins breeding in mixed coniferous forests during the shrub-seedling stage of succession (Sanderson, Bull and Edgerton, 1980).

Suggested Viewing Locations: Brushy foothill forests are good places in Wyoming to search for this species (Dorn and Dorn, 1990), which is widespread throughout all the major mountain ranges and the western Black Hills (Faulkner, 2010). Dense regional breeding populations probably occur in the Greater Yellowstone region, northeastern Wyoming, and over large parts of western Montana (*e.g.*, Lee Metcalf and Red Rock Lakes national wildlife refuges) and Idaho (Kootenai National Wildlife Refuge).

Population: National Breeding Bird Survey trend data indicate a nonsignificant annual population increase of 0.8%. Trend estimates for the Rocky Mountains & Plains States region (U.S.F.W.S. Region 6) indicate a significant annual increase of 2.9%. The North American population (north of Mexico) has been estimated at 3.56 million birds, of which 86% breed in the Intermountain West Avifaunal Biome (Rich *et al.*, 2004).

Gray Flycatcher *(Empidonax wrightii)*

Status: A local summer resident in the southwestern portions of the region, especially where pinyon pines and junipers form an open woodland habitat-type, often with interspersed sagebrush. Less frequently they occur in sagebrush scrub, or in dense brush near streams in semi-arid areas.

Habitats and Ecology: In Colorado these birds are almost exclusively found in pinyon–juniper woodlands, where they are a characteristic breeding species, nesting either in forks of junipers or sometimes in sagebrush. Their altitudinal range there is very limited, from 5,000–7,000 feet (Kingery, 1998).

Their nests are typically constructed of the weathered outside strands of juniper bark, which makes them very difficult to find.

Suggested Viewing Locations: In Wyoming, juniper woodlands and sagebrush areas such as Minnies Gap (Sweetwater County), Powder Wash (west of Baggs) and Alcova Reservoir (Natrona County) are good places to search. Relatively dense regional breeding populations probably occur in southwestern Idaho (*e.g.*, City of Rocks National Reserve, Curlew National Grassland) and northwestern Colorado (Dinosaur National Monument, Browns Park National Wildlife Refuge).

Population: National Breeding Bird Survey trend data indicate a significant annual population increase of 4.1%. The North American population has been estimated at 1.2 million birds, of which 96% breed in the Intermountain West Avifaunal Biome (Rich *et al.*, 2004).

Cordilleran Flycatcher (*Empidonax occidentalis*)

Status: A summer resident over much of the region, mainly near streams; probably present in all the montane parks, but a common breeder only in Rocky Mountain National Park, where it is the most common small flycatcher.

Habitats and Ecology: A widespread and adaptable small flycatcher (previously known as the western flycatcher), ranging from riparian woodlands through aspens into the coniferous forest zones, all the way to the upper spruce-fir zones. Shrubby riparian areas with cottonwoods are favored. In Colorado, nesting occurs from about 6,000–10,000 feet, in montane and subalpine forests. In the interior Northwest, this species begins breeding in mixed coniferous forests during the young forest stage of succession (Sanderson, Bull and Edgerton, 1980). Nests are often placed on ledges, including bridges and vacant buildings so they favor ravines with rock outcrops, steep banks, and similar sites suitable for such nest placement.

Suggested Viewing Locations: In Wyoming, this is a common species in woodlands above the juniper zone, and is usually found along watercourses. It is widespread throughout all the major mountain ranges and the western Black Hills (Faulkner, 2010), and is especially common in the Laramie Range (Scott, 1993). Dense regional breeding populations probably occur in Idaho (*e.g.*, Kootenai and Minidoka national wildlife refuges), north-central Colorado (Rocky Mountain National Park), and western Montana (Lee Metcalf National Wildlife Refuge).

Population: National Breeding Bird Survey trend data indicate a nonsignificant annual population decrease of 0.9% for the Pacific Slope/cordilleran species group. The North American population (north of Mexico) has been estimated at 2.21 million birds (Rich *et al.*, 2004).

Eastern Phoebe (*Sayornis phoebe*)

Status: A summer resident in eastern Alberta, and a migrant through the other eastern portions of the region east of the mountains. Rare in Banff National Park, and unreported for the other montane parks.

Habitats and Ecology: Associated with woodland edges, wooded ravines near water, woodlots, and lakes or streams in partially wooded areas. Often breeds close to humans, nesting on building ledges, or on the understructure of bridges. In southeastern Colorado canyons, the birds build nests under overhanging cliffs and boulders in shady grottos (Kingery, 1998).

Suggested Viewing Locations: In Wyoming, and elsewhere in the region this species is most likely to be found along its eastern boundary, mainly during migration. Breeding birds might be searched-for in riparian woodlands of plains rivers in northeastern Wyoming (breeding has not yet been proven for the state) as well as in southeastern Colorado (mainly Las Animas County), where the birds reach the western limits of their U.S. range.

Population: National Breeding Bird Survey trend data indicate a significant annual population increase of 0.8%. Trend estimates for the Rocky Mountains & Plains States region (U.S.F.W.S. Region 6) indicate a significant annual increase of 3.6%. The North American population has been estimated at 16 million birds (Rich *et al.*, 2004).

Say's Phoebe (*Sayornis saya*)

Status: A summer resident over most of the region, but infrequent in the mountains and generally rare or absent from the montane parks.

Habitats and Ecology: Generally associated with grasslands, sagebrush, and agricultural areas in the region, especially prairie coulees and steep, eroded riverbanks. They sometimes reach foothill areas, but do not breed in the wooded mountain zones. In Wyoming they are widespread throughout, except at high elevations. In Colorado the birds usually nest below 9,000 feet, with the highest densities under 8,000 feet on the Western Slope (Kingery, 1998).

Suggested Viewing Locations: In Wyoming, and elsewhere, this is an abundant species in open country. Good Wyoming birding sites include the badlands area NNW of Baggs, Fossil Butte National Monument (Lincoln County), the Newcastle area (Weston County), and around Boysen Reservoir (Fremont County)(Dorn and Dorn, 1990). Dense regional breeding populations probably occur in northwestern and southeastern Colorado (*e.g.*, Dinosaur National Monument), and in north-central Wyoming (Bighorn Basin).

Population: National Breeding Bird Survey trend data indicate a significant annual population increase of 1.3%. The North American population (north of Mexico) has been estimated at 3.37 million birds (Rich *et al.*, 2004).

Ash-throated Flycatcher (*Myiarchus cinerascens*)

Status: A summer resident in western Colorado, and a local breeder in southwestern Wyoming. Rare or absent from the montane parks.

Habitats and Ecology: A desert species associated in this region with open pinyon–juniper woodlands, grasslands with scattered trees, and gulches or riparian edges in dry country. In Colorado the birds are most abundant from 4,500–7,000 feet elevation, with two-thirds of the population in pinyon–ju-

nipers, and 16% in riparian woodlands. Nests are usually in natural cavities or old woodpecker holes (Kingery, 1998), especially in old-growth junipers (Faulkner, 2010).

Suggested Viewing Locations: In Wyoming, the juniper woodlands along the
south side of Powder Rim (Sweetwater County) are good places for finding this species (Dorn and Dorn, 1990). Besides the Sweetwater–Uinta population, breeding also occurs locally in southern Natrona County (Alcova
Reservoir) and south-central Fremont County (Faulkner, 2010). The densest
regional breeding populations probably occur in western Colorado (*e.g.*, Dinosaur and Colorado national monuments, Mesa Verde National Park), and
also in southeastern Colorado (mainly Los Amimas County).

Population: National Breeding Bird Survey trend data indicate a significant annual population increase of 0.9%. Trend estimates for the Rocky Mountains
& Plains States region (U.S.F.W.S. Region 6) indicate a significant annual
increase of 3.1%. The North American population (north of Mexico) has
been estimated at 6.67 million birds (Rich *et al.*, 2004).

Cassin's Kingbird (*Tyrannus vociferus*)

Status: A local summer resident in the high plains east of the mountains, from
central Montana southward.

Habitats and Ecology: Associated with open country, usually with scattered
trees, or with open woodlands, and extending out into grasslands and agricultural lands where there are locally available trees for nesting. However,
bushes and posts may at times also be used as nesting sites. In Colorado,
pinyon–junipers are the primary habitat, with riparian woodlands, shelterbelts and grasslands also used.

Suggested Viewing Locations: In Wyoming, riparian or open oak woodlands attract this species, such as along the North Platte River in Goshen County
and Lance Creek in Niobrara County (Dorn and Dorn, 1990). Breeding
mostly occurs east of a line from Sheridan to Cheyenne (Faulkner, 2010).
The densest regional breeding populations probably occur in southeastern
Colorado and south-central Montana. In Colorado the largest populations
are in Las Animas and Huerfano counties, along the Purgatorie River and
Piñon Canyon (Kingery, 1998), and in Montana they might be found around
Ashland and Birney (McEneaney, 1993).

Population: National Breeding Bird Survey trend data indicate a nonsignificant annual population decrease of 0.25%. The North American population
(north of Mexico) has been estimated at 170,000 birds (Rich *et al.*, 2004).

Western Kingbird (*Tyrannus verticalis*)

Status: A summer resident in most of the region, but infrequent in montane areas, and rare or absent from the northern montane parks.

Habitats and Ecology: This species is always associated with edge habitats near
open country, such as shelterbelts, hedgerows, margins of forests, tree-lined

residential districts, riparian forests, and the like. It occupies more open country and occurs at somewhat lower elevations than the Cassin's kingbird, in Colorado usually below 7,000 feet and most often in rural or residential areas, or in riparian woodlands (Kingery, 1998).

Suggested Viewing Locations: This is a nearly ubiquitous breeder in open country throughout the entire region, especially along riparian woodlands in otherwise tree-free habitats. In Wyoming the birds occur statewide at lower elevations, especially where flying insects are abundant (Faulkner, 2010). The densest regional breeding populations probably occur in eastern parts of Montana, Wyoming and Colorado, and in southern Idaho. Because of their loud and insistent calls, they can be readily found wherever they occur.

Population: National Breeding Bird Survey trend data indicate a significant annual population increase of 0.6%. Trend estimates for the Rocky Mountains & Plains States region (U.S.F.W.S. Region 6) indicate a significant annual increase of 0.4%. The North American population (north of Mexico) has been estimated at 18.24 million birds (Rich *et al.*, 2004).

Eastern Kingbird (*Tyrannus tyrannus*)

Status: A summer resident nearly throughout the region, except in the drier areas of the southwest. Present in all the montane parks, and probably breeding in all of them.

Habitats and Ecology: Associated with open areas with scattered trees or tall shrubs, such as forest edges, fencerows, riparian areas, agricultural lands and farmsteads. . In Wyoming the birds occur statewide at lower elevations, especially where scattered mature trees or utility poles occur in otherwise open landscapes (Faulkner, 2010). In Colorado they occupy many of the same habitats as western kingbirds, but are more likely to occur in woodlands adjacent to open water. Cottonwoods and elms are favored nesting trees there (Kingery, 1998).

Suggested Viewing Locations: This is a widespread and common species in Wyoming and elsewhere throughout the region, with a somewhat more easterly (more woodland adapted) distribution than the western kingbird. Like the western kingbird, riparian woodlands in eastern Wyoming attract large numbers of eastern kingbirds. The densest regional breeding populations probably occur in eastern Montana riparian woodlands, but very good populations are also present in northeastern Wyoming (*e.g.*, Black Hills, Devils Tower National Monument) and northeastern Colorado (South Platte valley).

Population: National Breeding Bird Survey trend data indicate a nonsignificant annual population decrease of 1.0%. The North American population has been estimated at 13 million birds (Rich *et al.*, 2004).

Shrikes

Northern Shrike (*Lanius excubitor*)

Status: A winter resident or migrant throughout the region, mainly at lower elevations. Reported from all the montane parks, but generally uncommon to rare in all.

Habitats and Ecology: Invariably associated with open landscapes such as agricultural lands or grasslands that have scattered observation points such as fence posts or small trees. Reported as high as 9,500 feet in Colorado, but primarily occurring at lower elevations. Usually solitary, and sometimes seen where small birds such as sparrows are likely to gather.

Suggested Viewing Locations: This is a mobile species that might turn up almost anywhere in open country, especially where wires and fenceposts provide perching sites.

Population: National Breeding Bird Survey trend data are not available. The North American population has been estimated at 210,000 birds (Rich *et al.*, 2004).

Loggerhead Shrike (*Lanius ludovicianus*)

Status: A summer resident and migrant nearly throughout the region, except for the mountainous portions of Alberta and northern Idaho, where rare or accidental. Probably breeds in most or all of the U.S. montane parks of the region, although specific records are lacking.

Habitats and Ecology: Like the northern shrike, this species is associated with open habitats having scattered perching sites, and ranges altitudinally from agricultural lands on the prairies to montane meadows. Sagebrush areas, desert scrub, and pinyon–juniper woodlands offer ideal nesting and foraging areas, but some nesting also occurs in woodland edge situations, farmlands, and similar habitats. In Wyoming these birds occur statewide at elevations up to the foothills zone (Faulkner 2010). In Colorado, rural areas account for most breeding-season sightings, with shortgrass prairies a distant second. Nests there are usually placed in small, scattered trees, at times at elevations as high as 8,000 feet, and possibly to 8,900 feet (Kingery, 1998).

Suggested Viewing Locations: This species can be found in the same open-country habitats as the northern shrike, and is usually seen perching on telephone or fence wires along country roads. The densest regional breeding populations probably occur in eastern Colorado (Pawnee National Grassland and South Platte valley grasslands), but very good populations are also present in eastern Wyoming (*e.g.*, Thunder Basin National Grassland) and northeastern Montana (grasslands in Missouri and Yellowstone valleys). Populations in all these regions have undergone sharp declines in recent years.

Population: National Breeding Bird Survey trend data indicate a significant annual population decrease of 3.7%. The North American population (north of Mexico) has been estimated at 3.7 million birds (Rich *et al.*, 2004).

Vireos

Plumbeous Vireo (*Vireo plumbeus*), **Cassin's Vireo** (*V. cassini*) and **Blue-headed Vireo** (*V. solitarius)*

Status: Until recently these three forms were considered conspecific, so the published regional records for each of them are still somewhat limited. The plumbeous vireo is a summer resident in forested areas over most of the region, breeding north to southern Idaho and south-central Montana. The Cassin's vireo is a migrant throughout the western Rockies, and breeds from the northern half of Idaho and the Montana Rocky Mountains north through the Canadian montane parks. The blue-headed vireo is a migrant along the eastern edge of the Rocky Mountains.

Habitats and Ecology: Open, coniferous or mixed forest with considerable undergrowth seem to be the plumbeous vireo's favored habitat, especially those that offer open branches for foraging at low to medium tree levels. Fairly dry and warm forests are favored by the plumbeous vireo over moist and cool ones, and breeding extends from the open oak or aspen and ponderosa pine zones upward through the lower coniferous communities, to about 8,000 feet in Colorado, or below the upper limits of the warbling vireo. The Cassin's is more inclined to favor low-altitude deciduous woods for nesting. The nests of both are usually in the lower branches of pines or oaks, often only a few feet above ground level.

Suggested Viewing Locations: Plumbeous vireos breed in all the major mountain ranges of Wyoming except for the Greater Yellowstone region, but are especially common in the western Black Hills, Laramie Mountains and eastern Bighorns (Dorn and Dorn, 1990; (Faulkner 2010). The densest regional breeding populations of the plumbeous vireo occur in the Black Hills, but good populations are also present in central and western Colorado. In Colorado the birds are mainly found in pinyon–juniper woodlands (such as at Dinosaur National Monument and Mesa Verde National Park) and open ponderosa pine forests, with more limited use of riparian woodlands, scrub oak and montane shrubs (Kingery, 1998). In Montana the Bighorn National Recreation Area provides good viewing opportunities. In Idaho they mainly occur in the intermountain semidesert region, such as at the City of Rocks National Reserve and Deer Flat National Wildlife Refuge. The densest populations of the Cassin's vireo are in northwestern Montana (*e.g.*, Lee Metcalf National Wildlife Refuge) and northern Idaho (Kootenai National Wildlife Refuge, Winchester Lake State Park). Migrant blue-headed vireos are only likely to be found in the eastern plains.

Population: National Breeding Bird Survey trend data for the plumbeous vireo indicate a nonsignificant annual population decrease of 0.2%, for the Cassin's vireo a significant annual population increase of 1.0%, and for the blue-headed vireo a significant annual population increase of 4.7%. The North American populations (north of Mexico) have been estimated at 2.16 million birds for the plumbeous, 4.6 million for the Cassin's and 6.9 million for the blue-headed vireo (Rich *et al.*, 2004).

Warbling Vireo (*Vireo gilvus*)

Status: A summer resident in forests throughout the entire region, and fairly common in nearly all the montane parks, and probably breeding in all of them. It is the most common and widespread vireo of the region.

Habitats and Ecology: Fairly open woodlands, especially of deciduous trees, are favored by this species. It is probably most common along riparian forests supporting tall trees, but also occurs in aspen groves and well-wooded residential or park areas, especially where tall cottonwoods are present. In coniferous forest areas the birds favor areas where single or clumped broad-leaved trees such as aspens or birches occur. Foraging is done near the crowns of fairly densely-leaved trees, and nests are sometimes located as high as 90 feet above ground in very tall forests. In Colorado, most breeders occupy aspen woodlands, and are more abundant on the Western Slope than eastwardly (Kingery, 1998). Relatively few nest in eastern riparian woodlands of the prairie rivers, whereas in Nebraska these constitute a primary nesting habitat.

Suggested Viewing Locations: In Wyoming, this species is likely to be found in almost every aspen stand in the mountains (Dorn and Dorn, 1990); at lower altitudes riparian willow and cottonwood stands are equally attractive. The densest regional breeding populations probably occur in the Greater Yellowstone region, the Black Hills region, and from central Idaho northeast to western Montana. The male's loud and distinctively syncopated songs, uttered even while he is sitting on the nest, makes locating the birds very easy.

Population: National Breeding Bird Survey trend data indicate a significant annual population increase of 0.9%. Trend estimates for the Rocky Mountains & Plains States region (U.S.F.W.S. Region 6) indicate a significant annual increase of 1.0%. The North American population (north of Mexico) has been estimated at 1.6 million birds (Rich *et al.*, 2004).

Philadelphia Vireo (*Vireo philadelphicus*)

Status: A local summer resident in central Alberta south to about Cold Lake and Sundre. Otherwise a spring and fall migrant east of the mountains; reported as a vagrant in Banff, Jasper, and Yoho National Parks.

Habitats and Ecology: Breeding habitats of this species include open deciduous woodlands such as regrowth areas of aspens and poplars, or secondary birch-poplar communities that have developed following logging of coniferous forests. The birds also occur in muskeg areas having willow or alder thickets around their edges.

Suggested Viewing Locations: In Wyoming, this is a very rare migrant, and is most likely to be seen in the eastern third of the state (Dorn and Dorn, 1990). Elsewhere in the region it is likewise rarely encountered, except along the eastern border.

Population: National Breeding Bird Survey trend data indicate a significant annual population increase of 2.5%. The North American population has been estimated at 4.3 million birds (Rich *et al.*, 2004).

Red-eyed Vireo (*Vireo olivaceous*)

Status: A summer resident over the northern half of the region, and local farther
south as a breeder or migrant. Largely limited to deciduous areas, and thus
rare or occasional in most of the montane parks except Glacier National
Park, where a common breeder.

Habitats and Ecology: This species is primarily associated with deciduous forests, especially those with semiopen canopies. In the Rocky Mountain region it is largely limited to broad-leaved riparian forests in prairie areas, especially where large, mature cottonwoods occur. It also occurs in planted
areas such as city parks and farmsteads, as well as in aspen groves or poplars growing among conifers.

Suggested Viewing Locations: In Wyoming, most breeding occurs east of a line
extending from Laramie to Sheridan (Faulkner 2010). Breeding birds are
likely to be found in the Tongue River Canyon (Sheridan County) and along
Sand Creek in Crook County (Dorn and Dorn, 1990). There are also good
northeastern Wyoming populations in the western Black Hills. The densest
regional breeding populations probably occur in northern Idaho (*e.g.*, Kootenai National Wildlife Refuge) and western Montana (Swan River National
Wildlife Refuge and National Bison Range) and British Columbia.

Population: National Breeding Bird Survey trend data indicate a significant annual population increase of 1.2%. The North American population has
been estimated at 140 million birds (Rich *et al.*, 2004).

Jays, Magpies & Crows

Gray Jay (*Perisoreus canadensis*)

Status: A resident in montane forests throughout the region; present in all the
montane parks and relatively common in all.

Habitats and Ecology: This species is associated with a wide variety of boreal
and montane coniferous forest types, and occasionally extends into aspens and riparian woodlands outside the breeding season. Nesting almost
always is in coniferous vegetation. In the interior Northwest, this species
breeds in mixed coniferous forests from the pole-sapling to the mature forest stage of succession (Sanderson, Bull and Edgerton, 1980). In Colorado,
nesting occurs from about 8,500 feet to timberline, usually in spruce–fir
forests, but sometimes in lodgepole pines. Territories are large and permanent, and almost everything edible is consumed, including a variety of
other birds (Kingery, 1998).

Suggested Viewing Locations: In Wyoming, gray jays occur in all the mountain
ranges, and are most often and easily seen at higher elevations at tourist-
frequented sites of the Teton, Wind River and Medicine Bow Ranges (Dorn
and Dorn, 1990). McEneaney (1988) judged the best observation areas of
the Yellowstone population to be around Yellowstone River and Bridge
Bay. Other montane parks in the region are also favorite places for finding this common species. The densest regional breeding populations prob-

ably occur in the central Rockies of Colorado (*e.g.*, Rocky Mountain National Park), and in the mountains of western Montana (Glacier National Park) and Idaho (Winchester Lake and McCrosky state parks).

Population: National Breeding Bird Survey trend data indicate a nonsignificant annual population decrease of 0.6%. The North American population has been estimated at 16 million birds (Rich *et al.*, 2004).

Steller's Jay (*Cyanocitta stelleri*)

Status: A resident in coniferous woodlands throughout the region. Present in all the montane parks, and breeding in most or all, but less common northwardly.

Habitats and Ecology: This species is centered in the ponderosa pine zone, but also extends down into the pinyon–juniper zone, and as high as the Douglas-fir zone. In the interior Northwest, this species breeds in mixed coniferous forests from the pole-sapling to the old-growth forest stage of succession (Sanderson, Bull and Edgerton, 1980). In Colorado it occurs in foothills and lower mountains from 6,000–9,000 feet, most often in ponderosa pines forests, but also in other conifers, and aspen forests. It occupies an ecological zone above that of blue jay and western scrub-jay, and below the gray jay's (Kingery, 1998).

Suggested Viewing Locations: This species is common throughout most of the coniferous forests of Wyoming, but are curiously absent from the Black Hills and are only rare vagrants in the Bighorns. More widely, they are generally common throughout the Rocky Mountains proper. The densest regional breeding populations probably occur in the central and southern Rockies of Colorado (*e.g.*, Rocky Mountain National Park), but good populations extend north to the mountains of western Montana (Glacier National Park, Lolo Pass) and northern Idaho (Kootenai National Wildlife Refuge).

Population: National Breeding Bird Survey trend data indicate a nonsignificant annual population increase of 0.15%. The North American population (north of Mexico) has been estimated at 3.74 million birds (Rich *et al.*, 2004).

Blue Jay (*Cyanocitta cristata*)

Status: A local resident in eastern and northern portions of the region, mainly in deciduous habitats. A rare vagrant in the montane parks, but gradually extending its range westwardly and perhaps becoming more common in the Rocky Mountain region. In Wyoming the birds now breed west approximately to a line extending from Park to Albany counties (Faulkner 2010).

Habitats and Ecology: Widely distributed in deciduous woodland, city parks, suburbs, and almost anywhere there is an intersection of woods and open grassy areas. Riparian woods, with large willows or cottonwoods, are favored habitats on the western plains. In Colorado, riparian riverbottoms are the most commonly used habitat, followed by urban and rural environments. They reach the foothills, but very rarely extend into montane habitats (Kingery, 1998).

Suggested Viewing Locations: In Wyoming, the North Platte River riparian woodlands, and the western Black Hills forests of Crook County are good places
for finding this conspicuous species, as are eastern towns such as Cheyenne and Laramie (Dorn and Dorn, 1990). The densest regional breeding
populations probably occur in eastern Montana (*e.g.,* Yellowstone and Missouri River valleys), eastern Wyoming (North Platte valley) and northeastern Colorado (South Platte valley).

Population: National Breeding Bird Survey trend data indicate a significant annual population decrease of 1.1%. The North American population has
been estimated at 22 million birds (Rich *et al.,* 2004).

Western Scrub-Jay *(Aphelocoma californica)*

Status: A local migrant or summer resident in the southwestern parts of the region, south of the Snake River in Idaho and in northwestern Colorado. Rare
or absent from the montane parks.

Habitats and Ecology: Associated with low arid woodlands, including pinyon–
juniper and oak–mountain mahogany, and less frequently extending into
the ponderosa pine zone where oaks are also present. It is often found along
brushy ravines or wooded creek-bottoms. In Colorado, pinyon–juniper and
scrub oak are primary habitats, especially if water is nearby (Kingery, 1998).

Suggested Viewing Locations: In Wyoming, the best chances of finding this
southwestern species are at Flaming Gorge and Powder Rim, in Sweetwater
County (Dorn and Dorn, 1990); they are regularly found only in Sweetwater and Uinta counties, and on the Power Rim of Carbon County (Faulkner
2010). The densest regional breeding populations probably occur the Front
Range foothills of south-central Colorado (Douglas to Huerfano counties),
but there are well-established populations in western Colorado, from Mesa
Verde National Park and Colorado National Monument northward through
Dinosaur National Monument and Browns Park National Wildlife Refuge to
the Wyoming border.

Population: National Breeding Bird Survey trend data indicate a nonsignificant annual population increase of 0.04%. The North American population (north of Mexico) has been estimated at 2.72 million birds (Rich *et al.,*
2004).

Pinyon Jay *(Gymnorhinus cyanocephalus)*

Status: A resident over much of the southern parts of the region, including
southern Idaho, southern Montana, and most of Wyoming, especially at
lower elevations. Rare in the montane parks of Wyoming and Colorado.

Habitats and Ecology: Generally associated with pine forests growing on dry
substrates, especially the pinyon–juniper association, but extending during the non-breeding period into oak–mountain mahogany, sagebrush, and
desert scrub habitats. In Colorado the birds strongly prefer pinyon–juniper
habitats, but where pinyon pines are lacking they use juniper woodlands
and consume juniper berries rather than pinyon nuts. On the eastern slope

of the Continental Divide only about two percent of the breeding-season observations were in ponderosa pines (Kingery, 1998).

Suggested Viewing Locations: In Wyoming, the best locations for finding this southwestern species are at Flaming Gorge in Sweetwater County, the lower eastern slope of the Bighorn Mountains, around Alcova Reservoir (Natrona County) and in the Guernsey area (Dorn and Dorn, 1990). In Colorado they occur commonly in the parks and refuges along the Utah border (see scrub-jay account), as well as in the woodlands of Las Animas and Huerfano counties. In Montana the birds are in pinyon–ponderosa woodlands (abundant at the C. R. Russell National Wildlife Refuge). In Idaho they may be found at the City of Rocks National Reserve, Curlew National Grassland, or the Juniper Rest Area (five miles north of the Utah border on I-84 (Svingen and Dumroese, 1997).

Population: The North American population has been estimated at 4.1 million birds, of which 92% breed in the Intermountain West Avifaunal Biome. Classified as a Partners-in-Flight Watch List species of continental conservation importance (Rich et al., 2004). National Breeding Bird Survey trend data indicate a significant annual population decrease of 4.3%.

Clark's Nutcracker (*Nucifraga columbiana*)

Status: A resident in wooded areas over much of the region, common in all the montane parks and breeding in all of them.

Habitats and Ecology: Widespread in coniferous habitats, from the ponderosa pine zone to timberline. In the interior Northwest, this species breeds in mixed coniferous forests in the mature and old-growth forest stages of succession (Sanderson, Bull and Edgerton, 1980). More common in the higher coniferous zones in summer, but descending during winter to the pinyon zone and sometimes out onto the plains areas. In Colorado the birds range from 8,000 feet to timberline during the breeding season, and are especially common in spruce–fir and mixed coniferous forests. During breeding they continue to eat pinyon seeds, obtained from stored underground caches (Kingery, 1998).

Suggested Viewing Locations: In Wyoming, this conspicuous species occurs in al the higher mountain ranges. High-altitude picnic areas and scenic viewpoints in national or state parks are good places to find nutcrackers, which often are very tame at such locations. McEneaney (1988) judged the population the best Yellowstone observation areas to be around Dunraven Pass and the Upper Terrace Drive. The densest regional breeding populations probably occur in central and southwestern Colorado (east locally in southern Colorado to Las Animas County), the Greater Yellowstone region, and the mountains along the Idaho-Montana border (*e.g.*, Red Rock Lakes and Kootenai national wildlife refuges, Pine Butte Swamp, Beartooth Mountains).

Population: National Breeding Bird Survey trend data indicate a significant annual population increase of 1.9%. The North American population (north of Mexico) has been estimated at one million birds, of which 89% breed in the Intermountain West Avifaunal Biome (Rich *et al.*, 2004).

Black-billed Magpie (*Pica pica*)

Status: A resident throughout the region, in most habitats. Present and a common breeder in all the montane parks.

Habitats and Ecology: Of widespread occurrence, but especially common in riparian areas with thickety vegetation, agricultural areas with scattered trees, sagebrush, aspen groves, and the lower levels of the coniferous forest zones. Small, thorny trees are especially favored nest sites, but junipers and similar trees are also used. In Colorado the birds range up to 10,000 feet of elevation during the breeding season, and tend to favor riparian, rural and pinyon–juniper areas for nesting. Conifers appear to be avoided except for open ponderosa woodlands (Kingery, 1998).

Suggested Viewing Locations: In Wyoming, this species might be seen almost anywhere in open country at lower elevations, especially in those areas with scattered small trees, and that are fairly close to water. The same rule applies elsewhere in the Rocky Mountain region. The densest regional breeding populations probably occur in southern Idaho (*e.g.*, Camas, Deer Flat and Minidoka national wildlife refuges) and southwestern Montana (Lee Metcalf National Wildlife Refuge and National Bison Range). National populations have not fully recovered from an outbreak of West Nile disease that devastated many corvid populations during the early years of this century.

Population: National Breeding Bird Survey trend data indicate a nonsignificant annual population decrease of 0.2%. The North American population has been estimated at 3.4 million birds (Rich *et al.*, 2004).

American Crow (*Corvus brachyrhynchos*)

Status: A summer or year-round resident in wooded habitats throughout the region, but more common at lower altitudes, and varying from common to rare in the montane parks, breeding in several.

Habitats and Ecology: Forested areas, wooded riverbottoms, orchards, woodlots, large parks, and suburban areas are all used by this species, but it is often replaced by common ravens in rocky canyons and higher montane areas. In Colorado, crows breed at elevations up to 10,000 feet, with rural areas and croplands favored habitats. Trees, shrub and utility poles may serve at nest sites (Kingery, 1998).

Suggested Viewing Locations: Crows are widespread in Wyoming wherever they are not displaced by ravens, and unlike most ravens will commonly frequent towns and villages. They are very common in the Black Hills, as well as around Lovell, Torrington and Riverton (Dorn and Dorn, 1990). The densest regional breeding populations probably occur in northeastern Montana, but there are also good populations in the Greater Yellowstone Ecosystem and the general boundary region of southwestern Montana (*e.g.*, National Bison Range, Lee Metcalf National Wildlife Refuge) and southeastern Idaho (Camas and Grays Lake national wildlife refuges).

Population: National Breeding Bird Survey trend data indicate a significant annual population increase of 0.6%. The North American population has been estimated at 31 million birds (Rich *et al.*, 2004). The West Nile epidemic that reached Wyoming in 2002 had serious effects on Nebraska crow and magpie populations as it moved west, but its impact on Wyoming is unknown (Faulkner 2010).

Common Raven (*Corvus corax*)

Status: A permanent resident throughout much of the region, mainly in mountainous regions. Present in all of the montane parks, and breeding in all of them.

Habitats and Ecology: Generally associated with wilderness areas of mountains and forests, especially where bluffs or cliffs are present for nesting. Where these are unavailable, tall coniferous trees are used for nesting, as in Jackson Hole, where over 90% of the nests are in trees. Often found all the way to timberline in late summer, or even in alpine tundra areas. They also extend out into sage and grassland areas, scavenging for road-killed mammals and birds. In Colorado ravens breed widely in mountains and forests, nesting most commonly in pine-juniper woodlands and upland coniferous forests, with foraging territories often covering several square miles, and often encompassing an entire valley, ridge to ridge. There they often nest on cliffs having a southern aspect (Kingery, 1998).

Suggested Viewing Locations: In Wyoming, ravens are almost entirely found west of a line extending from Sheridan to Casper and Laramie (Faulkner 2010), and are common in all the major mountain ranges. They are especially abundant in Jackson Hole, where they are attracted to carrion and wastes at places like the National Elk Refuge, or at wolf kills in the Yellowstone Park region. McEneaney (1988) judged the to best Yellowstone observation areas to be around Artist Point and Old Faithful, both heavily visited by humans. The densest regional breeding populations probably occur in the Greater Yellowstone region, and to a lesser degree in southern Idaho (Snake River valley). In the Idaho panhandle Kootenai National Wildlife Refuge also has a high raven population.

Population: National Breeding Bird Survey trend data indicate a significant annual population increase of 3.2%. Trend estimates for the Rocky Mountains & Plains States region (U.S.F.W.S. Region 6) indicate a significant annual increase of 3.3%. The North American population (north of Mexico) has been estimated at 9.22 million birds (Rich *et al.*, 2004).

Larks

Horned Lark (*Eremophila alpestris*)

Status: Present throughout the entire region, and resident in most areas, although migratory movements do occur. Present in nearly all the montane parks, and breeding in several.

Habitats and Ecology: Open-country habitats, ranging from shortgrass plains through agricultural lands such as pastures, desert scrub, mountain meadows, and alpine tundra, are the basic requirements for this species, which has an enormous ecological and geographic range in North America.

Suggested Viewing Locations: In Wyoming, horned larks occur statewide ands may easily be found during almost any season in open, grassy country up to the alpine zone. The same applies elsewhere in the Rocky Mountain region. The densest regional breeding populations probably occur in the mixed-grass prairies of northwestern Montana (e.g., Medicine Lake National Wildlife Refuge), and in the shortgrass plains of southeastern Colorado (Comanche National Grassland). There are also dense populations in central Montana (C. R. Russell National Wildlife Refuge) and southern Wyoming (Seedskadee and Hutton Lake national wildlife refuges).

Population: National Breeding Bird Survey trend data indicate a nonsignificant annual population decrease of 2.0%. Trend estimates for the Rocky Mountains & Plains States region (U.S.F.W.S. Region 6) indicate a significant annual decline of 1.4%. The North American population (north of Mexico) has been estimated at 98 million birds (Rich *et al.*, 2004).

Swallows

Purple Martin (*Progne subis*)

Status: Mostly limited as a summer resident to extreme northeastern Montana and the central portions of Alberta in the Rocky Mountain region, but breeds locally in western Colorado (Routt to Montezuma counties). Absent or accidental in all the montane parks.

Habitats and Ecology: Restricted in this region almost entirely to areas where "martin houses" have been erected. Like the other swallows, it obtains all of its food by aerial foraging for insects. Nesting in Colorado occurs at 8,000–9,000 feet, often in old-growth aspen stands close to water (Kingery, 1998).

Suggested Viewing Locations: The only known breeding colony in Wyoming is in an aspen stands on the west slope of the Sierra Madre Mountains (Albany County) (Faulkner 2010). In Colorado possible summer viewing locations include the Fourmile Road near Glenwood Springs and the Tabeguache Basin near Grand Junction (Kingery, 2007). The densest regional breeding populations probably occur in west-central Colorado, on the western plateaus such as Roan and Grand Mesa.

Population: National Breeding Bird Survey trend data indicate a nonsignificant annual population decrease of 0.1%. The North American population (north of Mexico) has been estimated at 9.9 million birds (Rich *et al.*, 2004).

Tree Swallow (*Iridoprocne bicolor*)

Status: A local summer resident throughout the region, but absent from treeless areas and the high montane communities. Present and breeding in all the montane parks.

Habitats and Ecology: Breeding in the region extends from riparian woodlands through the aspen zone, and into the lower levels of the coniferous forest zone. In the interior Northwest, this species breeds in mixed coniferous forests from the young to the old-growth forest stages of succession (Sanderson, Bull and Edgerton, 1980). Outside the breeding season often seen over lakes or rivers, as well as over other open habitats. Nesting is especially prevalent in the aspen stands where old woodpecker holes are available, but also occurs at times in bird houses erected for bluebirds. In some areas such as along plains rivers where aspens are lacking cottonwoods may serve for nesting; with either aspen or cottonwood nest sites, easy access to open foraging areas is important (Kingery, 1998).

Suggested Viewing Locations: Tree swallows in Wyoming and elsewhere are often associated with aspen stands having good woodpecker populations. They are also very common in the riparian woodlands of eastern Wyoming, such as along the North Platte River. They occur statewide, and are especially common in aspen groves near water (Faulkner, 2010). The densest regional breeding populations probably occur in the Greater Yellowstone region and western Montana (*e.g.*, National Bison Range; Lee Metcalf National Wildlife Refuge).

Population: National Breeding Bird Survey trend data indicate a nonsignificant annual population increase of 0.2%. Trend estimates for the Rocky Mountains & Plains States region (U.S.F.W.S. Region 6) indicate a significant annual increase of 2.5%. The North American population has been estimated at 20 million birds (Rich *et al.*, 2004).

Violet-green Swallow (*Tachycineta thalassina*)

Status: A summer resident in mountainous areas, occurring locally throughout the entire region; present and probably breeding in all of the montane parks.

Habitats and Ecology: Generally associated with open coniferous forests, such as ponderosa pines, but also breeds in aspen groves, in riparian woods, and sometimes also in urbanized areas. Nesting sites are rather variable, and include old woodpecker holes, natural tree or cliff cavities, and occasionally also birdhouses. Cliff nest sites in Colorado are second only to upland deciduous forest habitats in nest-choice frequency (Kingery, 1998).

Suggested Viewing Locations: In Wyoming, and elsewhere, this species occupies deep canyons during the breeding season. In Wyoming they occur in all the major mountain ranges and the Black Hills, and may be easily seen in the Wind River Canyon (Washakie County), Yellowstone Canyon, canyons in the Bighorn Mountains, as well as Beaver Rim (Fremont County), and LAK Reservoir in Weston County (Dorn and Dorn, 1990). The densest regional breeding populations probably occur in the plateau and canyon country western Colorado (e.g., Black Canyon National Park, Dinosaur National Monument). Good populations also occur in the Greater Yellowstone region, the western Black Hills, and the mountains bordering Idaho and Montana (National Bison Range, Swan River and Red Rock Lakes national wildlife refuges).

Population: National Breeding Bird Survey trend data indicate a nonsignificant annual population increase of 0.5%. Trend estimates for the Rocky Mountains & Plains States region (U.S.F.W.S. Region 6) indicate a significant annual increase of 1.6%. The North American population (north of Mexico) has been estimated at 8.25 million birds (Rich *et al.*, 2004).

Northern Rough-winged Swallow (*Stelgidopteryx ruficollis*)

Status: A local summer resident throughout the region, mainly at lower elevations and in open habitats. Present in all the montane parks, and breeds in most, but rare in the more southerly parks except at the lowest elevations.

Habitats and Ecology: Associated with open areas, including agricultural lands, rivers and lakes, and grasslands near water, and breeding almost exclusively in cavities dug in earthen banks of clay, sand, or gravel. Ready-made cavities in rock are also used. In Colorado the birds nest at altitudes up to 7,500 feet, mostly in streamside banks in prairies, foothills and deserts, and typically as solitary pairs (Kingery, 1998).

Suggested Viewing Locations: The swallow is found throughout Wyoming except at high elevations, and is especially common at Flaming Gorge Reservoir (Sweetwater County). Elsewhere in the Rocky Mountain region it is easily found, especially around water, The densest regional breeding populations probably occur in the Greater Yellowstone region, but good populations also exist in northwestern Montana (e.g., National Bison Range, Swan River National Wildlife Refuge) and southwestern Idaho (Deer Flat National Wildlife Refuge).

Population: National Breeding Bird Survey trend data indicate a nonsignificant annual population decrease of 0.5%. The North American population (north of Mexico) has been estimated at 5.1 million birds (Rich *et al.*, 2004).

Bank Swallow (*Riparia riparia*)

Status: A summer resident in suitable habitats throughout the region, mainly at lower elevations. Breeds in several of the montane parks, but not common in any.

Habitats and Ecology: Breeding almost always occurs near water, such as in steep banks along rivers, roadcuts near lakes, gravel pits, and similar areas with steep slopes of clay, sand, or gravel. Outside the breeding season the birds are of broader distribution, sometimes foraging over agricultural lands.

Suggested Viewing Locations: In Wyoming, this is a widespread species at lower elevations in summer, where it nests along cuts in clay, sand, or silt banks near water. Where these banks are unstable the colonies may last for only a year or two, and then disappear. The densest regional breeding populations probably occur in the Greater Yellowstone region, and along the common boundary of southeastern Idaho and southwestern Montana (e.g., Red Rock Lakes National Wildlife Refuge). Dense populations also exist in southern Idaho (Minidoka and Deer Flat national wildlife refuges) and part of northern Idaho (Kootenai National Wildlife Refuge).

Population: National Breeding Bird Survey trend data indicate a significant annual population decrease of 1.95%. The North American population (north of Mexico) has been estimated at 13.8 million birds (Rich *et al.*, 2004).

Cliff Swallow (*Petrochelidon pyrrhonota*)

Status: A summer resident throughout the region, including all of the montane parks, where it is a common to abundant breeder.

Habitats and Ecology: A wide variety of nesting areas are used by this species, but in the region under consideration vertical cliff-sides and the sides or undersides of bridges are perhaps most commonly used. The nests are gourd-like structures of dried mud, made of small mud globules that are gathered by the birds and carried back in the beak. In Colorado the birds nest as high as 12,000 feet, but are mostly found at much lower altitudes (Kingery, 1998).

Suggested Viewing Locations: In Wyoming, and elsewhere in the region this species occurs throughout the state except at high elevations, especially where there are sheer rock cliffs, concrete bridges or even large metal culverts. The densest regional breeding populations probably occur along the common boundary of southeastern Idaho and southwestern Montana (e.g., Red Rock Lakes National Wildlife Refuge), as well as over much of the Greater Yellowstone region.

Population: National Breeding Bird Survey trend data indicate a nonsignificant annual population increase of 0.65%. The North American population (north of Mexico) has been estimated at 80.1 million birds (Rich *et al.*, 2004).

Barn Swallow (*Hirundo rusticola*)

Status: A summer resident throughout the region, including the montane parks, where it is a common to occasional breeder in all.

Habitats and Ecology: Except for the purple martin, this species is the swallow that is most closely associated with humans in the Rocky Mountain region. Although it may still occasionally nest on cliff or cave walls, its normal current nesting sites are the horizontal beams or upright walls of buildings and similar structures. In Colorado the birds nest at elevations up to 10,000 feet, most often in rural sites such as farmhouses, barns and sheds (Kingery, 1998).

Suggested Viewing Locations: In Wyoming this species is abundant at elevations up to about 8,000 feet, especially around farmyards or abandoned buildings, as well as around bridges and culverts. The densest regional breeding populations probably occur in eastern Montana and eastern Idaho.

Population: National Breeding Bird Survey trend data indicate a significant annual population decrease of 0.9%. Trend estimates for the Rocky Mountains
& Plains States region (U.S.F.W.S. Region 6) indicate a significant annual
decline of 0.7%. The North American population (north of Mexico) has
been estimated at 51.3 million birds (Rich *et al.*, 2004).

Chickadees & Titmice

Black-capped Chickadee *(Poecile atricapillus)*

Status: Resident in deciduous and coniferous forests throughout the region.

Habitats and Ecology: Associated with a wide variety of wooded habitats, both
of coniferous and hardwood types, and breeding wherever suitable nesting
cavities exist. These typically consist of old woodpecker holes, but sometimes the birds excavate their own nest cavities in the rotted wood of dead
stumps. Bird houses are also occasionally used. Aspen groves and riparian woodlands are favored nesting areas in the Rocky Mountain region. In
Colorado most chickadees nest from 5,000–9,000 feet elevation, with large
aspens stands their favorite breeding habitat, and lowland riparian woods
with cottonwoods their second most frequent choice (Kingery, 1998).

Suggested Viewing Locations: In Wyoming, this eastern-oriented chickadee is
most common in eastern areas, such as the western Black Hills, but also the
Tongue River Canyon near Dayton, along the North Platte River, and also is
prevalent west to the Wind River near Dubois (Dorn and Dorn, 1990). It is
usually found below 8,500 feet, but may reach 10,000 feet along riparian
corridors (Faulkner 2010). The densest regional breeding populations probably occur in the western Black Hills, but good populations also occur in
northwestern Montana (*e.g.*, Lee Metcalf, Red Rock Lakes and Swan River
national wildlife refuges) and northern Idaho (Kootenai National Wildlife
Refuge).

Population: National Breeding Bird Survey trend data indicate a significant annual population decrease of 1.1%. The North American population has
been estimated at 34 million birds (Rich *et al.*, 2004).

Mountain Chickadee *(Poecile gambelii)*

Status: A resident in mountain forests of the region, occurring in all the montane parks and probably breeding in all. It is more montane in distribution
than the previous species.

Habitats and Ecology: Largely limited to montane coniferous forests, and usually absent from deciduous stands, although aspens are frequently used for
nesting. Prefers open coniferous forests, especially pines, including both
ponderosa pines and also pinyons. In the interior Northwest, this species
breeds in mixed coniferous forests from the pole-sapling to the old-growth
forest stages of succession (Sanderson, Bull and Edgerton, 1980). Woodpecker holes or self-excavated cavities in rotted wood are used for nesting.

In Colorado, this species ranges up to about 9,000 feet elevation, reaching its greatest abundance in old-growth spruce–fir, lodgepole pine and ponderosa pine forests, and is less common in pinyon–juniper woodlands (Kingery, 1998).

Suggested Viewing Locations: In Wyoming, this chickadee is common in all the mountain ranges except for the Black Hills (Dorn and Dorn, 1990), and largely limited to mature coniferous forests from 7,500 feet to timberline (Faulkner 2010). McEneaney (1988) judged the Yellowstone population to be in the thousands during the late 1980's, with the best observation areas being around Fishing Bridge and the Blacktail Plateau. The densest regional breeding populations probably occur in the Greater Yellowstone region, but large populations also occur in north-central Colorado, where they are abundant in the Front, Gore and Park ranges, as well as in the San Juan Mountains of southwestern Colorado.

Population: National Breeding Bird Survey trend data indicate a significant annual population decrease of 0.9%. The North American population (north of Mexico) has been estimated at 11.9 million birds (Rich *et al.*, 2004).

Boreal Chickadee (*Poecile hudsonica*)

Status: A permanent resident in coniferous forests of the northern part of the region, breeding south to west-central Montana and extreme northern Idaho, and occurring in all the more northern montane parks.

Habitats and Ecology: Associated with coniferous forests, including swampy areas and muskegs on lower elevations, as well as those of montane areas. Nesting occurs in trees and stumps, either in old woodpecker holes or in self-excavated holes.

Suggested Viewing Locations: Glacier National Park and Kootenai National Wildlife Refuge provide some likely locations for finding this boreal species. The densest regional breeding concentrations are probably in the Canadian montane parks.

Population: National Breeding Bird Survey trend data indicate a significant annual population decrease of 2.6%. The North American population has been estimated at 7.8 million birds (Rich *et al.*, 2004).

Chestnut-backed Chickadee (*Poecile rufescens*)

Status: A permanent resident in the wetter coniferous forests of the western slopes of the region, south to about central Montana.

Habitats and Ecology: Regionally this species is essentially limited to the western hemlock, western redcedar association, but it has a broader ecological range farther west, extending into some deciduous woodlands adjacent to coniferous forests. Areas along stream courses and other forest margins are favored for foraging, and nesting occurs in woodpecker holes or self-excavated cavities. In the interior Northwest, this species breeds in mixed coniferous forests from the pole-sapling to the old-growth forest stages of succession (Sanderson, Bull and Edgerton, 1980).

Suggested Viewing Locations: This species has a range that conforms to that of the moist Pacific Northwest coastal coniferous forest, is locally extended inland on the wettest west-facing slopes of the Rockies in Idaho, Montana and adjoining Canada. It is a common breeder in Glacier National Park. It is an uncommon breeder at Kootenai National Wildlife Refuge, and also summers at Heyburn and Winchester Lake state parks.

Population: National Breeding Bird Survey trend data indicate a nonsignificant annual population increase of 7.3%. The North American population has been estimated at 6.9 million birds (Rich *et al.*, 2004).

Juniper (Plain) Titmouse *(Baeolophus ridgwayi)*

Status: A summer or permanent resident in the southwestern portions of the region, from the Snake River southward in Idaho, and in the southwestern and western portions of Wyoming and Colorado, respectively.

Habitats and Ecology: In this region associated almost exclusively with the pinyon–juniper association; in some other areas also extending into oak woodlands. Nesting is usually done in cavities of partially decayed and split-open trunks of junipers. They sometimes also use woodpecker holes, and occasionally will nest in birdhouses. In Colorado, pinyon–juniper habitat accounted for 96% of breeding-season observations (Kingery, 1998). In Wyoming nesting is confined to Utah juniper woodlands in Sweetwater, Uinta and Carbon counties (Faulkner 2010).

Suggested Viewing Locations: In Wyoming, a good location for finding this highly localized species is the Minnie's Gap area in Sweetwater County, and among scrub junipers 5–10 miles west of Baggs. In Idaho the City of Rocks National Reserve, Curlew National Grassland, Massacre Lakes State Park, and Juniper Rest Area (five miles north of the Utah border along I-84) are all good locations. The densest regional breeding populations probably occur along the Utah border of western Colorado, from Mesa Verde National Park north to Dinosaur National Monument and Browns Park National Wildlife Refuge.

Population: National Breeding Bird Survey trend data indicate a nonsignificant annual population decrease of 0.7%. Trend estimates for the Rocky Mountains & Plains States region (U.S.F.W.S. Region 6) indicate a significant annual decline of 2.9%. The North American population has been estimated at 330,000 birds (Rich *et al.*, 2004).

Bushtits

Bushtit *(Psaltriparus minimus)*

Status: Limited to the southern edge of our area, where it is most likely to be found in pinyon pine-juniper woodlands, and similar scrub oak–mountain mahogany habitats. Within the geographic limits of this book this species is mainly a non-breeding vagrant, since the birds often move into ponderosa pines following breeding.

Habitats and Ecology: Commonest in open woodlands such as pinyon pine and juniper habitats, these birds also at times occur in sagebrush or even aspen-covered hillsides. Nests are usually in pinyon pines or junipers, and are beautiful soft, woven hanging structures woven of mosses, spider webs and hair or feathers, with lateral entrances and usually are less than ten feet above ground. In Colorado 85% of breeding bushtits were recorded in pinyon–juniper woodlands, with upland and riparian shrubland also used to a small degree (Kingery, 1998). In Wyoming known nesting is confined to Utah juniper woodlands in southern Sweetwater and Uinta counties, but a local population also exists and presumably nests around Coal Mountain, Natrona County (Faulkner 2010).

Suggested Viewing Locations: In Wyoming, Seedskadee National Wildlife Refuge and Flaming Gorge National Recreation Area of Sweetwater County is good locations for finding this tiny and elusive species. The densest regional breeding populations are in far-western Colorado (from Montezuma and La Plata counties north through the plateau and canyonland parks and monuments to Moffatt County; also Great Sand Dunes National Monument). There is also significant breeding in southeastern Colorado along the pinyon–juniper woodlands of the Arkansas River (Kingery, 1998). Southern Idaho also has a good population in its geographically limited pinyon–juniper habitats, such as at City of Rocks Natural Preserve.

Population: National Breeding Bird Survey trend data indicate a nonsignificant annual population decrease of 0.7%. The North American population (north of Mexico) has been estimated at 2.97 million birds (Rich *et al.*, 2004).

Nuthatches

Red-breasted Nuthatch (*Sitta canadensis*)

Status: Found in montane coniferous forests throughout the entire region, and a variably common breeding species in all the montane parks.

Habitats and Ecology: Limited largely but not entirely to coniferous forests, primarily those of relatively tall firs, and much of the foraging occurs at rather high portions of the trees. To a much more limited degree aspens and riparian woodlands are sometimes also used. Breeding occurs in the trunks of dead trees or the rotting portions of live trees, with the birds typically excavating their own nesting holes. In Colorado this species is associated with coniferous forests and aspens during the breeding season, at elevations from 6,500–11,500 feet (Kingery, 1998). In the interior Northwest, this species breeds in mixed coniferous forests from the young to the old-growth forest stages of succession (Sanderson, Bull and Edgerton, 1980).

Suggested Viewing Locations: In Wyoming, and elsewhere through the entire region this nuthatch is common in coniferous forests. It occurs statewide in all of Wyoming's mountain ranges, at middle to higher elevations (Faulkner 2010). The densest regional breeding populations probably occur in the western Black Hills, but good populations also occur in northwestern Montana and the Greater Yellowstone region.

Population: National Breeding Bird Survey trend data indicate a significant annual population increase of 1.3%. The North American population has been estimated at 18 million birds (Rich *et al.*, 2004).

White-breasted Nuthatch (*Sitta carolinensis*)

Status: A permanent resident of deciduous forests and woodlands nearly throughout the region; rather rare in the Canadian montane parks, and known to breed only from Glacier National Park southward.

Habitats and Ecology: Largely confined in eastern states to deciduous forests, but in the Rocky Mountains associated with lower elevation coniferous forests, especially the ponderosa pine zone and also the pinyon–juniper zone. Nesting occurs in old woodpecker holes or in self-excavated holes in rotted wood of dead or partially dead trees, often aspens. In Colorado, breeding birds extend up to 10,000 feet elevation (Kingery, 1998). In the interior Northwest, this species breeds in mixed coniferous forests from the mature to the old-growth forest stages of succession (Sanderson, Bull and Edgerton, 1980).

Suggested Viewing Locations: Almost any deciduous or coniferous woodland in the entire region is likely to support this nuthatch, especially open ponderous pine forests at lower altitudes. In Wyoming the species has a patchy distribution statewide (Faulkner, 2010). The densest regional breeding populations probably occur in the western Black Hills, but good populations also occur widely throughout the region. In Colorado, there were relatively few breeding n the deciduous riparian woodlands of the plains river valleys during breeding bird surveys, but were abundant along the Front Range and in the southwestern part of the state (Kingery, 1998).

Population: National Breeding Bird Survey trend data indicate a significant annual population increase of 1.9%. The North American population (north of Mexico) has been estimated at nine million birds (Rich *et al.*, 2004).

Pygmy Nuthatch (*Sitta pygmaea*)

Status: A local resident in some portions of the region, mainly in the drier areas of southern Idaho and western Colorado, but also local in western Montana and eastern Wyoming. Generally rare or absent in the montane parks, but common in Rocky Mountain National Park.

Habitats and Ecology: Primarily associated with the ponderosa pine zone, but also occurring locally in the pinyon–juniper zone. It generally forages fairly high in tall pines, but nests closer to the ground in snags or stubs that have rotted trunks providing excavation opportunities. In Colorado, two-thirds of breeding-season observations were in ponderosa pine woodlands, and most of the remainder were in other coniferous habitats. Mature trees, with snags or rotting portions, are preferred (Kingery, 1998). In the interior Northwest, this species breeds in mixed coniferous forests from the mature to the old-growth forest stages of succession (Sanderson, Bull and Edgerton, 1980).

Suggested Viewing Locations: In eastern and central Wyoming this nuthatch can be found in mature ponderosa pine forests, such as those in Curt Gowdy State Park (Laramie County), and along Middle Crow Creek (Albany County) in the Pole Mountain Division of Medicine Bow National Forest (Dorn and Dorn, 1990). The birds are rare and of uncertain breeding status in the Greater Yellowstone region (Faulkner, 2010). The densest regional breeding populations probably occur in the Rocky Mountain National Park and surrounding parts of central Colorado along the Front Range.

Population: National Breeding Bird Survey trend data indicate a nonsignificant annual population increase of 0.4%. The North American population (north of Mexico) has been estimated at 1.7 million birds (Rich *et al.*, 2004).

Creepers

Brown Creeper (*Certhia familiaris*)

Status: Resident in mature forested areas almost throughout the region. Present in all the montane parks and probably breeding in all.

Habitats and Ecology: Associated with forests throughout the year, including both deciduous and coniferous forests. Virtually all foraging is done on the trunks of fairly large trees, where the birds forage for insects in bark crevices and grooves. In Colorado, breeding birds most frequent occupy mature spruce–fir and lodgepole pine forests at 9,000–11,500 feet of elevation. Dense stands of mature trees are strongly favored, making them very sensitive for logging effects (Kingery, 1998). In the interior Northwest, this species breeds in mixed coniferous forests from the mature to the old-growth forest stages of succession (Sanderson, Bull and Edgerton, 1980).

Suggested Viewing Locations: In Wyoming, this inconspicuous bird occurs almost statewide and can often be found during winter in most towns and cemeteries. During summer it might be searched-for in high-altitude coniferous forests statewide, such as the Togwotee Pass area (Teton-Fremont county line) or along the west side of Blackhall Mountain in Carbon County (Dorn and Dorn, 1990). The apparently densest regional breeding populations probably occur in northern Idaho (*e.g.*, Kootenai National Wildlife Refuge), but they are also relatively common in western Montana (Lee Metcalf National Wildlife Refuge).

Population: National Breeding Bird Survey trend data indicate a nonsignificant annual population increase of 0.6%. Trend estimates for the Rocky Mountains & Plains States region (U.S.F.W.S. Region 6) indicate a significant annual increase of 5.0%. The North American population (north of Mexico) has been estimated at 5.02 million birds (Rich *et al.*, 2004).

Wrens

Rock Wren (*Salpinctes obsoletus*)

Status: Locally present throughout the region in rocky areas, especially in dry
   sagebrush- dominated localities. Present in all the U.S. montane parks and
   possibly breeding in all.

Habitats and Ecology: Closely associated with eroded slopes, badlands, rocky
   outcrops, cliff walls, talus slopes, and similar rock-dominated habitats at
   generally rather low elevations, but sometimes occurring in alpine areas at
   high as 13,000 feet on Colorado's Pike's Peak. Crannies in cliffs are favorite
   nesting sites, but the birds can nest among small as well at big rocks, which
   allows them a broader breeding distribution than is true of canyon wrens
   (Kingery, 1998).

Suggested Viewing Locations: This species occurs statewide in Wyoming, from
   low elevations to above timberline. Good Wyoming birding locations include the badlands NNW of Baggs, the lower western slope of the Bighorn
   Mountains, rocky areas around Boysen Reservoir (Fremont County), the
   foothills of the Laramie Range west of Wheatland, and rocky hills south of
   Rock Springs (Dorn and Dorn, 1990). The densest regional breeding populations probably occur in central and western Colorado. The birds are common to abundant in Dinosaur National Monument, Browns Park National
   Wildlife Refuge, Colorado National Monument, and Black Canyon and
   Mesa Verde national parks. They are also common in other rocky habitats,
   such as the mesas and canyons of Las Animas County.

Population: National Breeding Bird Survey trend data indicate a significant annual population decrease of 1.7%. The North American population (north
   of Mexico) has been estimated at 3.36 million birds (Rich *et al.*, 2004).

Canyon Wren (*Catherpes mexicanus*)

Status: Largely limited to arid canyons south of the Snake River in Idaho, western Wyoming, and central and western Colorado, with local breeding farther east. Absent from the montane parks except in Colorado.

Habitats and Ecology: Rocky canyons, river bluffs, cliffs, rockslides, and similar topographic sites are favored, especially those offering shady crevices.
   Wyoming populations include disjunct groups in the Black Hills and Devils
   Tower regions, a large region centered in the Laramie Mountains, and much
   of western Wyoming except for the Yellowstone–Teton region (Faulkner,
   2010). In Colorado these birds breed at altitudes of 3,900–8,500 feet, in
   varied topography such as sandstone outcrops, rimrocks, and vertical canyon cliffsides. They are often found in stream-fed canyons, but sometimes
   nest well away from water. At times old buildings, fences or other structures are also used for nest sites (Kingery, 1998).

Suggested Viewing Locations: In Wyoming, good localities for finding this wren
   include canyons in the Bighorn Mountains, Wind River Canyon (Washakie
   County), and the cliffs on the west side of Flaming Gorge Reservoir (Sweet-

water County) (Dorn and Dorn, 1990). The most extensive regional breed-ing populations probably occur in western and southeastern Colorado, at locations such as Mesa Verde and Black Canyon national parks, Dino-saur National Monument, and the mesa and canyon country of Las Animas County (Kingery, 1998).

Population: National Breeding Bird Survey trend data indicate a nonsignifi-cant annual population decrease of 1.8%. The North American population (north of Mexico) has been estimated at 330,000 birds (Rich *et al.,* 2004).

Bewick's Wren (*Thryomanes bewickii*)

Status: Limited to the drier areas of extreme southern Idaho, southwestern Wy-oming, and western Colorado, at fairly low elevations below 7,000 feet elevation.

Habitats and Ecology: Broken and rather low brushy areas, especially where heavier cover is present overhead, seems to be this species' favorite habitat. It occurs from riparian areas through sagebrush to pinyon–juniper and oak–mountain mahogany habitats in the region, but is perhaps most common in pinyon–juniper woodlands. In western Colorado over 80% of observed breeding areas were in pinyon–junper habitat, while on the Eastern Slope pinyon–juniper habitats were somewhat less important, and lowland ripar-ian forests were somewhat more frequently used (Kingery, 1998). Nesting is often done in natural tree cavities, but old woodpecker holes are some-times also used, as are cavities in manmade structures, such as birdhouses.

Suggested Viewing Locations: In Wyoming, Bewick's wrens are mostly confined to the southwestern counties (Uinta and Sweetwater) (Faulkner, 2010). They can be found in the Minnie's Gap area (Sweetwater County) and in ju-niper scrub 5–10 miles west of Baggs (Dorn and Dorn, 1990). The densest regional breeding populations probably occur in Colorado, mostly in brush-lands and woodlands along the Utah border (such as the parks and pre-serves from Moffatt to Montezuma counties), but also in the mesas and ri-parian woodlands of Las Animas and adjoining counties (Kingery, 1998).

Population: National Breeding Bird Survey trend data indicate a nonsignificant annual population decrease of 0.1%. The North American population (north of Mexico) has been estimated at 4.56 million birds (Rich *et al.,* 2004).

Marsh Wren (*Cistothorus palustris*)

Status: Locally present, mostly at lower altitudes, throughout the region. Rare or absent in the montane parks except Grand Teton National Park, where occasional.

Habitats and Ecology: Restricted to marshy or swampy areas having an abun-dance of emergent plants such as reeds and cattails. Slow-moving waters, such as the inlets of reservoirs, are sometimes also used. Nesting is always done over water, usually from 3 to 5 feet above the substrate.

Suggested Viewing Locations: In Wyoming, there are many possible viewing points for this species, which occurs nearly statewide. They include Hutton

Lake National Wildlife Refuge (Albany County), Table Mountain Wildlife Unit (Goshen County), Lake Camahweit (Fremont County) and Lyman Lake (Uinta County) (Dorn and Dorn, 1990). The densest regional breeding populations probably occur in southeastern Idaho (*e.g.*, Bear Lake, Camas and Grays Lake national wildlife refuges).

Population: National Breeding Bird Survey trend data indicate a significant annual population increase of 2.7%. The North American population has been estimated at 7.7 million birds (Rich *et al.*, 2004).

House Wren (*Troglodytes aedon*)

Status: Present almost throughout the region, although more common southwardly in the Rocky Mountain region, and rare or lacking in the montane parks of Canada.

Habitats and Ecology: Generally most common in the lower elevation forests, but occasionally reaching timberline. In this region the birds favor riparian woodlands, aspen groves, and the lower and more open coniferous forest zones, as well as areas of human habitations. In the interior Northwest, this species breeds in mixed coniferous forests from the pole-sapling to the oldgrowth forest stages of succession (Sanderson, Bull and Edgerton, 1980). Nesting occurs in natural tree cavities, old woodpecker holes, artificial cavities such as birdhouses, and the like. In Colorado, this species replaces the Bewick's wren above about 7,000 feet, and has been reported breeding at sites as high as 11,800 feet (Kingery, 1998).

Suggested Viewing Locations: In Wyoming, and elsewhere in the region this ubiquitous species can be found almost anywhere, from backyards where wren houses have been erected to brushy woods, wooded riparian edges and even alpine timberline. The densest regional breeding populations probably occur in eastern Montana, northern Idaho, and western Colorado.

Population: National Breeding Bird Survey trend data indicate a significant annual population increase of 0.4%. Trend estimates for the Rocky Mountains & Plains States region (U.S.F.W.S. Region 6) indicate a significant annual increase of 1.7%. The North American population (north of Mexico) has been estimated at 18.9 million birds (Rich *et al.*, 2004).

Winter Wren (*Troglodytes troglodytes*)

Status: A summer resident in the northern parts of the region, south to about central Montana, and rarely farther south. Generally a common breeder in the montane parks south to Glacier, but only a vagrant farther south.

Habitats and Ecology: This species is typically found in heavy forests, usually coniferous, and often occurs in moist and shady canyons where brushpiles and tangles of vegetation cover the ground. Root-tangles or cavities in or under logs are favorite nest locations, especially where there are undercut banks. In the interior Northwest, this species breeds in mixed coniferous forests from the shrub-seedling to the old-growth forest stages of succession (Sanderson, Bull and Edgerton, 1980).

Suggested Viewing Locations: This rather mouse-like bird might well be over-looked if it weren't for the marvelous song of the male. In Wyoming, the western slope of the Teton Range might turn up winter wrens in summer (Dorn and Dorn, 1990), but there is no evidence of breeding in the state (Faulkner, 2010). Johnston Creek is one of the areas where it can be read-ily seen in Banff National Park, while in Glacier National Park it is common along McDonald Creek. The densest regional breeding populations proba-bly occur in northern Idaho (*e.g.,* Kootenai National Wildlife Refuge, Hey-burn and McCrosky state parks).

Population: National Breeding Bird Survey trend data indicate a significant an-nual population increase of 1.3%. Trend estimates for the Rocky Mountains & Plains States region (U.S.F.W.S. Region 6) indicate a significant annual in-crease of 8.4%. The North American population (north of Mexico) has been estimated at 18 million birds (Rich *et al.,* 2004).

Dippers

American Dipper (*Cinclus mexicanus*)

Status: Found in suitable habitats throughout the region, and a relatively com-mon breeder in all the montane parks.

Habitats and Ecology: Rapidly flowing mountain streams, often with water-falls or cascades present, and ledges or crevices providing safe nesting sites, are this species' prime habitat. Nesting is sometimes done on rock walls or overhangs near or even sometimes behind waterfalls, but more often the nests are constructed under bridges that cross creeks or rivers. The birds are highly territorial, and pairs tend to be well separated. In Colorado breeding sites as high as 11,820 feet have been reported, and at elevations below 7,600 feet second nesting efforts may be attempted (Kingery, 1998).

Suggested Viewing Locations: In Wyoming, dippers can be found along streams in all the major mountains except for the Bear Lodge Mountains (but do oc-cur in several canyons in the adjacent South Dakota Black Hills). McEne-aney (1988) judged the best observation areas in Yellowstone National Park to be along the Gibbon River, Firehole Canyon, and the Gardiner River. The most extensive regional breeding populations probably occur in northern Idaho (*e.g.,* Kootenai National Wildlife Refuge), western Montana (Red Rock Lakes National Wildlife Refuge, Glacier National Park) and central Colorado (Rocky Mountain National Park).

Population: National Breeding Bird Survey trend data indicate a significant an-nual population increase of 0.02%. The North American population (north of Mexico) has been estimated at 586,000 birds (Rich *et al.,* 2004).

Kinglets

Golden-crowned Kinglet (*Regulus satrapa*)

Status: Present in coniferous forests throughout the region, mainly at higher elevations. An abundant to rare breeder in all the montane parks.

Habitats and Ecology: During the breeding season primarily associated with spruce-fir forests, but otherwise generally present in the coniferous zones and sometimes extending out into riparian woodlands. Nesting occurs in dense and fairly tall coniferous trees, usually spruces, and nests usually are placed rather high in the tree. In Colorado the birds breed from 7,600–11,600 feet, with old-growth spruce–fir their most favored nesting habitat (Kingery, 1998). In the interior Northwest, this species breeds in mixed coniferous forests from the young to the old-growth forest stages of succession (Sanderson, Bull and Edgerton, 1980).

Suggested Viewing Locations: In Wyoming, breeding occurs widely across the state in the higher elevation coniferous forests, including the Medicine Bow and Park ranges, and presumably but not certainly also the Bear Lodge, Bighorn and Laramie ranges (Faulkner, 2010). During summer, these birds can be found along Pacific Creek in Grand Teton National Park, and along North Mullen Creek in the Medicine Bow Mountains (Dorn and Dorn, 1990). The densest regional breeding populations probably occur in northern Idaho (*e.g.*, Kootenai National Wildlife Refuge) and western Montana.

Population: National Breeding Bird Survey trend data indicate a significant annual population decrease of 1.0%. Trend estimates for the Rocky Mountains & Plains States region (U.S.F.W.S. Region 6) indicate a significant annual decline of 4.5%. The North American population has been estimated at 34 million birds (Rich *et al.*, 2004).

Ruby-crowned Kinglet (*Regulus calendula*)

Status: Present in coniferous forests throughout the region, more widespread and generally more numerous than the golden-crowned kinglet. Present in all the montane parks, and probably breeding in all of them.

Habitats and Ecology: Breeding occurs in coniferous forests from the lower zones almost to timberline in the subalpine zone, but is usually in taller and denser forests of medium altitude. In Colorado they breed exclusively in conifers, but are not as closely associated with dense sands as it true of the golden-crested. Like that species, they have been seen as high as 11,000 feet during the breeding season (Kingery, 1998). In the interior Northwest, this species breeds in mixed coniferous forests from the mature to the old-growth forest stages of succession (Sanderson, Bull and Edgerton, 1980). During winter the birds often move toward lower elevations, including prairie riverbottom woodlands and sometimes even move into cities.

Suggested Viewing Locations: In Wyoming, this species occurs in all the major mountain ranges, middle to higher elevations, and sometimes also among conifers at lower elevations (Dorn and Dorn, 1990; Faulkner, 2010). The densest and most extensive regional breeding populations probably occur in Colorado, but the species breeds commonly in all of the region's mountain ranges, including the Black Hills. In Colorado it has been estimated to be the most common breeding bird species in the mountains (Kingery, 1998).

Population: National Breeding Bird Survey trend data indicate a nonsignificant annual population decrease of 1.1%. The North American population (north of Mexico) has been estimated at 72 million birds (Rich *et al.*, 2004).

Gnatcatchers

Blue-gray Gnatcatcher (*Polioptila caerulea*)

Status: Limited to the southern and eastern parts of the region, mainly western Colorado and adjacent southern Wyoming, north to southern Idaho. Absent or a vagrant in the montane parks.

Habitats and Ecology: Breeding in the region occurs in pinyon–juniper and also adjacent oak woodland or sagebrush areas, up to at least 7,000 feet elevation. Arid park-like areas, with scattered thickets, are preferred for foraging, and nests are usually placed in low junipers. In Colorado, breeding birds selectively use pinyon–juniper woodlands, but also occupy scrub oak thickets, and mountain mahogany or serviceberry scrublands (Kingery, 1998).

Suggested Viewing Locations: In Wyoming, gnatcatchers are most common in the southwest, but have expanded their range and now breed east to the Nebraska and South Dakota borders (Faulkner, 2010). Sweetwater County populations occur in the Flaming Gorge area and Powder Rim. Junipers around Alcova Reservoir in Natrona County may also support breeding birds, and they are known to nest north to Hot Springs County. Their status in the Wyoming Bighorns is unknown, but they are known to breed across the Montana line (Faulkner, 2010). In Idaho, Craters of the Moon National Monument, City of Rocks Natural Preserve, and Curlew National Grasslands are good birding sites. The densest regional breeding populations probably occur in the woodlands of far-western Colorado, from the Wyoming to the New Mexico borders. The birds are especially common on the Roan Plateau.

Population: National Breeding Bird Survey trend data indicate a nonsignificant annual population increase of 0.3%. The North American population (north of Mexico) has been estimated at 42.2 million birds (Rich *et al.*, 2004).

Thrushes & Solitaires

Eastern Bluebird (*Sialia sialis*)

Status: Limited to deciduous woodlands along rivers on the plains of eastern Montana, Wyoming, and Colorado. Generally absent from the montane parks, but rare during migration in Rocky Mountain National Park.

Habitats and Ecology: Generally associated with open deciduous woods that are close to grass, lands, such as riparian forests, shelterbelts, farmsteads, and city parks. Nesting occurs in old woodpecker holes, natural cavities of dead trees, dead limbs, or sometimes utility poles. Birdhouses are also frequently used, especially where natural cavities are relatively rare. In a Colorado survey nearly 90% of breeding-season sightings were in lowland riparian habitats, often among cottonwoods. Edges of cottonwood forests were sometimes also used (Kingery, 1998).

Suggested Viewing Locations: Riverbottom woodlands along the eastern border of Wyoming provide good viewing possibilities, and Wyoming breeding records are known from Crook County (Faulkner, 2010). A small regional breeding populations occurs in eastern parts of Montana, Wyoming and Colorado. In eastern Montana this uncommon species might be found around Fort Union National Historic Site, or around the town of Fort Peck. In Colorado the highest concentration has been found along the Republican River near the Kansas border (Kingery, 1998).

Population: National Breeding Bird Survey trend data indicate a significant annual population increase of 2.2%. Trend estimates for the Rocky Mountains & Plains States region (U.S.F.W.S. Region 6) indicate a significant annual increase of 5.0%. The North American population (north of Mexico) has been estimated at eight million birds (Rich *et al.*, 2004).

Western Bluebird (*Sialia mexicana*)

Status: A local summer resident in northwestern and southwestern parts of the region, primarily in western Colorado and the northern parts of Idaho. Generally rare or absent from the montane parks, but an uncommon breeder in Rocky Mountain National Park.

Habitats and Ecology: Rather open timberlands, either of deciduous or coniferous trees, seem to be this species' favored habitats. It breeds regionally in both aspens and ponderosa pine woodlands, but in Colorado pine woodlands from 5,000–8,000 feet elevation are the most favored breeding habitat (Kingery, 1998). Breeding also extends to the level of mountain meadows, sometimes to about 10,000 feet above sea level. In the interior Northwest, this species breeds in mixed coniferous forests from the young to the old-growth forest stages of succession (Sanderson, Bull and Edgerton, 1980). During the non-breeding season it moves out into the woodlands of pinyon–juniper, oak–mountain mahogany, and some agricultural or desert scrub habitats. A combination of timberlands having dead trees with natural cavities or living trees with woodpecker holes, and nearby open grassy areas for foraging seem to provide optimum habitats.

Suggested Viewing Locations: Confirmed breeding records from Wyoming are lacking (Faulkner, 2010)., but mature woodlands along the southern edge of Wyoming are most likely to support breeding western bluebirds. The densest regional breeding populations probably occur in central and southern Colorado. There, the San Juan Basin and from Fort Collins south to the Wet

Mountains (especially the Black Forest) have the highest densities. Good birding areas include Rocky Mountain National Park, Castlewood Canyon State Park, Genesee Mountain Park, and Navajo State Park (Kingery, 2007).

Population: National Breeding Bird Survey trend data indicate a nonsignificant annual population decline of 0.06%. However, trend estimates for the Rocky Mountains & Plains States region (U.S.F.W.S. Region 6) indicate a significant annual increase of 6.9%. The North American population (north of Mexico) has been estimated at 1.22 million birds (Rich *et al.*, 2004).

Mountain Bluebird (*Sialia currucoides*)

Status: The most common and widespread of the bluebirds of the region, especially in rather open woodlands. Fairly common in all of the montane parks, and breeding in all.

Habitats and Ecology: Breeding occurs in open woodlands and forest-edge habitats from mountain meadows downward through the ponderosa pine zone, the aspen zone, and into the pinyon–juniper zone. Typically the birds favor nesting where either dead trees are available for nest cavities, or where rock crevices or other suitable sites are present. In Colorado pinyon–juniper woodlands are the most favored breeding habitat, followed closely by aspen forest and mountain grasslands. Nests there have been observed as high as 13,500 feet elevation (Kingery, 1998)! In the interior Northwest, this species breeds in mixed coniferous forests from the young to the old-growth forest stages of succession (Sanderson, Bull and Edgerton, 1980).

Suggested Viewing Locations: Mature forests at all altitudes support mountain bluebirds throughout Wyoming and elsewhere in the region. McEneaney (1988) judged the best observation areas in Yellowstone National Park to be around Roosevelt Lodge and the Upper Terrace Drive, and the birds are abundant at Montana's C. R. Russell National Wildlife Refuge during summer. The densest regional breeding populations probably occur in the coniferous forests of western Colorado, where they are nearly ubiquitous (Kingery, 1998).

Population: National Breeding Bird Survey trend data indicate a significant annual population increase of 1.2%. The North American population has been estimated at 5.2 million birds, of which 76% breed in the Intermountain West Avifaunal Biome (Rich *et al.*, 2004).

(Rich *et al.*, 2004).

Townsend's Solitaire (*Myadestes townsendi*)

Status: Widespread in wooded mountainous areas of the region during the breeding season, and extending into lower pinyon–juniper woodlands during winter. Present and probably breeding in all the montane parks.

Habitats and Ecology: Forested mountain slopes that provide snow-free areas for nesting on or near the ground, and which also offer sources of berries for food, are favored for nesting. In Colorado, closed canopy forests, especially upland coniferous forests are favored by breeding birds, but deciduous forests, pinyon–juniper communities and other habitats including brushlands are also used, at altitudes of 7,000–11,000 feet (Kingery, 1998). Nesting occurs in root tangles, rock crevices, or other cavities, including even hollows in road cuts. In the winter the birds feed almost entirely on junipers or other kinds of berries, but while breeding the usual thrush diet of insects is the most important source of food.

Suggested Viewing Locations: In summer the loud songs of this species make it easy to find in all the mountain ranges of the entire region. In Wyoming it breeds statewide in montane forests (Faulkner, 2010). The densest regional breeding populations probably occur in western Montana (*e.g.*, Red Rock Lakes and Lee Metcalf national wildlife refuges) and the Black Hills of Wyoming and South Dakota. They are also common through the Greater Yellowstone Ecosystem and the mountains of southeastern Wyoming. In Colorado they occur in nearly every wooded habitat above 7,000 feet (Kingery, 1998).

Population: National Breeding Bird Survey trend data indicate a significant annual population decrease of 1.0%. The North American population (north of Mexico) has been estimated at 731,000 birds (Rich *et al.*, 2004).

Veery (*Catharus fuscescens*)

Status: Widespread in wooded areas of the region, especially near water. Variably common in the montane parks, but less common in the northern ones, and also rather rare in Rocky Mountain National Park.

Habitats and Ecology: In this region the favored habitats consist of wooded river valleys and canyons that range from deciduous gallery forests along prairie areas of Alberta, through aspen forests of the foothills, and willow-lined mountain streams up to about 8,000 feet at the southern end of the region. Areas with heavy and thickety undergrowth that are difficult for humans to penetrate are this species' favorite habitats, and most of its foraging is done on the ground. In Colorado the birds favor moist and dense riparian thickets along mountain streams, at altitudes up to 10,000 feet (Kingery, 1998).

Suggested Viewing Locations: In Wyoming the veery breeds in all the higher mountain ranges. Some good locations for veerys include the lower canyons on the east slope of the Bighorns, the western Black Hills (Crook County), and overgrown aspen groves in the southern Laramie Range of Albany County (Dorn and Dorn, 1990). The densest regional breeding populations probably occur in western Montana (*e.g.*, Swan River and Red Rock Lakes national wildlife refuges) and northern Idaho (Kootenai National Wildlife Refuge, Heyburn State Park).

Population: National Breeding Bird Survey trend data indicate a significant annual population decrease of 1.4%. Trend estimates for the Rocky Mountains & Plains States region (U.S.F.W.S. Region 6) indicate a significant annual decline of 3.9%. The North American population has been estimated at 14 million birds (Rich *et al.*, 2004).

Gray-cheeked Thrush (*Catharus minimus*)

Status: A rare migrant in the region, primarily east of the mountains. Reported as a summer vagrant at Banff National Park; the nearest known breeding areas are extreme northern Saskatchewan or northern Alberta.

Habitats and Ecology: In our region these birds are likely to be seen foraging on or near the ground in almost any dense woodland, but their breeding habitats are typically scrub willows, alders, and dwarf birches near arctic timberline. While on migration the birds are often seen in association with Swainson's thrushes or other forest thrushes; at this time they are relatively quiet and elusive, as they move about through the shaded woodland floor searching for ground-dwelling insects and worms.

Suggested Viewing Locations: In Montana, Wyoming and Colorado, look for migrants along the eastern edge of the states in riverbottom forests. This thrush is not known to breed in the region.

Population: National Breeding Bird Survey trend data are not available. The North American population (north of Mexico) has been estimated at 10.8 million birds (Rich *et al.*, 2004).

Swainson's Thrush (*Catharus ustulatus*)

Status: Widespread during summer through the forested areas of the region, and relatively common in the montane parks, probably breeding in all of them.

Habitats and Ecology: On migration these birds are likely to be found in almost any fairly dense woodlands, but during the breeding season the birds are likely to be found at higher and cooler elevations. There they use shaded canyons where there are also fairly large areas of tangled brushy undergrowth, permitting ground-level foraging. Riparian thickets, often of willows or alders, and moist mountain slopes supporting aspens, are usually used for nesting in this region. In the interior Northwest, this species breeds in mixed coniferous forests from the shrub-seedling to the old-growth forest stages of succession (Sanderson, Bull and Edgerton, 1980).

Suggested Viewing Locations: In Wyoming, where it breeds in all the major mountain ranges, these birds might be found in streamside thickets and aspen groves with dense understories, such as occur at Fossil Butte National Monument (Lincoln County)(Dorn and Dorn, 1990). In Colorado, the birds prefer moist mountain valleys, typically breeding in riparian thickets of willow and alder, but also nest in aspen woodlands, mountain shrublands, and upper-level conifer forests (Kingery, 1998). The densest regional breeding populations probably occur in western Montana (*e.g.*, Red Rock Lakes and Swan River national wildlife refuges) and adjacent Idaho (Grays Lake and Kootenai national wildlife refuges, Farragut, Heyburn, McCrosky and Ponderosa state parks).

Population: National Breeding Bird Survey trend data indicate a nonsignificant annual population decrease of 0.6%. The North American population has been estimated at 100 million birds (Rich *et al.*, 2004).

Hermit Thrush (*Catharus guttatus*)

Status: A summer resident in wooded areas almost throughout the region; present in all the montane parks and probably breeding in all of them.

Habitats and Ecology: Moist woodlands, especially of coniferous or mixed hardwoods and conifers, are preferred for breeding. Spruces, ponderosa pines, and higher zones of coniferous forests almost all the way to timberline are sometimes used. Shady and leaf-littered forest floors are favored for foraging, and the altitudinal range of breeding often spans several thousands of feet. In Colorado, they have been found to nest from 6,200–11,100 feet, and perhaps to 12,000 feet. There they prefer coniferous forests over deciduous forests, where they mostly nest in spruce-fir communities, and are strongly impacted by clear-cut logging (Kingery, 1998). In the interior Northwest, this species breeds in mixed coniferous forests from the mature to the old-growth forest stages of succession (Sanderson, Bull and Edgerton, 1980).

Suggested Viewing Locations: In Wyoming, this species may be found during summer in all the major mountain ranges except the Bear Lodge Mountains and Black Hills (Dorn and Dorn, 1990; Faulkner, 2010). Dense regional breeding populations probably occur in western Montana (Glacier National Park), western Wyoming (Greater Yellowstone region) and central Colorado (e.g., Rocky Mountain National Park).

Population: National Breeding Bird Survey trend data indicate a significant annual population increase of 1.1%. The North American population has been estimated at 56 million birds (Rich *et al.*, 2004).

American Robin (*Turdus migratorius*)

Status: Widespread throughout the entire region, especially in open woodland areas. Fairly common to abundant in all the montane parks, and breeding in all.

Habitats and Ecology: Open woodlands, whether natural or artificial, such as suburbs, city parks, and farmsteads, are typical habitats. The birds tend to occur almost anywhere there are at least scattered trees, soft ground suitable for probing for insects and worms, and where mud can be gathered for the nest. Nesting on human-made structures seems to be preferred over natural nest sites such as trees, at least in protected areas. In Colorado, they nest from farmsteads though forested vegetation up to timberline, in nearly every habitat type, but most often in mountain conifers and aspen woodlands (Kingery, 1998). In the interior Northwest, this species breeds in mixed coniferous forests from the shrub-seedling to the old-growth forest stages of succession (Sanderson, Bull and Edgerton, 1980).

Suggested Viewing Locations: Robins breed commonly and ubiquitously throughout the entire region. Notably dense regional breeding populations probably occur in western Colorado, the Greater Yellowstone region, and in northwestern Montana and adjacent Idaho.

Population: National Breeding Bird Survey trend data indicate a significant annual population increase of 0.4%. The North American population (north of Mexico) has been estimated at 307 million birds (Rich *et al.*, 2004).

Varied Thrush (*Ixoreus naevius*)

Status: A permanent or summer resident in northern and western portions of the region, south to about west-central Montana. A common breeder in the montane parks from Glacier National Park north, but only a vagrant farther south.

Habitats and Ecology: In this region the varied thrush is associated with mature coniferous forests, especially rather wet forests that have completely shaded floors and a relatively open understory vegetation that permits ground foraging. In the interior Northwest, this species breeds in mixed coniferous forests from the pole-sapling to the old-growth forest stages of succession (Sanderson, Bull and Edgerton, 1980). During winter the birds turn to berries and fruits such as those of mountain ash and Russian olive, as well as frozen apples.

Suggested Viewing Locations: This thrush is somewhat robin-like in its habitat preferences and, like robins, often appears in cities and towns. There are no Wyoming breeding records (Faulkner, 2010). The densest regional breeding populations probably occur in northern and central Idaho (*e.g.*, McCrosky and Ponderosa state parks, Bear Basin) and adjacent Montana.

Population: National Breeding Bird Survey trend data indicate a nonsignificant annual population decrease of 0.45%. The North American population has been estimated at 26 million birds (Rich *et al.*, 2004).

Mockingbirds and Thrashers

Gray Catbird (*Dumetella carolinensis*)

Status: Present in wooded habitats nearly throughout the entire region, but rarer to the north, and absent or only a vagrant in the Alberta-British Columbia montane parks. Breeding probably occurs in all the U.S. montane parks.

Habitats and Ecology: Dense thickets, ranging from riverine forests or prairie coulees, city parks and suburbs, orchards, woodland edges, shrubby marsh borders, and similar overgrown areas that provide a combination of dense vegetation and transitional "edge" situations are the ideal habitats of this species. Coniferous forests are avoided, although aspen groves are used, as are other natural vegetational habitats that offer rich sources of insects and berries. In Colorado catbirds usually nest in dense shrubbery long streamsides at fairly low elevations, but sometimes nest in mountain shrublands and willow thickets (Kingery, 1998).

Suggested Viewing Locations: In Wyoming, where breeding occurs statewide below 8,000 feet (Faulkner, 2010), good locations for finding catbirds include brushy North Platte river woodlands between Torrington and Casper, and at Fort Steele (Carbon County), as well as woodlands along the Bighorn River (Dorn and Dorn, 1990). The densest regional breeding populations probably occur in eastern and central Montana (*e.g.*, C. R. Russell, Medicine Lake and Ninepipe national wildlife refuges).

Population: National Breeding Bird Survey trend data indicate a nonsignificant annual population increase of 0.06%. Trend estimates for the Rocky Mountains & Plains States region (U.S.F.W.S. Region 6) indicate a significant annual increase of 2.0%. The North American population has been estimated at ten million birds (Rich *et al.,* 2004).

Northern Mockingbird *(Mimus polyglottos)*

Status: A local permanent resident in eastern Colorado and a highly local and infrequent breeder in eastern Alberta; in other areas a migrant or vagrant, primarily east of the mountains, but local in southwestern Colorado and southwestern Idaho. Absent from the montane parks except for Rocky Mountain National Park, where it is a vagrant.

Habitats and Ecology: In Colorado and the other Rocky Mountain states mockingbirds are not so much city birds as they are associated with windbreaks, shrubs and scattered trees in grasslands. Lowland riparian habitats, with scattered cottonwoods or other isolated trees, represent typical habitat. Nesting up to 8,500 feet has been reported in Colorado, but breeding under 6,000 feet is more typical (Kingery, 1998).

Suggested Viewing Locations: In Wyoming, breeding occurs sparsely across the southeastern and southwestern corers of the state at lower elevations (Faulkner, 2010). Good locations for finding mockingbirds include the Laramie River bottoms at Ft. Laramie and the rim of Goshen Hole (Goshen County)(Dorn and Dorn, 1990). The densest regional breeding populations probably occur in southeastern Colorado, especially Las Animas County, where pinyon–juniper woodlands provide the region's best habitat (Kingery, 1998).

Population: National Breeding Bird Survey trend data indicate a significant annual population decrease of 0.6%. The North American population (north of Mexico) has been estimated at 36.9 million birds (Rich *et al.,* 2004).

Sage Thrasher *(Oreoscoptes montanus)*

Status: A local summer resident in sagebrush habitats almost throughout the region, north as far as north-central Montana. Largely limited to lower altitudes, and absent or rare in the montane parks except Rocky Mountain National Park, where uncommon and not known to breed, and Yellowstone National Park, where a rare breeder.

Habitats and Ecology: This species is closely associated with sage-dominated grasslands and to a much lesser extent other shrub lands dominated by shrubs of similar growth-forms such as rabbitbrush and greasewood. In Colorado, over 80% of breeding season observations were in shrubland habitats, with sagebrush accounting for more than half (Kingery, 1998). Most foraging is done on the ground, but nests are placed in shrubs. A greater array of shrublands is used in other seasons.

Suggested Viewing Locations: In Wyoming, where breeding is statewide in sagebrush habitats, look for this species in sage-dominated places such as

Hutton Lake National Wildlife Refuge (Albany County), below Fontenelle Dam in Lincoln County, the north side of Pathfinder Reservoir (Natrona County), and in valley-bottoms south of Fossil Butte in Lincoln County (Dorn and Dorn, 1990). The densest regional breeding populations probably occur in southern Idaho (*e.g.*, Craters of the Moon National Monument, Curlew National Grassland) and western Colorado (North Park, San Luis Valley, Gunnison Basin and the northwestern corner of the state) (Kingery, 1998).

Population: National Breeding Bird Survey trend data indicate a nonsignificant annual population decrease of 0.6%. The North American population has been estimated at 7.9 million birds, of which 99% breed in the Intermountain West Avifaunal Biome (Rich *et al.*, 2004).

Brown Thrasher (*Toxostoma rufum*)

Status: A local summer resident east of the mountains throughout the entire region; present in the montane parks only as a rare migrant or vagrant.

Habitats and Ecology: Associated with open, brushy woodlands, scattered clumps of woodland in open environments, shelterbelts, woodlots, and shrubby residential areas. At the western edge of their range in Colorado breeding birds are most often found in rural plantings of trees and shrubs (Kingery, 1998). In grassland areas the birds are mostly confined to shrubby coulees or to riparian forests that provide sources of berries, as well as foraging locations in open grasslands.

Suggested Viewing Locations: In Wyoming, breeding generally occurs east of a line from Sheridan to Cheyenne, and also in the Bighorn Basin south to Lander (Faulkner, 2010). The species is most common in brushy areas along eastern rivers, such as the North Platte River from Torrington to Casper, around Sheridan, and at LAK Reservoir (Weston County)(Dorn and Dorn, 1990). The densest regional breeding populations probably occur in eastern Montana (Missouri and Yellowstone River valleys).

Population: National Breeding Bird Survey trend data indicate a significant annual population increase of 1.2%. The North American population has been estimated at 7.3 million birds (Rich *et al.*, 2004).

Starlings

European Starling (*Sturnus vulgaris*)

Status: Present throughout the entire region, usually as a year-round resident, although substantial migration does occur. A fairly common breeder in all the montane parks.

Habitats and Ecology: Largely associated with humans, and most abundant in cities, farm areas, and suburbs, but also utilizing natural woodlands with woodpecker holes or other nest sites, such as aspen groves, where it competes effectively with native hole-nesting birds.

Suggested Viewing Locations: Of ubiquitous occurrence, this species breeds abundantly throughout the region.

Population: National Breeding Bird Survey trend data indicate a significant annual population increase of 0.9%.

Pipits

American Pipit (*Anthus rubescens*)

Status: A breeding summer resident in high montane areas over much of the region, and present on open grasslands or beaches during migration. Present in all the montane parks and probably breeding in all. In Colorado these birds only rarely breed in subalpine areas, on montane grasslands, but nest on nearly every patch of alpine tundra in the state, including both wet and dry tundra (Kingery, 1998).

Habitats and Ecology: During the breeding season this species is found on alpine tundra and high meadows, while at other seasons it occurs on similar very open terrain, usually with only sparse vegetation, and often a moist substrate. Shorelines, flooded fields, river edges, and similar habitats are commonly used by migrants or wintering birds.

Suggested Viewing Locations: In Wyoming this species is widespread during summer in the alpine zone of the Greater Yellowstone region, and also occurs in the alpine zones of the Beartooth and Medicine Bow–Sierra Madre ranges (Faulkner, 2010). Roads over alpine tundra provide for easy summer viewing, such as along the Beartooth Plateau in Park County, Libby Flats and Brooklyn Ridge in the Medicine Bow Range and near Medicine Wheel in Big Horn County (Dorn and Dorn, 1990). No dense breeding centers occur in the region; roads into alpine areas of Montana (Beartooth Mountains, Glacier National Park), Idaho (Henrys Lake State Park) and adjacent Canada probably reach the most accessible birds. In Colorado good birding locations include Rocky Mountain National Park, Guanella Pass, Loveland Pass, Independence Pass and Shrine Ridge (Kingery, 2007).

Population: National Breeding Bird Survey trend data are not available. The North American population (north of Mexico) has been estimated at 19.8 million birds (Rich *et al.*, 2004).

Sprague's Pipit (*Anthus spragueii*)

Status: A summer resident on native grasslands east of the mountains in Alberta and Montana, and a rather rare migrant in plains areas farther south. Absent or at most a rare vagrant in the montane parks.

Habitats and Ecology: Associated during the breeding season with native grasslands of short to moderate stature. Breeding also occurs in alkaline meadows and around the edges of alkaline lakes. Outside the breeding season the birds are also associated with grassy habitats.

Suggested Viewing Locations: This very elusive pipit might be searched-for in the shortgrass or mixed-grass prairies of northeastern Wyoming, where it is no yet known to breed. Such sites include the Lake DeSmet (Johnson County) area, and the lower grassy slopes of the Bighorn Mountains (Dorn and Dorn, 1990, Scott, 1993). The densest regional breeding populations probably occur in northeastern Montana (*e.g.*, Bowdoin National Wildlife Reservation, Blackfeet Indian Reservation).

Population: The North American population has been estimated at 870,000 birds. Classified as a Partners-in-Flight Watch List species of continental conservation importance (Rich *et al.*, 2004). National Breeding Bird Survey trend data indicate a significant annual population decrease of 3.9%. Trend estimates for the Rocky Mountains & Plains States region (U.S.F.W.S. Region 6) indicate a significant annual increase of 1.7%.

Waxwings

Bohemian Waxwing (*Bombycilla garrulus*)

Status: A breeding summer resident in montane forests as far south as Glacier National Park, and a migrant or winter resident farther south, including the other montane parks.

Habitats and Ecology: During the breeding season this species is associated with coniferous and mixed forests, often nesting as loosely associated groups in conifer groves. Outside the breeding season the birds move about opportunistically, seeking out sources of berries and small fruits in trees and hedges, such as mountain ash, crab apples, pyracantha, and the like.

Suggested Viewing Locations: Bohemian waxwings are most likely to be seen among flocks of cedar waxwings, often during migration or winter when the birds gather at fruit-or berry-bearing trees. The densest regional breeding populations probably occur in the Canadian montane parks.

Population: National Breeding Bird Survey trend data are not available. The North American population (north of Mexico) has been estimated at 2.4 million birds (Rich *et al.*, 2004).

Cedar Waxwing (*Bombycilla cedrorum*)

Status: Widespread over the wooded areas of the region, present in all the montane parks and probably breeding in all of them except Rocky Mountain National Park, where apparently only a rare summer visitor.

Habitats and Ecology: Somewhat open woodlands, primarily of broad-leaved species, are used for nesting, including riparian forests, farmsteads, parks, cedar groves, shelterbelts, and brushy edges of forests. Areas that have abundant growths of berry-bearing bushes are especially favored, although insects, buds, and other food sources are also consumed. In Colorado they nest rather sparingly, in deciduous riparian forests below 7,500 feet (Kingery, 1998).

Suggested Viewing Locations: In Wyoming, cedar waxwings throughout the state at elevations up to about 8,000 feet, especially in riparian deciduous forests. They can often be found during summer in the northwestern mountains, or at places such as Battle Creek Campground of the Sierra Madre Mountains (Carbon County)(Dorn and Dorn, 1990). The densest regional breeding populations probably occur in northern Idaho (*e.g.*, Kootenai National Wildlife Refuge, Farragut State Park) and adjacent western Montana (Glacier National Park, National Bison Range). They are also common in central Montana (Benton Lake and Bowdoin national wildlife refuges).

Population: National Breeding Bird Survey trend data indicate a significant annual population increase of 0.5%. The North American population has been estimated at 15 million birds (Rich *et al.*, 2004).

Wood Warblers

Tennessee Warbler (*Vermivora peregrina*)

Status: A summer resident and local breeder in the northwestern part of the region, and a variably common spring and fall migrant east of the mountains elsewhere in the region. An occasional breeder in Banff and Jasper National Park, and perhaps some of the other Canadian montane parks. Reported during summer in Clearwater National Forest (Idaho), but not proven to breed there.

Habitats and Ecology: The usual breeding habitat consists of coniferous boggy areas such as those of spruce and tamarack or white cedar, usually where sphagnum mosses are abundant. It also occurs on brushy hillsides, along forest clearings, and deciduous forests, and in Alberta favors deciduous or mixed woods that have poplars or aspens present. Foraging is done rather high up in the crown foliage, although nesting is on the ground, usually in sphagnum-covered hummocks.

Suggested Viewing Locations: This warbler is a common migrant in eastern parts of Montana, Wyoming and Colorado. No known breeding populations occur south of Canada.

Population: National Breeding Bird Survey trend data indicate a nonsignificant annual population decrease of 0.9%. The North American population has been estimated at 62 million birds (Rich *et al.*, 2004).

Orange-crowned Warbler (*Vermivora celata*)

Status: A summer resident in most wooded areas throughout the region, at least at lower elevations. It is present and probably breeding in all the montane parks, but common only in the more northerly ones.

Habitats and Ecology: A variety of woodland and brushy habitats are used by this species for breeding, ranging from riparian woodlands, pinyon–juniper habitats, and aspen groves. In montane areas they favor willow or alder thickets near streams, or willow thickets at treeline. At lower elevations they tend to breed along riverine woods or in brushy vegetation surrounding beaver ponds in northern coniferous woodlands. In Colorado they breed from 6,500–9,500 feet, in montane and subalpine vegetation, largely in scrub oak, aspen and mountane shrubs, including mountain mahogany and willows. On migration the birds are found in a wide variety of brushy or wooded habitats, but favor the brushy areas of riverbottoms.

Suggested Viewing Locations: During migration this species is common across all of Wyoming. Breeding occurs locally at higher altitudes, over all the major mountain ranges except for the Black Hills (Faulkner, 2010). They often may be found in moist deciduous woodlands, such as the headwaters of Coral Creek on Muddy Mountain (south of Casper Mountain in the Laramie Range) and near Battle Lake in the Sierra Madre range (Scott, 1993). The densest regional breeding populations probably occurs in northwestern Montana (Glacier National Park). Another region of relatively dense breeding occurs in western Colorado (e.g., San Juan Mountains; Umcompahgre Plateau, and White River and Routt national forests (Kingery, 1998).

Population: National Breeding Bird Survey trend data indicate a significant annual population decrease of 1.0%. The North American population has been estimated at 76 million birds (Rich *et al.*, 2004).

Nashville Warbler (*Vermivora ruficapilla*)

Status: A local summer resident in the northwestern part of the region, south to about west-central Montana and adjacent northern Idaho, but a vagrant or rare migrant in the montane parks.

Habitats and Ecology: Moderately open deciduous woods, or the deciduous portions of mixed woods, are the primary breeding habitats of this species. It seems limited to those woodlands sufficiently open to allow for the growth of shrubbery under which nesting occurs. In the interior Northwest, this species breeds in mixed coniferous forests during the shrub-seedling stage of succession (Sanderson, Bull and Edgerton, 1980). Areas that allow for feeding at heights of 2–40 feet seem preferred, although some foraging in shrubbery also occurs. On migration a wider array of habitats are used, but riparian woodlands are apparently favored.

Suggested Viewing Locations: Generally an uncommon migrant in Wyoming, it has been most frequently seen in the Green River valley (Scott, 1993). The densest regional breeding populations probably occur in northern Idaho (e.g., Kootenai National Wildlife Refuge, Salmon River drainage, and adjacent Montana (Swan River National Wildlife Refuge).

Population: National Breeding Bird Survey trend data indicate a nonsignificant annual population increase of 0.9%. The North American population has been estimated at 34 million birds (Rich *et al.*, 2004).

Virginia's Warbler (*Vermivora virginiae*)

Status: A local summer resident in the southernmost part of the region, including extreme southern Idaho and northwestern Colorado. A common breeder in Rocky Mountain National Park, and in Dinosaur National Monument, but absent from the more northern montane parks. In Rocky Mountain National Park, nesting occurs at lower elevations of the eastern slope, such as in Moraine Park.

Habitats and Ecology: In Idaho this species is essentially limited to mountainsides that are covered with dense thickets of mountain mahogany; more generally it is associated with scrubby oak, open pinyon–juniper woodlands, and similar semi-arid and brush-dominated habitats. Nesting in Colorado often occurs where scrub oaks meet the ponderosa pine zone. In Colorado breeding surveys, scrub oaks has the highest number of nestings, with pinyon–juniper habitats in second place (Kingery, 1998). Trees and taller shrubs provide singing and foraging posts, while nests are located at ground-level under bushes.

Suggested Viewing Locations: In Wyoming, this bird breeds locally in southern Sweetwater County, and extends north in the Laramie Mountains and along the North Platte River to the vicinity of Casper (Faulkner, 2010). It is often found in dry gulches and canyons, such as Linwood Canyon in Sweetwater County (Dorn and Dorn, 1990), near Minnies Gap, along the Little Firehole, and in Lesser Jackson Canyon on the west side of Casper Mountain (Scott, 1993). In Idaho the City of Rocks National Preserve is a good location r seeing this species. The densest regional breeding populations probably occur in western Colorado from the Wyoming to the New Mexico borders, and along the Eastern Slope foothills (Kingery, 1998). Good birding sites include Mesa Verde and Black Canyon national parks, and Castlewood Canyon and Golden Gate Canyon state parks (Kingery, 1993).

Population: Classified as a Partners-in-Flight Watch List species of continental conservation importance (Rich et al., 2004). National Breeding Bird Survey trend data indicate a nonsignificant annual population increase of 1.3%. However, trend estimates for the Rocky Mountains & Plains States region (U.S.F.W.S. Region 6) indicate a significant annual decline of 0.9%. The North American population has been estimated at 410,000 birds, of which 62% breed in the Intermountain West Avifaunal Biome (Rich *et al.*, 2004).

Northern Parula (*Parula americana*)

Status: A rare migrant in the eastern plains; accidental farther west. Absent from the montane parks except Rocky Mountain National Park, where it is a vagrant.

Habitats and Ecology: On migration these birds are likely to be seen in riverine forests or other deciduous forest areas, but on the breeding grounds the birds are closely associated with swampy woodlands, especially those draped with moss-like lichens (*Usnea*) or "Spanish moss" (*Tillandsia*).

Suggested Viewing Locations: This species is too rare in the region to suggest possible viewing locations.

Population: National Breeding Bird Survey trend data indicate a significant annual population increase of 0.8%. The North American population has been estimated at 7.3 million birds (Rich *et al.*, 2004).

Yellow Warbler (*Dendroica petechia*)

Status: A local summer resident throughout the region in suitable habitats, including the montane parks, where abundant to uncommon, and probably breeding in all.

Habitats and Ecology: Generally moist habitats, such as riparian woodlands and brush, the brushy edges of marshes, swamps, or beaver ponds, and also drier areas including roadside thickets, hedgerows, orchards, and forest edges. A combination of open areas and dense shrubbery seem to be important for breeding, although migrant birds are rather more widely distributed. In Colorado studies, deciduous streambottoms provided the most typical nesting habitat during breeding bird surveys, but the birds used a wide variety of other deciduous sites, with only two percent occurring in conifers (Kingery, 1998).

Suggested Viewing Locations: This species breeds commonly throughout the entire region. It occurs throughout Wyoming below 8,000 feet, especially in riverbottoms with mature cottonwoods and an understory of willows or Russian olives (Faulkner, 2010). In Colorado these birds are known to occur in every county and every latilong, but are most abundant in the north-central and west-central regions, plus North Park (Kingery, 1998).The most extensive regional breeding populations may occur in western and central Montana (common in nearly all the national wildlife refuges).

Population: National Breeding Bird Survey trend data indicate a nonsignificant annual population increase of 0.7%. Trend estimates for the Rocky Mountains & Plains States region (U.S.F.W.S. Region 6) indicate a significant annual increase of 0.9%. The North American population (north of Mexico) has been estimated at 33.15 million birds (Rich *et al.*, 2004).

Chestnut-sided Warbler (*Dendroica pensylvanica*)

Status: A rare migrant in the eastern plains, and a vagrant farther west. Local breeding occurs in the Front Range region of Colorado.

Habitats and Ecology: During the breeding season this species is generally associated with low shrubbery, forest edges and clearings, briar thickets, overgrown pastures, and similar rather open and dry areas having scattered trees and shrubs. In Alberta the birds inhabit fairly open but mature deciduous woodlands with loose understories of cranberries and dogwoods.

Suggested Viewing Locations: Migrants are rare in Wyoming, so no reliable birding locations can be suggested. The only breeding known to occur in the region has been in the foothills of the Front Range in Colorado, near Colorado Springs, Boulder, Golden, and perhaps a few other locations (Kingery, 1998).

Population: National Breeding Bird Survey trend data indicate a nonsignificant annual population increase of 0.7%. The North American population has been estimated at 9.4 million birds (Rich *et al.*, 2004).

Magnolia Warbler (*Dendroica magnolia*)

Status: A local summer resident in the northwestern corner of the region, south to the vicinity of Banff National Park, although not yet proven to breed there. It is a rare migrant farther south to the east of the mountains, and a vagrant in Rocky Mountain National Park.

Habitats and Ecology: In Alberta breeding occurs in open coniferous and mixed forests, especially areas of young spruce and pines only about 6–8 feet high and which are not too dense. In some areas open coniferous bogs that are dominated by white cedars or other species are preferred, and likewise coniferous forest edges, second-growth following logging, and other habitats dominated by bush and saplings are also used.

Suggested Viewing Locations: Like many other eastern warblers, this species is too rare to predict where it might turn up in the region. The only regional breeding populations probably occur in Jasper and Banff National Parks.

Population: National Breeding Bird Survey trend data indicate a significant annual population increase of 1.2%. The North American population has been estimated at 32 million birds (Rich *et al.*, 2004).

Yellow-rumped Warbler (*Dendroica coronata*)

Status: A common and widespread species in wooded areas throughout the region; perhaps the commonest breeding warbler. A common to abundant breeder in all the montane parks.

Habitats and Ecology: This species breeds in a wide array of coniferous forests, from the ponderosa pine zone upwards, and also breeds in riparian forests with conifers present. Habitats range from open, park-like ponderosa pine communities through dense montane forests to timberline species, foraging from low branches to the highest crown levels. In the interior Northwest, this species breeds in mixed coniferous forests from the young to the old-growth forest stages of succession (Sanderson, Bull and Edgerton, 1980). In Colorado studies, these birds bred in conifer and aspen forests from 7,000–11,500 feet, with two-thirds of the observations in coniferous forests (Kingery, 1998). During winter the habitats used are more varied, and may range from berry-eating and nectar-drinking to aerial flycatching.

Suggested Viewing Locations: This species breeds commonly throughout the entire region. In Wyoming it occurs widely in coniferous forests of all of the state's mountain ranges and in some low elevation habitats, such as cottonwood-lined riparian corridors (Faulkner, 2010). Extensive regional breeding populations probably occur in north-central and west-central Colorado, the Greater Yellowstone region, and the mountains of Idaho and western Montana. The Canadian montane parks support populations of the eastern (Myrtle) race. In Colorado, this species is the most common of the state's breeding warblers (with over a million estimated breeding pairs), with yellow warblers in second place. They are almost entirely limited to the western half of the state, but are rare in the state's northwestern sector and some of the drier valleys and plateaus (Kingery, 1998).

Population: National Breeding Bird Survey trend data indicate a nonsignificant annual population increase of 0.1%. The North American population (north of Mexico) has been estimated at 127.4 million birds (Rich *et al.*, 2004).

Black-throated Gray Warbler *(Dendroica nigrescens)*

Status: Limited to the southernmost portions of the region, including southern Idaho, southwestern Wyoming and northwestern Colorado.

Habitats and Ecology: During the breeding season this species is closely associated with pinyon–juniper woodlands; in Colorado they have been found nesting at about 7,000 feet elevation or higher, almost exclusively in mature pinyon–juniper woodlands. In Idaho they nest on low ridges covered by large, gnarled junipers. Elsewhere they have been found breeding in oak woodlands, and in general they seem to prefer trees with dense and stiff foliage, of relatively low stature, and in dry environments. The birds forage in dense terminal foliage, and nest at medium heights.

Suggested Viewing Locations: In Wyoming this species breeds only in Utah juniper woodland from the southeastern corner of the state north through central Wyoming at least to Washakie County, and perhaps to Big Horn County (Faulkner, 2010). Good Wyoming birding sites for this warbler during the breeding season include Flaming Gorge National Recreation Area and Powder Rim, both in Sweetwater County (Dorn and Dorn, 1990). Besides the juniper scrublands of the Rock Springs–Green River region, they have been seen in junipers south of Casper (Scott, 1993). In Idaho, the City of Rocks National Preserve is a good birding area for this species. The densest regional breeding populations probably occur in western Colorado (Dinosaur and Colorado national monuments, Black Canyon and Mesa Verde national parks; they are also common at Alamosa National Wildlife Refuge)

Population: National Breeding Bird Survey trend data indicate a nonsignificant annual population increase of 0.2%. However, trend estimates for the Rocky Mountains & Plains States region (U.S.F.W.S. Region 6) indicate a significant annual decline of 2.9%. The North American population (north of Mexico) has been estimated at 2.84 million birds (Rich *et al.*, 2004).

Townsend's Warbler *(Dendroica townsendi)*

Status: A summer resident in the northwestern part of the region, south at least to west-central Montana and possibly farther. Reported as an occasional breeder in Yellowstone National Park by some early observers, but currently believed to be only a migrant. Also a confirmed breeder in Clearwater National Forest, Idaho, which possibly represents its southernmost breeding limits.

Habitats and Ecology: This is a crown-level forager in tall conifers, favoring dense and mature montane forests. This general adaptation seems to be true during the non-breeding season as well as when nesting. Nesting appears to be in spruces and firs, sometimes within 15 feet of the ground, but rather few nests of this species have been described. In Alberta the nesting habitat consists of dense stands of spruce or fir, often with a stream or a willow-lined swamp nearby. In the interior Northwest, this species breeds in mixed coniferous forests from the mature to the old-growth forest stages of succession (Sanderson, Bull and Edgerton, 1980).

Suggested Viewing Locations: During migration these birds are more common in western parts of Wyoming, and are often plentiful in the Green River valley (Scott, 1993). The densest regional breeding populations probably occur in northern Idaho (*e.g.*, Farragut, McCrosky and Winchester Lake state parks, Upper Salmon River drainage) and adjacent western Montana (Glacier National Park, Swan River National Wildlife Refuge).

Population: National Breeding Bird Survey trend data indicate a nonsignificant annual population increase of 0.15%. The North American population has been estimated at 12 million birds (Rich *et al.*, 2004).

Blackburnian Warbler (*Dendroica fusca*)

Status: A rare migrant in the eastern plains, and a vagrant elsewhere, including the montane parks.

Habitats and Ecology: In Alberta breeding birds are associated with heavy stands of spruce and fir in mixed forests, while farther east the birds occur in a variety of coniferous and deciduous forest habitats. Mature coniferous forests, especially those in swampy areas and with *Usnea* lichens present, seem to be the favored breeding habitats. In any habitat, the birds tend to forage high in trees, and nests are typically placed very high in tall trees as well.

Suggested Viewing Locations: This species is too rare in the region to predict good birding areas.

Population: National Breeding Bird Survey trend data indicate a significant annual population increase of 0.7%. The North American population (north of Mexico) has been estimated at 5.9 million birds (Rich *et al.*, 2004).

Palm Warbler (*Dendroica palmarum*)

Status: A rare migrant over the eastern plains; accidental in the western areas and the montane parks. There have some been sightings from Wyoming, Montana and Idaho, and many from Colorado. The nearest breeding area is in central Alberta, south to about Grande Prairie, Elk Island National Park, and Cold Lake.

Habitats and Ecology: During the breeding season these birds are associated with dense boggy areas dominated by larch, spruce, and white cedar, and with alders, willows, and cranberry thickets present. Nesting is done on the ground, in fairly dry sphagnum mosses. Outside of the breeding season and on migration the birds are seen in various habitats such as roadside shrubbery and deciduous trees.

Suggested Viewing Locations: This species is too rare in the region to predict good birding areas.

Population: National Breeding Bird Survey trend data indicate a significant annual population increase of 2.8%. The North American population has been estimated at 23 million birds (Rich *et al.*, 2004).

Bay-breasted Warbler (*Dendroica castanea*)

Status: A rare migrant in the eastern plains; accidental in western areas and in the montane parks. The nearest breeding areas are in central Alberta, south to about Cold Lake and perhaps to Jasper National Park, where it is rare.

Habitats and Ecology: During the breeding season this species is associated with coniferous forests, especially those in rather swampy areas and with birches or maples present, and also with mixed forests having clearings or edge areas. In Alberta it is found in extensive stands of mature spruce or mixed spruce, larch, and pine. It may also occur in mixed forests, but only those that are dominated by conifers.

Suggested Viewing Locations: This species is too rare in the region to predict good birding areas.

Population: The North American population has been estimated at 3.1 million birds. Classified as a Partners-in-Flight Watch List species of continental conservation importance (Rich *et al.*, 2004). National Breeding Bird Survey trend data indicate a nonsignificant annual population decrease of 1.8%.

Blackpoll Warbler (*Dendroica striata*)

Status: A local summer resident in the montane forest of the northwestern corner of the region, south at least through Banff National Park and Bragg Creek; a migrant in the eastern plains.

Habitats and Ecology: In Alberta the favored breeding habitat consists of mixed woodland- especially spruce and aspen or alder regrowth on previously burned areas. In northern Alberta they inhabit deciduous shrubs as frequently as young coniferous growth. Outside the breeding season they occur over a much broader array of wooded habitats.

Suggested Viewing Locations: In Wyoming, locations from about Lander east are most likely to support migrants, as are eastern parts of Montana and Colorado. The only known regional breeding populations probably occur in the Canadian montane parks.

Population: National Breeding Bird Survey trend data indicate a nonsignificant annual population decrease of 2.6%. The North American population has been estimated at 21 million birds (Rich *et al.*, 2004).

Black-and-white Warbler (*Mniotilta varia*)

Status: A local summer resident in the northeastern parts of the region, and a rare migrant east of the mountains throughout the region. An accidental vagrant in the montane parks, except for Banff, where occasionally seen, but breeding is unproven.

Habitats and Ecology: During the breeding season this species inhabits deciduous or mixed woods bordering lakes and streams, or in shrubbery around muskeg areas. It also breeds in immature or scrubby trees on hillsides or ravines, and in riverside forests in grassland areas.

Suggested Viewing Locations: This species is too infrequent in the region to predict good birding areas. The only possible regional breeding populations might occur in the Canadian montane parks.

Population: National Breeding Bird Survey trend data indicate a significant annual population decrease of 0.8%. The North American population has
been estimated at 14 million birds (Rich *et al.*, 2004).

American Redstart (*Setophaga ruticilla*)

Status: A relatively common summer resident in woods over most of the region, including montane forests. Present and variably common in all the
montane parks, but rarer southwardly, and apparently absent from Rocky
Mountain National Park during the summer (but breeding in the nearby
foothills).

Habitats and Ecology: Breeding habitats of this species include moist bottomland woodlands, the margins or openings of mature forests, young or second-growth stands of various types of forests, and especially deciduous
forests. Nearby water and of a brush layer seem to be important habitat
components. In Colorado, redstarts typically nest in streamside habitats,
with undergrowth shrubs, but may also nest in either deciduous or mixed
deciduous–coniferous forests (Kingery, 1998).

Suggested Viewing Locations: Breeding in Wyoming occurs in the Black Hills
National Forest, the Bighorn Mountains, the Laramie Mountains, and the
Snowy Range (Faulkner, 2010). Good summer birding sites in Wyoming
include the Black Hills, Tongue River Canyon in Sheridan County, Sand
Creek in Crook County and LAK Reservoir in Weston County (Dorn and
Dorn, 1990). The densest regional breeding populations probably occur in
the Black Hills, and also in northern Idaho (*e.g.*, Kootenai National Wildlife
Refuge) and adjacent western Montana (Glacier National Park, Red Rock
Lakes National Wildlife Refuge).

Population: National Breeding Bird Survey trend data indicate a nonsignificant
annual population decrease of 0.7%. Trend estimates for the Rocky Mountains & Plains States region (U.S.F.W.S. Region 6) indicate a significant annual decline of 3.9%. The North American population has been estimated
at 25 million birds (Rich *et al.*, 2004).

Ovenbird (*Seiurus aurocapillus*)

Status: A local summer resident east of the mountains, and a rare migrant or vagrant in the mountains, including the montane parks.

Habitats and Ecology: During the breeding season in eastern North America this species occupies well-drained, bottomland deciduous forests, and
well-shaded and mature upland forests, especially on north-facing slopes
or shady ravines. In Alberta the birds favor deciduous or mixed woods in
which the undergrowth is not too dense for ground foraging, but avoid the
darkest coniferous forests. In Colorado they favor habitats with a scrub oak
understory and a canopy of trees varying from young to mature ponderosa
pine to mixed forests of pines and Douglas-firs (Kingery, 1998).

Suggested Viewing Locations: Breeding in Wyoming occurs in the Black Hills
National Forest, the Bighorn Mountains, and in a small area of aspen forest in the Laramie Mountains near Esterbrook (Faulkner, 2010). Good sum-

mer birding sites in Wyoming include the Sand Creek drainage of the Black Hills (Crook County), and the Story Fish Hatchery in Sheridan County (Dorn and Dorn, 1990). They also breed at Devils Tower National Monument and throughout the Bear Lodge Mountains (Scott, 1993). The densest regional breeding populations probably occur in northeastern Wyoming (e.g., the Bighorn Mountains) and adjacent southeastern Montana. Other good Montana sites include the Beartooth Mountains, the Blackfeet Indian Reservation, and the Snowy Mountains Special Recreation Management Area near Judith (McEneaney, 1993).

Population: National Breeding Bird Survey trend data indicate a significant annual population increase of 0.3%. Trend estimates for the Rocky Mountains & Plains States region (U.S.F.W.S. Region 6) indicate a significant annual decline of 2.0%. The North American population has been estimated at 24 million birds (Rich *et al.,* 2004).

Northern Waterthrush (*Seiurus noveboracensis*)

Status: A local summer resident in the northwestern portions of the region, south to about the Wyoming-Montana border. Breeding occurs in the montane parks south at least to Glacier National Park, and locally perhaps to Beaverhead-Deer Lodge National Forest or Red Rock Lakes National Wildlife Refuge in Montana. Also breeds locally in north-central Colorado (North Park).

Habitats and Ecology: During the breeding season this species inhabits woodlands with ponds, lakes and streams, especially those with brushy bogs and swampy areas of forest. Standing-water habitats are favored over those with moving streams, and in Alberta the birds are found in deciduous forests having heavy underbrush and are often partially or recently flooded. In Colorado they have been found nesting in wet middle-elevation willow and alder thickets that are nearly impenetrable (Kingery, 1998).

Suggested Viewing Locations: Breeding in Wyoming is apparently limited to the Medicine Bow Mountains near Laramie, but possible breeding activities have also been reported from the Bighorn Mountains and Jackson (Faulkner, 2010). In Wyoming, eastern parts of the state attract the largest number of migrants, at sites such as Sloan's Lake in Cheyenne's Lion Park (Scott, 1993). The densest regional breeding populations probably occur in northwestern Montana (*e.g.* Glacier National Park, Swan River National Wildlife Refuge, Blackfeet Indian Reservation) and adjacent Canada.

Population: National Breeding Bird Survey trend data indicate a nonsignificant annual population decrease of 0.7%. The North American population has been estimated at 13 million birds (Rich *et al.,* 2004).

MacGillivray's Warbler (*Oporornis tolmiei*)

Status: Widespread in woodlands and brushy areas throughout the region, including montane areas; relatively common and a probable breeder in all the montane parks.

Habitats and Ecology: Generally associated with brushy thickets, especially riparian woodlands. Less often it occurs in dense deciduous woods or mixed woodland on upland slopes, or in mature riverbottom forests. In Colorado, 80% of breeding birds were found in montane willow and alder thickets, montana shrublands, riparian deciduous forests, and scrub oaks (Kingery, 1998). In the interior Northwest, this species breeds in mixed coniferous forests from the shrub to the mature forest stages of succession (Sanderson, Bull and Edgerton, 1980). In Alberta the birds are usually found close to water in thick brushy growth, in prairie coulees, mountain slopes with dense shrubbery, or along forest clearings.

Suggested Viewing Locations: In Wyoming, this species breeds in lower to middle-level zones of all the major mountain ranges (Faulkner 2010). It is a common species in brushy streambottoms throughout the state, from Sand Creek in the Black Hills (Crook County) to Jenny's Lake in Grand Teton National Park (Scott, 1993). In Colorado it is common in Western Slope shrubby habitats from the Wyoming to the New Mexico borders. The densest regional breeding populations probably occur in northern Idaho (*e.g.*, Kootenai National Wildlife Refuge, Heyburn & McCrosky state parks) and adjacent western Montana (Glacier National Park, National Bison Range, Red Rock Lakes National Wildlife Refuge).

Population: National Breeding Bird Survey trend data indicate a nonsignificant annual population decrease of 0.8%. The North American population (north of Mexico) has been estimated at 5.35 million birds (Rich *et al.*, 2004).

Common Yellowthroat (*Geothlypis trichas*)

Status: A summer resident throughout the region, including montane areas, and a common to uncommon breeder in all of the montane parks except Rocky Mountain National Park, where rare.

Habitats and Ecology: Moist to wet ground, with associated vegetation such as tall grasses, shrubs, and small trees, are the primary breeding habitat, although at times the birds extend to upland thickets of shrubbery and low trees. Willow thickets around beaver ponds, the edges of muskegs, and scrub alders are among its favorite nesting areas. In Colorado studies, cattail and other emergent wetlands accounted for more than half of the breeding records, and streamside locations accounted for a further 40%. Few were found above 8,000 feet elevation and most were in marshes at the eastern and western edges of the mountains, and along major stream corridors (Kingery, 1998).

Suggested Viewing Locations: This is a common breeder throughout the region in wetland habitats, especially in cattail marshes. In Wyoming breeding occurs up to about 8,000 feet (Faulkner, 2010). Some good Wyoming birding sites include Yellowtail Reservoir (Big Horn County), Rawhide Wildlife Unit (Goshen County), LAK Reservoir (Weston County) and Currant Creek (Sweetwater County)(Dorn and Dorn, 1990). The most extensive regional breeding populations probably occur in Montana (*e.g.*, Bowdoin, Ninepipe, Lee Metcalf and Red Rock Lakes national wildlife refuges).

Population: National Breeding Bird Survey trend data indicate a significant annual population increase of 0.5%. The North American population has been estimated at 32 million birds (Rich *et al.*, 2004).

Wilson's Warbler (*Wilsonia pusilla*)

Status: A widespread summer resident in woodlands throughout most of the region, including the montane parks, where generally common and probably breeding in all.

Habitats and Ecology: On their breeding grounds these birds inhabit willow, alder thickets along rivers or beaver ponds, brushy edges of lakeshores, the edges of mountain meadows, timberline areas of low shrubby vegetation, and sometimes aspen thickets. In Colorado they regularly breed at altitudes of 9,000–10,500 feet, especially in willow thickets around high mountain lakes, beaver ponds or other montane wetlands (Kingery, 1998). In the interior Northwest, this species breeds in mixed coniferous forests from the pole-sapling to the mature forest stages of succession (Sanderson, Bull and Edgerton, 1980).

Suggested Viewing Locations: In Wyoming, this species breeds in most higher mountain ranges, including those in the Greater Yellowstone region, the Bighorns, and the Medicine Bows (Faulkner, 2010). It most often nests in montane willow thickets, such as those in Grand Teton and Yellowstone National Parks, as well as in the Bighorn Mountains (Scott, 1993). The densest regional breeding populations probably occur in central Colorado (nearly all the higher mountains and plateaus of the central Rockies and San Juan Mountains) and northwestern Montana (*e.g.*, Red Rock Lakes and Swan River national wildlife refuges, Glacier National Park).

Population: National Breeding Bird Survey trend data indicate a significant annual population increase of 0.5%. The North American population has been estimated at 36 million birds (Rich *et al.*, 2004).

Yellow-breasted Chat (*Icteria virens*)

Status: A summer resident at lower altitudes over most of the drier portions of the region. A vagrant only in the montane parks south of Canada, but a possible breeder at Dinosaur National Monument and Colorado National Monument.

Habitats and Ecology: During the breeding season this species occurs along the shrubby coulee areas of the plains, the oak and mountain mahogany woodlands of the foothills, along alder and willow-lined creeks of the prairies, brushy forest edges, and in shrubby overgrown pasturelands. In Colorado, most nesting occurs below 7,000 feet, but sometimes is as high as 8,000 feet, with nearly 95% in dense riparian thickets (Kingery, 1998).

Suggested Viewing Locations: In Wyoming, this species occurs statewide at elevations up to about 7,500 feet (Faulkner, 2010). During the breeding season it can be easily found in the Laramie River bottoms, the Yellowtail Wildlife Unit near Lovell, the LAK Reservoir (Weston County) and Richard's Gap (Sweetwater County)(Dorn and Dorn, 1990). The densest regional breeding populations probably occur in eastern Montana (*e.g.*, Benton Lake, Bowdoin, and Medicine Lake national wildlife refuges), and in southern Idaho (Camas and Deer Flat national wildlife refuges).

Population: National Breeding Bird Survey trend data indicate a nonsignificant annual population decrease of 0.1%. The North American population (north of Mexico) has been estimated at 10.44 million birds (Rich *et al.*, 2004).

Towhees & Sparrows

Green-tailed Towhee (*Pipilo chlorurus*)

Status: A local summer resident in the southern parts of the region, breeding north to central Montana, and a variably common summer resident in the montane parks north to Yellowstone National Park.

Habitats and Ecology: During the breeding season this species occurs in brushy foothills areas dominated by sagebrush, scrub oaks, saltbush and greasewood flats, scrubby riparian woodlands, and similar open and semi-arid habitats. Forested areas are avoided, but scattered trees in brushlands are used as singing posts. Spreading shrubs that allow for easy movement and foraging on the ground surface below are favored vegetation types. In Colorado the birds breed at an average altitude of 7,300 feet, most often in montane shrublands such as snowberry, serviceberry, chokecherry, mountain mahogany, scrub oaks and sagebrush (Kingery, 1998).

Suggested Viewing Locations: In Wyoming, this towhee breeds mostly west of a line from Sheridan to Laramie, especially in mountain-foothills shrublands and juniper-woodland shrubsteppe under 8,000 feet (Faulkner, 2010). Grand Teton National Park is a prime location for seeing green-tailed towhees. They are uncommon in Yellowstone National Park, where there is less sagebrush. They are common in nearly all sage-dominated areas of Wyoming, but are absent from the Black Hills. The densest regional breeding populations probably occur in the foothills, low mountains and mesas of western Colorado (*e.g.*, Dinosaur National Monument, Roan and Umcompahgre plateaus) and western Wyoming (Greater Yellowstone region, Seedskadee National Wildlife Refuge, Lewis and Clark Caverns State Park).

Population: National Breeding Bird Survey trend data indicate a nonsignificant annual population increase of 0.4%. The North American population has been estimated at 4.1 million birds, of which 92% breed in the Intermountain West Avifaunal Biome (Rich *et al.*, 2004).

Spotted Towhee (*Pipilo maculatus*)

Status: A summer resident over most of the region, becoming rarer in the Canadian mountains and also in the more arid southwestern areas. It is common in Colorado's western canyon parks, but uncommon to rare in the montane parks.

Habitats and Ecology: Breeding occurs in brushy fields, thickets, woodland openings or edges, second-growth forests, city parks, and well-planted suburbs. Habitats that have a good accumulation of litter and humus, and a protective screen of shrubby foliage above the ground, are highly favored by these birds.

Suggested Viewing Locations: Wyoming is at the western edge of the hybrid zone with eastern towhees, and spotted towhees breed widely over the eastern three-fourths of the state (Faulkner, 2010). Brushy areas almost statewide are good birding sites for this species, such as around LAK Reservoir (Weston County), along Birdseye Creek (Fremont County), and the North Platte River bottomlands from Torrington to Casper (Dorn and Dorn, 1990). Along the eastern edges of the region in Colorado hybrids or intergrades with the eastern towhee (*Pipilo erythropthalmus*) are common. The densest regional breeding populations probably occur in western and central Colorado, in juniper and scrub oak habitats along the state's western border and in the foothills of the Front Range.

Population: National Breeding Bird Survey trend data indicate a nonsignificant annual population increase of 0.6%. The North American population (north of Mexico) has been estimated at 12.6 million birds (Rich *et al.*, 2004).

Cassin's Sparrow (*Aimophila cassinii*)

Status: A summer resident in eastern Colorado; casual and irregular in southeastern Wyoming.

Habitats and Ecology: This southwestern sparrow is adapted to desert grasslands and semiarid prairies, and irregularly moves north to nest in southeastern Wyoming and western Nebraska during drought years in the Southwest. In Colorado, where they regularly nest, nearly half of the breeding habitat records came from shortgrass prairies, especially those sprinkled with sandsage, yuccas, shrubs, cacti and mesquites, usually at elevations under 5,200 feet (Kingery, 1998).

Suggested Viewing Locations: Wyoming's breeding records of Cassin's sparrows are few, and are concentrated in Goshen County, in a sandsage area near Torrington (Faulkner, 2010). In Colorado, Comanche and Pawnee national grasslands are prime birding spots, as well as Bonny Lake State Park, John Martin Reservoir State Park and Two Buttes State Wildlife Area (Kingery, 1993, 1998).

Population: National Breeding Bird Survey trend data indicate a significant annual population decrease of 1.5%. The North American population (north of Mexico) has been estimated at ten million birds (Rich *et al.*, 2004).

American Tree Sparrow (*Spizella arborea*)

Status: An overwintering migrant virtually throughout the entire region, including both montane areas and plains, but rarer in mountains farther north, and unreported from Kootenay National Park.

Habitats and Ecology: While in the Rocky Mountain region this species occu-

pies brushy prairie areas, roadside thickets, farmsteads, old orchards, overgrown and weedy pastures, and similar relatively open habitats. The birds
often occur in company with juncos and other gregarious and hardy sparrows, and feed about on the ground or snow surface, industriously searching out small seeds. During the breeding season they are associated with
arctic timberline habitats.

Suggested Viewing Locations: During winter this species is likely to appear in
open to somewhat brushy areas at lower altitudes throughout the entire region. No breeding occurs in the region.

Population: National Breeding Bird Survey trend data are not available. The
North American population has been estimated at 26 million birds (Rich *et
al.*, 2004).

Chipping Sparrow (*Spizella passerina*)

Status: A widespread and common summer resident throughout the region in
all wooded areas; possibly the most common breeding sparrow in the montane parks, and very probably breeding in all of them.

Habitats and Ecology: Breeding in this species is done in open deciduous or
mixed forests, the margins of forest clearings, the edges of muskegs, in timberline scrub, riparian woodlands, pinyon–juniper or oak–mountain mahogany woodlands, and similar diverse habitats. Generally scattered trees,
an unshaded forest floor, and a sparse ground covering of herbaceous
plants seem to be the kinds of habitat considerations that are important.
In Colorado the birds nest in diverse habitats, but most often in coniferous
woodlands, especially those dominated ponderosa pine or pinyon–juniper
(Kingery, 1998). In the interior Northwest, this species breeds in mixed coniferous forests from the shrub-seedling to the old-growth forest stages of
succession (Sanderson, Bull and Edgerton, 1980).

Suggested Viewing Locations: This species can be found throughout the region
in most wooded habitats, at low to moderate altitudes. In Wyoming, birds
breed statewide, mostly in open coniferous forests at middle altitudes, but
with some use of junipers, aspens and other wooded habitats (Faulkner,
2010). The densest regional breeding populations probably occur in the
Black Hills, the Greater Yellowstone region, and in central and western
Montana (*e.g.*, Glacier National Park; Red Rock Lakes and C. R. Russell national wildlife refuges).

Population: National Breeding Bird Survey trend data indicate a nonsignificant annual population decrease of 0.1%. The North American population (north of Mexico) has been estimated at 89.1 million birds (Rich *et al.*,
2004).

Clay-colored Sparrow (*Spizella pallida*)

Status: A local summer resident in the northern part of the region, mainly east
of the mountains and north of Wyoming, but locally breeding north to Banff
and Jasper National Parks, and probably locally south to southern Montana.

Habitats and Ecology: Favored breeding habitats consist of brushy thickets in prairies, fenceline shrubbery along pastures or meadows, mixed-grass prairies with scattered shrubs or low trees, brushy woodland margins, early successional stages of forests following logging or fires, and retired croplands. Nesting sometimes also occurs in city parks or residential areas.

Suggested Viewing Locations: In Wyoming, eastern parts of the state support the largest number of clay-colored sparrows, where they often mix with migrating chipping sparrows. The densest regional breeding populations probably occur in northern Montana (*e.g.*, Bowdoin, C. R. Russell and Medicine Lake national wildlife refuges).

Population: National Breeding Bird Survey trend data indicate a nonsignificant annual population decrease of 0.1%. The North American population has been estimated at 23 million birds (Rich *et al.*, 2004).

Brewer's Sparrow (*Spizella breweri*)

Status: A summer resident in nearly the entire region, except for the plains of Alberta, where replaced by the clay-colored sparrow. Variably common in most and probably all of the montane parks, and known to breed in several. Usually found in semi-desert scrub habitats, but also breeds at alpine timberline in the northern Rockies, at least in Colorado, Montana and southern Canada.

Habitats and Ecology: In the Rocky Mountain region this species breeds in two very different habitats. The first is in short-grass prairies with sage or other semi-arid shrubs present in varying densities. In Wyoming it breeds in sage shrubsteppe statewide, but is not yet known to have an alpine-breeding population (Faulkner, 2010). In Colorado, sage shrubland accounts for most breeding habitats, but mountain mahogany or currants growing in brushy hillsides or mesa edges are sometimes used (Kingery, 1998). In Idaho the birds have been found breeding on both sagebrush flats as well as in serviceberry-covered slopes of mountain ridges. In southern Alberta the birds also breed on short-grass plains with scattered sage and cacti, as well as along timberline in Banff and Jasper parks, in stunted spruces, firs, willows, and alders. This alpine adapted population has at times been proposed as constituting a new species, the timberline sparrow (*S. taverneri*).

Suggested Viewing Locations: Sagebrush is an ideal habitat for Brewer's sparrows, such as in localities around Rock Springs, along the Green River northwest of Daniel, the Upper Birdseye Pass Road at Boysen Reservoir, or along the Laramie Range east of Pole Mountain in Albany County (Dorn and Dorn, 1990). The densest regional breeding populations probably occur in southern Idaho (*e.g.*, Craters of the Moon National Monument; Minidoka National Wildlife Refuge) and northwestern Colorado (Dinosaur National Monument; Browns Valley National Wildlife Refuge).

Population: National Breeding Bird Survey trend data indicate a significant annual population decrease of 2.1%. Trend estimates for the Rocky Mountains & Plains States region (U.S.F.W.S. Region 6) indicate a significant annual decline of 1.4%. The North American population has been estimated at 16 million birds, of which 94% breed in the Intermountain West Avifaunal Biome (Rich *et al.*, 2004).

Field Sparrow (*Spizella pusilla*)

Status: A local summer resident in northeastern Montana, and a local migrant farther south. Also breeds locally in extreme eastern Colorado and northeastern Wyoming. A vagrant in some montane parks, such as Glacier National Park.

Habitats and Ecology: Breeding occurs in brushy, open woodlands, brushy ravines or coulees, sagebrush flats, abandoned hayfields, forest clearings, and similar habitats having a combination of low grassy areas and scattered shrubs or trees. Very similar habitats are used by the chipping sparrow, but that species tolerates a greater tree density and a later vegetational succession stage.

Suggested Viewing Locations: The only significant regional breeding populations probably occur locally in eastern Montana, and breeding probably also occurs locally in Campbell County, northeastern Wyoming (Faulkner, 2010). Migrants may at times be found along the eastern edge of the region in habitats such as abandoned and overgrown fields.

Population: National Breeding Bird Survey trend data indicate a significant annual population decrease of 2.8%. The North American population has been estimated at 8.2 million birds (Rich *et al.*, 2004).

Vesper Sparrow (*Pooecetes gramineus*)

Status: A summer resident throughout the region in grassland areas; variably common in all the montane parks and probably breeding in all.

Habitats and Ecology: During the breeding season this species is found in overgrown fields, prairie edges, grasslands with scattered shrubs and small trees, sagebrush areas where the plants are scattered and stunted, and similar open habitats, but not extending to mountain meadows or tundra zones. In Colorado nesting occurs widely, but is most common in middle- to higher-elevation sagebrush, but extends to lower sagebrush stands locally, especially where the sage is interspersed with a good grass cover (Kingery, 1998).

Suggested Viewing Locations: In Wyoming this species breeds statewide, from low elevations grasslands to mountain meadows. Open grasslands, especially where some shrubs and bare ground are present, are good places to look for this large sparrow. In Colorado dense populations exist in the Gunnison Basin and South Park. It is widespread through the region, with the densest regional breeding populations in southwestern and central Montana (*e.g.*, Red Rock Lakes National Wildlife Refuge, National Bison Range).

Population: National Breeding Bird Survey trend data indicate a significant annual population decrease of 1.0%. Trend estimates for the Rocky Mountains & Plains States region (U.S.F.W.S. Region 6) indicate a significant annual decline of 0.9%. The North American population has been estimated at 30 million birds (Rich *et al.*, 2004).

Lark Sparrow (*Chondestes grammacus*)

Status: A summer resident over most of the region in grassland habitats, but rarer northwardly and absent from the montane areas of Alberta. Rare to occasional in the montane parks farther south.

Habitats and Ecology: This species favors grasslands that have scattered trees, shrubs, large forbs, or adjoin such vegetation; thus weedy fencerows near grasslands, open brushland on slopes, sagebrush flats, scrubby and open oak woodlands, orchards, and similar habitats are all suitable. Generally, open views and a variety of plants, including scattered woody vegetation some grasses and herbs are preferred. In Colorado, grasslands with junipers, greasewood, or yucca are commonly used, as are plains grasslands within cottonwood stands, and shortgrass prairie with cholla cacti present (Kingery, 1998).

Suggested Viewing Locations: Lark sparrows are widespread regionally, and in Wyoming they occur statewide except for higher elevations in the Greater Yellowstone region, the Bighorns, and other major ranges (Faulkner, 2010). Some good viewing localities include the Belle Fourche River from the South Dakota line to near Colony, the Table Mountain Wildlife Unit in Goshen County, around Superior, and junipers west of Baggs (Dorn and Dorn, 1990). The densest regional breeding populations probably occur in central and eastern Montana (*e.g.,* C. R. Russell and Medicine Lake national wildlife refuges).

Population: National Breeding Bird Survey trend data indicate a significant annual population decrease of 1.6%. The North American population has been estimated at 27 million birds (Rich *et al.,* 2004).

Black-throated Sparrow (*Amphispiza lineata*)

Status: A local summer resident in the southwestern part of the region, mainly in Idaho south of the Snake River, and also in southwestern Wyoming (Green River valley). Breeds locally in extreme western and southeastern Colorado. Of uncertain breeding status at Dinosaur National Monument, but known to breed at the Colorado National Monument.

Habitats and Ecology: Breeding habitats consist of thinly grassed pastures with scattered cactus, yucca, or mesquite. Desert uplands with much exposed ground, but with hiding places in thick and woody twig growth or cactus plants, are especially favored. Nests are also often placed in cactus or dense shrub growth. In Colorado they have been most frequently found in pinyon–juniper, woodland, followed by tall desert shrubs, lowland sagebrush, and low desert shrubs (Kingery, 1998).

Suggested Viewing Locations: This species is rare in southwestern Wyoming, but might be searched-for in sagebrush flats between Farson and the Green River (Scott, 1993). The densest regional breeding populations probably occur in southern Idaho (*e.g.,* Bruneau Dunes State Park, Curlew National Grassland, City of Rocks National Reserve, Snake River Birds of Prey Natural Area).

Population: National Breeding Bird Survey trend data indicate a significant annual population decrease of 2.9%. Trend estimates for the Rocky Mountains & Plains States region (U.S.F.W.S. Region 6) indicate a significant annual decline of 2.7%. The North American population (north of Mexico) has been estimated at 13.5 million birds (Rich *et al.*, 2004).

Sage Sparrow (*Amphispiza belli*)

Status: A summer resident in sage areas in the southern parts of the region, north to the Snake River in Idaho and north-central Wyoming. Generally rare or absent from the montane parks.

Habitats and Ecology: The species is closely associated with fairly dense to sparse and scrubby sagebrush vegetation during the breeding season, but also breeds at times in similar semi-desert vegetation types, such as in saltbush. Foraging is done on rather bare ground areas of gravel or alkali soil around the bushes, and escapes are made by fleeing into the shrubbery.

Suggested Viewing Locations: In Wyoming, these birds are sagebrush obligates, and are most common in tall, dense stands of sage. They are most common in the southwestern parts of the state, especially in Carbon, Lincoln, Sweetwater and Uinta counties, but expand north in central Wyoming at least as far north as Hot Springs and Washakie counties (Faulkner 2010). The densest regional breeding populations probably occur in southern Idaho (*e.g.*, Bruneau Dunes State Park, Snake River Birds of Prey Natural Area, Camas and Bear Lake national wildlife refuges).

Population: National Breeding Bird Survey trend data indicate a nonsignificant annual population decrease of 0.1%. The North American population (north of Mexico) has been estimated at 3.87 million birds, of which 99% breed in the Intermountain West Avifaunal Biome (Rich *et al.*, 2004).

Lark Bunting (*Calamospiza melanocorys*)

Status: A summer resident in the eastern half of the region, mainly on plains and foothill grasslands; a rare migrant or vagrant in the montane parks, probably not breeding in any.

Habitats and Ecology: This species favors mixed-grass prairies for nesting, but also can be found in short-grass and tallgrass prairies, as well as sage grasslands, retired croplands, alfalfa fields, and stubble fields. Areas with abundant shrubs are avoided, but fence posts or scattered trees may be used as song posts. In Colorado, over 40% of breeding birds were found in short-grass prairie, and taller prairies plus croplands comprise an additional 42% of the total (Kingery, 1998).

Suggested Viewing Locations: In Wyoming, breeding occurs nearly statewide, except for the Greater Yellowstone region, the Bighorns, and other major mountain ranges (Faulkner 2010). They are most likely to be found on drier grasslands of the eastern plains, but their local distribution varies greatly from year to year, depending on local rainfall or (in Nebraska at least) irrigation activities. Irrigated alfalfa fields seem to be especially favored. They are common at Hutton Lake National Wildlife Refuge. The densest regional breeding populations probably occur in the eastern parts of Montana, Wyoming and Colorado. In Montana, lark buntings are common to abundant at Benton Lake, Bowdoin and Medicine Lake national wildlife refuges. In Colorado an estimated 1.6 million breeding pairs were present in the late 1980's and early 1990's, with the densest concentrations in east-central Colorado (*e.g.,* Two Buttes State Wildlife Area) and northeastern Colorado (Pawnee National Grassland; Bonny Lake State Park)(Kingery, 2007).

Population: National Breeding Bird Survey trend data indicate a significant annual population decrease of 1.6%. Trend estimates for the Rocky Mountains & Plains States region (U.S.F.W.S. Region 6) also indicate a significant annual decline of 1.6%. The North American population has been estimated at 27 million birds (Rich *et al.,* 2004).

Savannah Sparrow (*Passerculus sandwichensis*)

Status: A summer resident throughout the region, mainly at lower altitudes, but occurring commonly in all the montane parks and probably breeding in all.

Habitats and Ecology: During the breeding season this species is closely associated with moist but low-stature prairies, the wet meadow zones around marshes or other wetlands, and the moist and open areas of mountain meadows. A growth of dense ground cover, preferably only a few inches tall, with scattered bushes or clumps of taller vegetation for song perches, are typical aspects of nesting habitats. In Colorado, moist mountain meadows are this species favored breeding habitat, with montane grasslands, croplands and emergent marshes representing 58% of all habitat reports (Kingery, 1998). Nests are placed on the ground, under thick herbaceous cover of grasses or sedges, and are usually hidden from above by overhanging leaves.

Suggested Viewing Locations: In Wyoming, breeding occurs statewide at lower to mid-elevations in moderately moist herbaceous habitats (Faulkner 2010). Good viewing locations during the breeding season include wet meadows along Chicken Creek in Fossil Butte National Monument, and in meadows along Billy Creek Road in the Bighorn Mountains of Johnson County. Similar meadows 12–13 miles south of Boulder in Sublette County, and along Spring Lane northwest of Laramie (Dorn and Dorn, 1990) are also good possibilities. The densest regional breeding populations probably occur in north-central Montana, such as at Freezeout Lake Wildlife Management Area, but the birds are also common at Benton Lake, Bowdoin and Medicine Lake national wildlife refuges.

Population: National Breeding Bird Survey trend data indicate a significant annual population decrease of 1.0%. Trend estimates for the Rocky Mountains & Plains States region (U.S.F.W.S. Region 6) also indicate a significant annual decline of 1.0%. The North American population (north of Mexico) has been estimated at 79.5 million birds (Rich *et al.,* 2004).

Baird's Sparrow (*Ammodramus bairdii*)

Status: A summer resident in grasslands of eastern Alberta and northeastern Montana, and a migrant east of the mountains farther south. Reported in the montane parks only as a vagrant.

Habitats and Ecology: Closely associated during the breeding season with native prairie areas, including ungrazed or lightly grazed mixed-grass prairies, wet meadows, and various disturbance habitats such as hayfields, stubble fields, and retired croplands. A dense but low vegetation over the soil and a few scattered shrubs for singing posts seem to be desirable aspects of the habitat.

Suggested Viewing Locations: In Wyoming, this uncommon migrant is confined to eastern parts of the state, such as the grasslands around Van Tassel, Cheyenne and Laramie (Scott, 1993). The densest regional breeding populations probably occur in northeastern Montana, such as at Bowdoin and Medicine Lake national wildlife refuges.

Population: The North American population has been estimated at 1.2 million birds. Classified as a Partners-in-Flight Watch List species of continental conservation importance (Rich et al., 2004). National Breeding Bird Survey trend data indicate a significant annual population decrease of 3.6%. Trend estimates for the Rocky Mountains & Plains States region (U.S.F.W.S. Region 6) indicate a significant annual decline of 5.0%.

Grasshopper Sparrow (*Ammodramus savannarum*)

Status: A local summer resident, mainly at lower altitudes, in grasslands throughout the region. Generally absent from the montane parks except as a vagrant, but reported as a rare breeder at Yellowstone National Park.

Habitats and Ecology: During the breeding season this species is mostly associated with mixed-grass prairies, but also occurs in short-grass and tall-grass areas, as well as on sage grasslands and disturbed grasslands such as hayfields, stubble fields, and retired croplands. Mountain meadows are not used, nor are grassland areas that have largely grown up to shrubs. In Colorado, prairie grasslands, or grasslands with rabbitbrush or saltbush are often used; over 80% of breeding-season observations occurred on various types of grasslands (Kingery, 1998).

Suggested Viewing Locations: In Wyoming, this species is mainly confined to eastern parts of the state, east of a line from Sheridan to Cheyenne (Faulkner 2010). There are good viewing opportunities in grasslands along the lower North Platte valley (Dorn and Dorn, 1990). It also occurs in grasslands of the Black Hills, south of Van Tassel, and along Bird Farm Road south of Big Horn (Scott, 1993). The densest regional breeding populations probably occur in the eastern parts of Montana, Wyoming and Colorado. In southeastern Montana, Medicine Rocks State Park and Ekalaka Park (near the North Dakota–South Dakota border) offer fine birding for grassland sparrows such as the grasshopper sparrow, and it is also common at Bowdoin and Medicine Lake national wildlife refuges. In Colorado these birds are most common along the easternmost counties. Pawnee National Grassland, and Mesa Trail near Boulder are good birding areas for this species (Kingery, 1993).

Population: National Breeding Bird Survey trend data indicate a significant annual population decrease of 3.6%. Trend estimates for the Rocky Mountains & Plains States region (U.S.F.W.S. Region 6) indicate a significant annual decline of 3.3%. The North American population (north of Mexico) has been estimated at 13.95 million birds (Rich *et al.,* 2004).

LeConte's Sparrow *(Ammodramus leconteii)*

Status: A local summer resident in southeastern Alberta east of the mountains. Probable breeding in Montana is limited to a few areas, such as meadows in Glacier National Park. Breeding in Glacier National Park is unproven, as is also true of Watertown Lakes National Park.

Habitats and Ecology: Breeding in this species is largely limited to hummocky bogs with alder or willows present, but it also nests in the wet meadows around prairie ponds or marshes, in moist tallgrass prairies, and in moist hayfields or retired croplands. A favorite nesting cover is cordgrass *(Spartina)*; usually the nest is on a dry hummock surrounded by dense grass or a shrub.

Suggested Viewing Locations: Visiting McGee Meadows overlook in Glacier National Park during June may provide the best regional opportunities south of Canada. The densest regional breeding populations probably occur in the grasslands of southern Alberta.

Population: National Breeding Bird Survey trend data indicate a nonsignificant annual population decrease of 0.3%. The North American population has been estimated at 2.9 million birds (Rich *et al.,* 2004).

Fox Sparrow *(Passerella iliaca)*

Status: A summer resident in wooded areas almost throughout the region; present and variably common in all the montane parks, and probably breeding in all.

Habitats and Ecology: During the breeding season dense brushy thickets, and the brushy margins of thick forests, are the favored habitats. Riparian thickets of willows or alders, alder clumps on mountain slopes, and the twisted and stunted conifers near timberline all serve to attract this species. Thickets that provide sufficient space underneath for ground foraging, and have a carpet of leaves and litter for scratching towhee-like for food, are particularly favored. In Colorado, riparian willow shrublands and wet, meadows overgrown with willows at elevations of 7,500–11,000 feet are preferred breeding habitats (Kingery, 1998).

Suggested Viewing Locations: In Wyoming, most breeding occurs in the Greater Yellowstone region (Faulkner, 2010). Fox sparrows may readily be found in wet thickets and marshy areas during summer in Grand Teton National Park (*e.g.,* Willow Flats and marshland around Two Ocean Lake)(Scott, 1993). Other likely sites include Swift Creek east of Afton, and Burbank

Creek near Teton Pass (Dorn and Dorn, 1990). The densest regional breeding populations probably occur in northwestern Montana (Glacier National Park) and northern Idaho (western Selkirk Mountains, Heyburn State Park).

Population: National Breeding Bird Survey trend data indicate a nonsignificant annual population decrease of 0.9%. The North American population has been estimated at 16 million birds (Rich *et al.*, 2004).

Song Sparrow (*Melospiza melodia*)

Status: A summer or permanent resident in suitable habitats throughout the region; present in all the montane parks and probably breeding in all.

Habitats and Ecology: Breeding habitats include such woodland edge types as the brushy margins of forest openings, the edges of ponds or lakes, shelterbelts, farmsteads, coulees on prairies, aspen groves, and the like. Foraging occurs mostly on the ground, both in open areas and leaf-covered ones, where the birds scratch to expose foods.

Suggested Viewing Locations: Song sparrows are common to abundant summer residents throughout the lower altitudes of the Wyoming mountains, as well as breeding in virtually all the state's riparian areas and in brushlands near marshes or beaver ponds, up to as high as 10,000 feet (Faulkner, 2010). The same is true throughout the region, but the densest regional breeding populations probably occur in northern Idaho (*e.g.,* Kootenai National Wildlife Refuge) and western Montana (National Bison Range).

Population: National Breeding Bird Survey trend data indicate a significant annual population decrease of 0.5%. However, trend estimates for the Rocky Mountains & Plains States region (U.S.F.W.S. Region 6) indicate a significant annual increase of 3.3%. The North American population (north of Mexico) has been estimated at 52.9 million birds (Rich *et al.*, 2004).

Lincoln's Sparrow (*Melospiza lincolnii*)

Status: A summer resident in wooded areas almost throughout the region, from lowland bogs to alpine timberline. Present in all the montane parks, and probably breeding in all of them.

Habitats and Ecology: In Alberta this species is mainly associated with marshes and bogs having extensive growths of willows and alders. Willow thickets along slow-moving streams are also utilized, as are the brushy borders of muskeg pools. In mountainous areas the birds favor boggy mountain meadows, especially those fringed with willow thickets and supporting a fairly tall growth of grasses, sedges and herbs. In Colorado these birds usually breed at medium to high altitudes (above 8,000 feet) in dense willow thickets, with aspen groves providing a secondary habitat choice (Kingery, 1998).

Suggested Viewing Locations: In Wyoming this sparrow breeds throughout the Greater Yellowstone region, and in other higher ranges such as the Bighorns, the Medicine Bows, and the Laramie Range (Faulkner, 2010). The abundant beaver ponds in and around Grand Teton National Park and elsewhere support good breeding populations of Lincoln's sparrows. Other good birding areas include the Medicine Bow Mountains in Albany County, willow thickets along the Powder River Pass in Johnson County, and the Uinta Mountains in Uinta County (Dorn and Dorn, 1990). The densest regional breeding populations probably occur in central Colorado along the main spine of the Rocky Mountain Range, where they are common in all of the montane parks from the Wyoming to the New Mexico borders (Kingery, 1998).

Population: National Breeding Bird Survey trend data indicate a nonsignificant annual population increase of 0.6%. Trend estimates for the Rocky Mountains & Plains States region (U.S.F.W.S. Region 6) indicate a significant annual increase of 2.3%. The North American population has been estimated at 39 million birds (Rich *et al.*, 2004).

Swamp Sparrow (*Melospiza georgiana*)

Status: A local summer resident in Alberta east of the mountains and south to the vicinity of Red Deer. Generally absent from the montane parks, but a rare visitor (May, June, September) to Banff National Park.

Habitats and Ecology: This species is strongly associated with wetlands during the breeding season, especially areas that are well grown with cattails, phragmites, shrubs, or small trees. In Alberta muskeg-like woodland swamps that have willows, alders, birches, and sometimes black spruces are favored areas, but the birds also nest in dense shrubbery along woodland streams or pools.

Suggested Viewing Locations: Migrants are mostly likely to be seen in marshes along the eastern edge of the region; in Wyoming the Black Hills have provided many records (Scott, 1993).

Population: National Breeding Bird Survey trend data indicate a nonsignificant annual population decrease of 0.7%. The North American population has been estimated at nine million birds (Rich *et al.*, 2004).

White-throated Sparrow (*Zonotrichia albicollis*)

Status: A local summer resident in northern and central Alberta south as far as Banff National Park and the area west of Calgary. Elsewhere a migrant or overwintering visitor throughout the region, but rarer in the mountains and generally only a vagrant in the montane parks.

Habitats and Ecology: In Alberta this species nests in deciduous and mixed woodlands, particularly woodland edge habitats such as lake shores, riverbanks, previously burned areas, logged areas, and roadsides. Outside the breeding season the birds are often observed foraging on the ground in somewhat brushy situations.

Suggested Viewing Locations: Migrants are common throughout the region in thickets and brushy habitats, and sometimes are attracted to suburban bird feeders. Local breeding occurs in the Canadian montane parks.

Population: National Breeding Bird Survey trend data indicate a significant annual population decrease of 0.6%. The North American population has been estimated at 140 million birds (Rich *et al.*, 2004).

Golden-crowned Sparrow (*Zonotrichia atricapilla*)

Status: A local summer resident in the Alberta mountains south to Banff National Park, and a migrant from that area south to northern Idaho, but a vagrant elsewhere in the region.

Habitats and Ecology: During the breeding season these birds seek out low coniferous or deciduous growth at or even above tree line. Thickets of stunted willows, alders, and conifers growing in high meadows or on scree slopes provide the nesting habitats; the nests are placed in low woody vegetation or on the ground. While on migration and on wintering areas the birds prefer interrupted brushlands, where leafy litter provides for ground-foraging opportunities.

Suggested Viewing Locations: Banff and Jasper parks are among the few places in the region where one might readily find this species during summer.

Population: National Breeding Bird Survey trend data indicate a significant annual population decrease of 3.7%. The North American population has been estimated at 5.2 million birds (Rich *et al.*, 2004).

White-crowned Sparrow (*Zonotrichia leucophrys*)

Status: A summer or permanent resident in suitable habitats almost throughout the region, including the montane parks, where it is a common to abundant breeder in all.

Habitats and Ecology: During the breeding season this species occurs in riparian brush, in coniferous forests with well developed wooded undergrowth, in aspen groves with a shrubby understory, willow thickets around beaver ponds or marshes, and on mountain meadows with alders or similar low and thick shrubbery, often to timberline. Damp, grass-covered ground and nearby shrubbery seem to be important habitat components. In Colorado these birds were most often found to breed in willow thickets at medium to high altitudes, with a secondary use of krummholz (stunted, often wind-shaped, timberline conifers)(Kingery, 1998). On migration and during winter the birds are found in a variety of lower elevation habitats that offer a combination of brushy cover and open ground for foraging.

Suggested Viewing Locations: In Wyoming, this sparrow breeds commonly at higher elevations (above 7,500 feet in the Greater Yellowstone region; above 9,000 feet farther east) in the major mountain ranges, and is absent from the Black Hills (Faulkner, 2010). Roads over high passes during summer, such as over the Snowy Range west of Laramie, or across Libby Flats in the Medicine Bow Range, are good places to find this species (Dorn and Dorn, 1990; Scott, 1993). The birds are common breeders in the Greater Yellowstone ecosystem, especially around Jackson Hole. The densest regional breeding concentrations are probably in the montane forests of western Montana (*e.g.*, Glacier National Park, Red Rock Lakes National Wildlife Refuge) and eastern Idaho (Grays Lake National Wildlife Refuge).

Population: National Breeding Bird Survey trend data indicate a nonsignificant annual population decrease of 1.0%. Trend estimates for the Rocky Mountains & Plains States region (U.S.F.W.S. Region 6) indicate a significant annual decline of 1.2%. The North American population has been estimated at 72 million birds (Rich *et al.*, 2004).

Harris' Sparrow (*Zonotrichia querula*)

Status: A seasonal or overwintering migrant over much of the region, but far more common on the plains east of the mountains, and rare or accidental in the montane parks.

Habitats and Ecology: During the breeding season this species is associated with the edges of the spruce forest adjoining arctic tundra, especially rather damp and open areas where the trees are low and scattered. Outside the breeding season the birds are much like the other *Zonotrichia* sparrows, foraging on the ground in areas close to thickets, which are used for protection and roosting. At that time, hedgerows, orchards, farmsteads, riparian thickets, woodland edges, and even sagebrush and desert scrub habitats are often utilized.

Suggested Viewing Locations: In this region Harris's sparrows are most often found along its eastern boundaries of Montana, Wyoming and Colorado during migration, where they often mix with other *Zonotrichia* sparrows in edge or brushy habitats. During inclement weather they may be attracted to urban bird feeders.

Population: National Breeding Bird Survey trend data are not available. The North American population has been estimated at 3.7 million birds. Classified as a Partners-in-Flight Watch List species of continental conservation importance.

Dark-eyed Junco (*Junco hyemalis*)

Status: A seasonal or permanent resident in wooded habitats throughout the region, including all the montane parks, where the species is a common to abundant breeder. Winters at lower altitudes, including the entire plains region. Breeding populations of slate-colored juncos (*J. h. hyemalis*) occur in the Canadian montane parks and winter widely southwardly through the western mountains and plains. The breeding juncos of this region include a "white-winged" form (*J. h. aikeni*) that is endemic to and semi-residential in the Black Hills region and surrounding areas. The complex of up to eight intergrading northwestern races collectively known as the "Oregon junco" breeds east from the Pacific Coast to the Pacific slope of Montana. The most interior "pink-sided" variant of the Oregon complex (*J. h. mearnsi*) breeds from Alberta to northern Wyoming. The "gray-headed" form (*J. h. caniceps*) of the southern Rockies breeds north to northern Colorado and adjacent southern and western Wyoming, intergrading with the pink-sided form in southwestern Wyoming. In the interior Northwest, dark-eyed juncos breed in mixed coniferous forests from the grass-forb to the old-growth forest stages of succession (Sanderson, Bull and Edgerton, 1980).

Habitats and Ecology: Breeding habitats include open coniferous forests, especially pinyon–juniper woodlands, ponderosa pine forests, mixed forests, aspen woods, forest clearings, the edges of muskegs or jackpine-covered ridges,
and similar habitats that offer ground-foraging and ground-nesting opportunities as well as tree or brush cover for escape. In Colorado, 70% of breeding
birds were found in coniferous forests, with most of the remainder in aspens,
and a few in other deciduous or shrubby habitats (Kingery, 1998).

Suggested Viewing Locations: Collectively, four races of juncos occur throughout Wyoming and most are present throughout the year. Three of these
subspecies breed in Wyoming, occupying all of the state's montane forests (Faulkner, 2010). All of the races are likely to appear commonly at
bird feeders during winter. Pink-sided juncos are abundant breeders in the
Greater Yellowstone region and the mountains of eastern Idaho and western Montana. The densest regional breeding populations of gray-headed
juncos occur in central Colorado (Front, Flat Tops, Gore, Park and San Juan
ranges)(Kingery, 1998), and white-winged juncos are largely confined to the
Black Hills region, where overwintering is regular.

Population: National Breeding Bird Survey trend data indicate a significant annual population decrease of 1.0%. The North American population has
been estimated at 260 million birds (Rich *et al.*, 2004).

McCown's Longspur (*Calcarius mccownii*)

Status: A summer resident on the plains east of the mountains from southern
Alberta southward through the region. It is a rare migrant in the mountains,
and a vagrant in the montane parks.

Habitats and Ecology: During the breeding season this species is mostly limited to short-grass prairies and grazed mixed-grass prairies, but also breeds
to some degree on stubble fields or newly sprouting grainfields. In Colorado the birds breed almost exclusively in very sparse grasses, with a large
amount of exposed bare soil and a low diversity of other plants, including
prickly pear cactus, lupine, and locoweed (Kingery, 1998). While on migration and during the winter period the birds occur on open grasslands, low
sage prairies, mountain meadows, and similar open habitats.

Suggested Viewing Locations: In Wyoming, breeding mostly occurs east of a line
from Sheridan to the western edge of the Laramie plains in Albany County
(Faulkner, 2010). Very good numbers of McCown's longspurs may be seen
in places such as the native grasslands south of Van Tassel, the high grasslands around Cheyenne, and in and around Hutton Lake National Wildlife Refuge (Albany County) (Dorn and Dorn, 1990; Scott, 1993). The densest regional breeding populations probably occur in southeastern Wyoming
and northeastern Colorado (Pawnee National Grassland).

Population: The North American population (north of Mexico) has been estimated at 1.1 million birds. Classified as a Partners-in-Flight Watch List species of continental conservation importance (Rich *et al.*, 2004). National
Breeding Bird Survey trend data indicate a significant annual population
decrease of 1.0%.

Lapland Longspur (*Calcarius lapponicus*)

Status: A seasonal or overwintering migrant over much of the region, primarily at lower altitudes and in grassland habitats, and generally rare to accidental in the montane parks.

Habitats and Ecology: Associated during the breeding season with arctic tundra. While on migration and on wintering areas they are typically found on open habitats, such as snow-covered grasslands, mud flats, and the like. Shortly after their arrival in fall the birds often are found in the lower mountain parks, but as the weather becomes more severe the birds move to the foothills and plains, where they usually occur in fairly large flocks.

Suggested Viewing Locations: Migrants of this arctic-breeding longspur can be found in eastern grasslands of the region during fall, winter and spring, often foraging with horned larks or other longspurs (Scott, 1993).

Population: National Breeding Bird Survey trend data are not available. The North American population has been estimated at 75 million birds (Rich *et al.*, 2004).

Chestnut-collared Longspur (*Calcarius ornatus*)

Status: A summer resident on the plains areas east of the mountains over most of the region, and a local migrant somewhat farther west of the breeding range. Generally uncommon to accidental in the montane parks.

Habitats and Ecology: Primary breeding habitats consist of grazed or hayed mixed-grass prairies, short-grass plains, the meadow zones of salt grass around alkaline ponds or lakes, mowed hayfields, heavily grazed pastures, and the like. In Colorado, this longspur breeds in taller and damper grasslands than the previous species, on sites having less bare ground and more singing posts provided by tall forbs (Kingery, 1998). Outside the breeding season the birds often are found in cultivated fields rich in weed seeds, especially of such species as amaranth.

Suggested Viewing Locations: In Wyoming, breeding occurs almost entirely along the easternmost tier of counties, but extends west to Albany County at the southern end of the state (Faulkner, 2010). In common with the McCown's longspur, grasslands of the eastern plains are also most likely to support migrating longspurs. During the breeding season they may be found on grasslands south of Van Tassel, but the birds favor mixed-grass prairies rather than the short-stature grasslands that are used by McCown's longspurs. Only a few breed in Colorado (Pawnee National Grassland). The densest regional breeding populations probably occur in northern and eastern Montana (*e.g.*, Benton, Bowdoin and Medicine Lake national wildlife refuges). *Population:* National Breeding Bird Survey trend data indicate a significant annual population decrease of 2.7%. Trend estimates for the Rocky Mountains & Plains States region (U.S.F.W.S. Region 6) indicate a significant annual decline of 2.5%. The North American population has been estimated at 5.6 million birds (Rich *et al.*, 2004).

Snow Bunting (*Plectrophenax nivalis*)

Status: An overwintering migrant throughout the region, mainly at lower altitudes on grasslands. Generally uncommon to rare in the more southern montane parks, but fairly common in the Canadian parks. The nearest breeding areas are in northern Canada.

Habitats and Ecology: While in the Rocky Mountain region these birds are usually found along snow-free roads, in partially snow-free weedy fields, stubble fields, snow-free hilltops on cultivated lands, and similar areas where grain or weed seeds are likely to be found.

Suggested Viewing Locations: In Wyoming, and elsewhere in the region snow buntings are uncommon to irregular winter visitors, probably depending on the severity of the winter. Wyoming areas that often have them include the Shirley Basin and Sweetwater Plateau (Scott, 1993).

Population: National Breeding Bird Survey trend data are not available. The North American population (north of Mexico) has been estimated at 19.5 million birds (Rich *et al.*, 2004).

Cardinals & Grosbeaks

Scarlet Tanager (*Piranga olivacea*)

Status: An accidental vagrant or rare migrant in the region; there is at least one Montana record, about 20 from Wyoming, and many records from Colorado. The nearest breeding areas are in the western Dakotas and Nebraska.

Habitats and Ecology: Breeding typically occurs in mature hardwood forests growing in river valleys, slopes, and bottomlands. Less often it occurs in coniferous forests and in city parks or orchards.

Suggested Viewing Locations: This species is too rare in the region to predict good birding areas.

Population: National Breeding Bird Survey trend data indicate a nonsignificant annual population decrease of 0.1%. The North American population has been estimated at 2.2 million birds (Rich *et al.*, 2004).

Western Tanager (*Piranga ludoviciana*)

Status: A summer resident in coniferous forests throughout the region, including all the montane parks, where it is variably common and probably a breeder in all. Its status is uncertain in the Cypress Hills of Alberta, where singing males have been observed.

Habitats and Ecology: Breeding occurs in various habitats, including riparian woodlands, aspen groves, ponderosa pine forests, and occasionally in Douglas-fir forests and pinyon–juniper or oak–mountain mahogany woodlands. It is usually found in areas having a predominance of coniferous trees, preferably those that are fairly open, but occasionally extends into fairly dense forests. In Colorado nesting occurs from 5,000–8,000 feet, especially in ponderosa pine and aspen woodlands, and in various coniferous forests ranging from foothills to mid-elevations (Kingery, 1998). In the interior Northwest, this tanager breeds in mixed coniferous forests from the young forest to the old-growth forest stages of succession (Sanderson, Bull and Edgerton, 1980).

Suggested Viewing Locations: In Wyoming, this bird breeds from the western Black Hills to the Greater Yellowstone region, including the Bighorn, Laramie and Medicine Bow ranges (Faulkner, 2010). It is notably common in the Black Hills and around Moose and Jenny's Lake in the Tetons. The densest regional breeding populations probably occur in north-central Idaho *(e.g.,* McCrosky and Ponderosa state parks) and adjacent western Montana (Glacier National Park, National Bison Range, Red Rock Lakes National Wildlife Refuge).

Population: National Breeding Bird Survey trend data indicate a significant annual population increase of 1.2%. Trend estimates for the Rocky Mountains & Plains States region (U.S.F.W.S. Region 6) indicate a significant annual increase of 2.0%. The North American population (north of Mexico) has been estimated at 88.1 million birds (Rich *et al.,* 2004).

Rose-breasted Grosbeak *(Pheucticus ludovicianus)*

Status: A local summer resident in the extreme northern portion of the region (Red Deer and Rocky Mountain House area, possibly south to Bottrel and the Porcupine Hills), and a rare to occasional migrant farther southeast of the mountains, with vagrants occurring in the montane parks.

Habitats and Ecology: During the breeding season this species is found in deciduous woodlands or the deciduous portions of mixed forests on floodplains, slopes, and bluffs. Forests where the undergrowth is tall but not too dense are apparently preferred, although a variety of undergrowth conditions are utilized.

Suggested Viewing Locations: This species is not a regular breeder in the region, but probably breeds locally in the Canadian montane parks. Hybridization with the black-headed grosbeak obscures breeding limits. The rose-breasted grosbeak is most likely to occur in deciduous riparian woodlands along eastern rivers.

Population: National Breeding Bird Survey trend data indicate a nonsignificant annual population decrease of 0.7%. The North American population has been estimated at 4.6 million birds (Rich *et al.,* 2004).

Black-headed Grosbeak *(Pheucticus melanocephalus)*

Status: A summer resident and variably common breeder over most of the region in wooded areas excepting the northernmost parks in Alberta, where only a vagrant.

Habitats and Ecology: During the breeding season this species is associated with open deciduous woodlands having fairly well developed shrubby understories, and usually on floodplains or upland areas. It extends into wooded coulees and riparian forests of cottonwoods and similar vegeta-

tion in the plains, and sometimes also nests in orchards, oak–mountain mahogany woodlands, and aspen groves. In Colorado, most nesting occurs
from 5,000–8,000 feet, in a variety of habitats that especially include ponderosa pine, aspen and riparian foothill forests, as well as pinyon–juniper
and scrub oak woodlands. The presence of tick clover, either as a canopy
or undergrowth cover plant, might be a common positive selection factor
(Kingery, 1998). In the interior Northwest, this species breeds in mixed coniferous forests from the young forest to the old-growth forest stages of succession (Sanderson, Bull and Edgerton, 1980).

Suggested Viewing Locations: In Wyoming, this grosbeak breeds nearly statewide at elevations below 8,000 feet (Faulkner, 2010). I t is most likely to be
found in deciduous riparian woodlands across the state, but especially in
the eastern half where these habitats are more common. Favorable birding
areas include oak woodlands of the western Black Hills (Crook County), aspen groves in Fossil Butte National Monument (Lincoln County), and lower
parts of Grand Teton National Park (Dorn and Dorn, 1993). The most extensive regional breeding populations probably occur in Colorado (along the
Front Range foothills and Western Slope plateaus) and northern Idaho (*e.g.*,
Kootenai National Wildlife Refuge.

Population: National Breeding Bird Survey trend data indicate a significant annual population increase of 0.8%. Trend estimates for the Rocky Mountains
& Plains States region (U.S.F.W.S. Region 6) indicate a significant annual increase of 3.6%. The North American population (north of Mexico) has been
estimated at 3.92 million birds (Rich *et al.*, 2004).

Blue Grosbeak (*Passerina caerulea*)

Status: A local summer resident in the southern parts of the region, west in Colorado to the Front Range, and in southwestern Colorado north possibly to
Dinosaur National Monument. Breeding also occurs in southeastern and
southwestern Wyoming, and southern Idaho.

Habitats and Ecology: During the breeding season these birds are found in
brushy and weedy pastures, old fields with scattered saplings, forest edges,
hedgerows, and streamside thickets. The presence of large seeds, such
as sunflowers, seems to favor its occurrence, and the birds are also often
found near water. In Colorado, the breeding birds associate with the edges
of lowland and foothills riparian forest, as well as with a variety of rural environments, usually at elevations no higher than 6,000 feet (Kingery, 1998).

Suggested Viewing Locations: In Wyoming, blue grosbeaks are mostly limited
to the southeastern corner of the state, from Niobrara to Laramie counties. They also extend northwest along the North Platte River to Natrona
County and breed locally in Fremont County (Faulkner, 2010). They are
most are likely to be seen in summer along the North Platte River between
Torrington and Casper (Dorn and Dorn, 1990), or farther upstream from
Alcova to Casper (Scott, 1993). Elsewhere in Wyoming they are less common to rare. Colorado also has dense populations, especially in the San

Juan Mountains, and also on both slopes of the mountain ranges extending down the center of the state. Good birding sites include the Carrizo Unit of Comanche National Grassland and Navajo State Park (Kingery, 1998).

Population: National Breeding Bird Survey trend data indicate a significant annual population increase of 1.0%. The North American population (north of Mexico) has been estimated at 6.16 million birds (Rich *et al.*, 2004).

Lazuli Bunting (*Passerina amoena*)

Status: A summer resident in suitable habitats nearly throughout the region, but becoming rarer eastwardly and northwardly; reported in all the montane parks except Banff/Jasper, and probably breeding in most.

Habitats and Ecology: In the mountain areas these birds breed along the edges of deciduous forests on gentle valley slopes, such as aspen groves, or thickets of willow or alder. On the foothills and plains the birds are usually found in riparian woodlands supporting a mixture of shrubs, low trees, and herbaceous vegetation. Plant diversity and discontinuity of cover seem to be important habitat characteristics for this species. In Colorado the birds mostly breed between 5,500 and 7,000 feet, especially in riparian habitats with an abundance of shrubs. Mountain shrublands of mountain mahogany or serviceberry were also found to be important breeding habitats (Kingery, 1998).

Suggested Viewing Locations: Lazuli buntings breed statewide in Wyoming, in shrub-dominated habitats up to about 9,500 feet (Faulkner, 2010). Streamside thickets in Wyoming provide ideal habitats, such as along Tongue River at Dayton, and in brushlands such as above LAK Reservoir (Weston County), Richard's Gap south of Rock Springs, and near Battle Creek campground in the Sierra Madre Mountains (Carbon County)(Dorn and Dorn, 1990). The densest regional breeding populations probably occur in north-central and southern Idaho (*e.g.*, Deer Flat and Grays Lake national wildlife refuges).

Population: National Breeding Bird Survey trend data indicate a significant annual population increase of 0.6%. The North American population (north of Mexico) has been estimated at 2.28 million birds (Rich *et al.*, 2004).

Indigo Bunting (*Passerina cyanea*)

Status: A local summer resident primarily east of the major mountain ranges, but extending west locally to the upper Missouri drainage, the Bighorn Mountains of Wyoming, and central Colorado. It is a vagrant in some montane parks.

Habitats and Ecology: This species typically breeds in relatively open hardwood forests on floodplains or uplands. Open woodlands, with a high density of shrubs and an open canopy, are favored, and thus forest edges, second-growth areas, orchards, overgrown pastures, and similar habitats are typically utilized.

Suggested Viewing Locations: In Wyoming, this species is mostly confined to the easternmost tier of counties as breeders (Faulkner, 2010). The northwestern Black Hills (Crook County) and the lower eastern slope of the Big-horn Mountains provide good birding sites for this species (Dorn and Dorn, 1990). Regional breeding populations also occur in Colorado (mostly along the Front Range, but also locally in Comanche National Grasslands and west to some western mesas) (Kingery, 1998). They are local in eastern Montana (easternmost counties), but hybridization with the lazuli bunting obscures both species' breeding limits where they come into contact. In one Wyoming study about one-third of the pairs were of mixed-species compo-sition, but the hybrids showed low viability (Baker and Boylan, 1999).

Population: National Breeding Bird Survey trend data indicate a significant an-nual population decrease of 0.5%. However, trend estimates for the Rocky Mountains & Plains States region (U.S.F.W.S. Region 6) indicate a signifi-cant annual increase of 0.9%. The North American population has been es-timated at 28 million birds (Rich *et al.*, 2004).

Dickcissel (*Spiza americana*)

Status: A local summer resident in the Great Plains east of the mountains, breeding in extreme eastern Montana and very locally in northeastern Wy-oming, and more extensively in eastern Colorado. A vagrant in montane ar-eas, and not yet reported from any of the montane parks.

Habitats and Ecology: This is a prairie-adapted species that breeds in grasslands having a combination of tall forbs, grasses, and shrubs, or in grassy mead-ows having nearby hedges or brushy fencerows.

Suggested Viewing Locations: In Wyoming, breeding mostly occurs east of a line from Sheridan to Torrington, and the amount of breeding may vary from year to year (Faulkner, 2010). The best populations occur in the western Black Hills (Crook County) and in the lower North Platte valley from Fort Laramie eastward (Scott, 1993). Localized regional breeding populations probably occur in eastern Colorado (easternmost counties), eastern Wyo-ming and eastern Montana.

Population: Classified as a Partners-in-Flight Watch List species of continental conservation importance (Rich *et al.*, 2004). National Breeding Bird Survey trend data indicate a nonsignificant annual population decrease of 0.2%. The North American population has been estimated at 22 million birds (Rich *et al.*, 2004).

Icterids (Blackbirds and their Relatives)

Bobolink (*Dolichonyx oryzivorus*)

Status: An uncommon summer resident in lower altitude grasslands and wet meadows throughout the region, and a rare migrant in most of the montane parks.

Habitats and Ecology: Breeding occurs in tallgrass prairies, ungrazed or lightly grazed mid-grass prairies, wet meadows, hayfields, retired croplands, and similar habitats. Scattered bushes or other singing posts in the territory add to its attractiveness.

Suggested Viewing Locations: In Wyoming, bobolinks are only local breeders, They are most common along the eastern slopes of the Bighorns, and in Creek County, but small populations occur elsewhere, such as at the National Elk Refuge (Faulkner, 2010).   They might be found in fields southwest of Sheridan, and on the Wagon Box Road west of Fort Phil Kearny (Scott, 1993). They have also been reported between Ranchester and Dayton, and between Dayton and the Tongue River canyon (Dorn and Dorn, 1990). They are also very local in Idaho, at locations such as around the west side of Cascade Reservoir, at Camas Prairie Centennial Marsh Wildlife Management Area near Fairfield, and in Teton Valley (Svingen and Dumroese, 1997). In northern Colorado they are found locally in irrigated hayfields and meadows, and breed locally west to Moffatt, Routt and Rio Blanco counties (Kingery, 1998).   The densest regional breeding populations probably occur in the native grasslands of eastern and central Montana (*e.g.*, Medicine Lake and Lee Metcalf national wildlife refuges).

Population: National Breeding Bird Survey trend data indicate a significant annual population decrease of 1.8%. However, trend estimates for the Rocky Mountains & Plains States region (U.S.F.W.S. Region 6) indicate a significant annual increase of 1.9%. The North American population has been estimated at 11 million birds (Rich *et al.*, 2004).

Red-winged Blackbird (*Agelaius phoeniceus*)

Status: A seasonal or permanent resident in suitable habitats throughout the region, including both lowlands and montane areas, and a relatively common breeder in all the montane parks.

Habitats and Ecology: Typical breeding habitats are wetlands ranging from deep marshes or the emergent vegetation zones of lakes and reservoirs through variably drier habitats including wet meadows, ditches, brushy patches in prairies, hayfields, and weedy croplands or roadsides. Wetlands with bulrushes or cattails are especially favored for nesting, but sometimes shrubs or other woody vegetation are used for nest sites. Outside the breeding season the birds often stray far from water, and seek grainfields, city parks, pasturelands, and other habitats offering food sources.

Suggested Viewing Locations: Perhaps the most abundant songbird in North America, this species can be found breeding in marshlands, ditches, and agricultural lands anywhere in the region. The most extensive dense regional breeding populations probably occur in the moister grasslands and marshes of eastern Montana.

Population: National Breeding Bird Survey trend data indicate a significant annual population decrease of 2.0%. The North American population (north of Mexico) has been estimated at 193.2 million birds (Rich *et al.*, 2004).

Western Meadowlark (*Sturnella neglecta*)

Status: A seasonal or permanent resident almost throughout the region, becoming rarer northwardly, and not known to nest in the higher mountain areas of Alberta, although common on the adjacent plains and foothills. Eastern meadowlarks are unreported for the region.

Habitats and Ecology: During the breeding season this species occupies mixed-grass to tallgrass prairies, wet meadows, hayfields, the weedy borders of croplands and retired croplands. to some extent short-grass prairies and sage prairies from about 3,000 feet elevation to mountain meadows as high as about 10,000 feet are used in southern parts of the region. In Colorado shortgrass prairies are the most common breeding habitat, followed closely by croplands (Kingery, 1998).

Suggested Viewing Locations: Probably Wyoming's most abundant bird, this meadowlark is abundant in all grasslands at lower altitudes. It is less common in the grasslands of western Wyoming, and also occurs in the juniper scrublands of southwestern Wyoming (Scott, 1994). The densest regional breeding populations probably occur in eastern Montana, eastern Wyoming and eastern Colorado.

Population: National Breeding Bird Survey trend data indicate a significant annual population decrease of 0.9%. The North American population (north of Mexico) has been estimated at 29,44 million birds (Rich *et al.*, 2004). Although still abundant, the species is declining nationally (0.9% annually), as are nearly all other North American grassland birds (Johnsgard, 2001).

Yellow-headed Blackbird (*Xanthocephalus xanthocephalus*)

Status: A summer resident in wetland habitats throughout almost the entire region; rarer in montane areas and not known to breed in most of the Canadian montane parks, although common on the nearby prairie marshes.

Habitats and Ecology: Restricted during the breeding seasons to relatively permanent marshes, the marsh zones of lakes, and the shallows of river impoundments where there are good stands of cattails, bulrushes, or phragmites. Although sometimes breeding in the same areas as red-winged blackbirds, yellow-headed blackbirds occupy the deeper areas adjacent to open water. In Colorado they breed at elevations up to 9,000 feet (Kingery, 1998).

Suggested Viewing Locations: Nearly all of Wyoming's cattail marshes that are deep enough to have some open water are likely to support this conspicuous blackbird. Large colonies occur in places such as Hutton Lake National Wildlife Refuge, Table Mountain Wildlife Unit, Ocean Lake and Loch Katrine (Dorn and Dorn, 1990). The most extensive dense regional breeding populations probably occur in eastern Montana (*e.g.,* Benton, Bowdoin. Medicine Lake and Ninepipe national wildlife refuges).

Population: National Breeding Bird Survey trend data indicate a significant annual population increase of 0.5%. The North American population has been estimated at 23 million birds (Rich *et al.*, 2004).

Rusty Blackbird (*Euphagus carolinus*)

Status: A local summer resident in woodland wetlands near the northern limits of the region, breeding south to Jasper National Park; rare in summer and breeding unproven for Banff National Park. Otherwise an irregular to uncommon migrant over much of the region, mainly east of the mountains.

Habitats and Ecology: During the breeding season this species is largely limited to wooded wetlands including alder-willow bogs, the brushy borders of lakes and slow-moving streams, receding muskegs, forest edges, and the borders of beaver ponds. Nests are usually placed over water, either in bushes or low conifers. On migration and during winter they use a wider variety of habitats, but typically roost in marshy or swampy areas.

Suggested Viewing Locations: In Wyoming, migrants are rare, and are mostly likely to be seen in eastern parts of the state. They are often found in flooded woodlands, sometimes walking over ice, apparently in search of frozen insects. Local breeding occurs in Jasper National Park and possibly other Canadian montane parks.

Population: The North American population has been estimated at two million birds. Classified as a Partners-in-Flight Watch List species of continental conservation importance (Rich et al., 2004). National Breeding Bird Survey trend data indicate a significant annual population decrease of 0.95%.

Brewer's Blackbird (*Euphagus cyanocephalus*)

Status: A summer resident virtually throughout the entire region, breeding in most habitats from plains to mountain meadows, and a fairly common breeder in nearly all the montane parks.

Habitats and Ecology: Low-stature grasslands are the primary breeding habitats of this species, including mowed or burned areas, farmsteads and residential areas, the edges of marshes, especially where scattered shrubs are present, aspen groves, the brushy banks of prairie creeks, and similar locations. Nesting occurs on the ground or in low shrubs such as sage and shrubs or fenceposts also serve as singing posts where they are available. Outside the breeding season a wider array of open habitats are used, including grainfields, orchards, berry farms, and similar agricultural lands. In Colorado, juxtaposed cropland/rural areas appear to be the ideal breeding habitats. They breed at altitudes from the plains up about 10,000 feet, but mostly in the foothills, intermountain valleys and along the Western Slope (Kingery, 1998). In the interior Northwest, this species breeds in mixed coniferous forests from the shrub-seedling to the old-growth forest stages of succession (Sanderson, Bull and Edgerton, 1980

Suggested Viewing Locations: This species is a common to abundant summer resident across Wyoming, in sage-dominated scrub, ranchlands, and agricultural lands up to about 9,500 feet (Faulkner, 2010). In Colorado they are especially common in the North Park region. The most extensive dense regional breeding populations are in Wyoming (*e.g.*, Hutton Lake National Wildlife Refuge, Greater Yellowstone region), Montana (National Bison

Range, C. R. Russell National Wildlife Refuge) and southern Idaho (Camas, Deer Flat and Grays Lake national wildlife refuges).

Population: National Breeding Bird Survey trend data indicate a significant annual population increase of 1.3%. The North American population (north of Mexico) has been estimated at 34.65 million birds (Rich *et al.*, 2004).

Common Grackle *(Quiscalus quiscula)*

Status: A summer resident in suitable plains or foothills habitats over most of the region west to eastern Idaho, but usually rare or absent from the montane parks.

Habitats and Ecology: Breeding habitats consist of woodland edges, areas partially planted to trees such as residential areas, farmsteads, shelterbelts, coniferous or deciduous woodlands of an open nature, woody shorelines around lakes, and riparian woodlands. Junipers, spruces, and other small and dense conifers are preferred for nesting, although hardwoods, shrubs, buildings, birdhouses, and even cattails are sometimes also used. In Colorado, about 80% of breeding birds used rural habitats, and the remainder used wooded riparian habitats, of various deciduous trees including cottonwoods (Kingery, 1998).

Suggested Viewing Locations: Probably all Wyoming's lower elevation cities, towns and villages have resident flocks of grackles, where they can be almost as much a nuisance as starlings. The densest regional breeding populations probably occur in eastern Montana, eastern and central Wyoming and eastern Colorado.

Population: National Breeding Bird Survey trend data indicate a nonsignificant annual population decrease of 1.0%. Trend estimates for the Rocky Mountains & Plains States region (U.S.F.W.S. Region 6) indicate a significant annual decline of 1.1%. The North American population has been estimated at 3.7 million birds (Rich *et al.*, 2004).

Great-tailed Grackle *(Quiscalus mexicana)*

Status: An expanding species now well established in most of Colorado at lower altitudes, and also present in southeastern Wyoming and southern Idaho.

Habitats and Ecology: Wet meadows, feedlots, and other areas where waste grain or other diverse foods are abundant attract these birds; less often they are attracted to cities and suburbs. In Colorado, breeding has occurred in marshes, evergreen trees, windbreaks, and scattered trees, often close to water or to agricultural activities (Kingery, 1998).

Suggested Viewing Locations: This expanding southern species was first reported from Colorado in 1970, and began nesting there by 1973. After arriving in Wyoming in 1989, the great-tailed grackle has become well established only in the southeastern corner of the state, where since 1998 it has bred locally in the Table Mountain marshes of Goshen County (Faulkner, 2010). By 2004 the densest regional breeding populations had developed in Colorado, but the species has exhibited a notably dynamic distribution as it moves west and north into the Pacific Northwest. Feed-lots are a good place to look for them, as are fast-food sites where waste food is often accessible.

Population: National Breeding Bird Survey trend data indicate a significant an-nual population decrease of 3.5%. The North American population (north of Mexico) has been estimated at 7.75 million birds (Rich *et al.,* 2004).

Brown-headed Cowbird (*Molothrus ater*)

Status: A summer resident throughout the region in most habitats; variably common in the montane parks and probably breeding in all.

Habitats and Ecology: Breeding occurs in a variety of woodland edge habitats, including brushy thickets, forest clearings, brushy creek-bottoms in prai-ries, aspen groves, sagebrush, desert scrub, agricultural lands, and open coniferous forests at lower altitudes (up to about 7,000 feet in southern parts of the region). In the interior Northwest, this species parasitizes birds breeding in mixed coniferous forests from the shrub-seedling to the mature forest stages of succession (Sanderson, Bull and Edgerton, 1980).

Suggested Viewing Locations: Wyoming's ubiquitous cowbird population is cen-tered at lower altitudes where cattle-grazing is common. Cowbirds can of-ten be seen feeding on insects around the feet of cattle and other ungulates. The densest and most extensive regional breeding populations probably oc-cur in eastern Montana, but the birds are nearly ubiquitous. In Colorado breeding birds were reported from 46 different habitat types, and have been implicated as parasitizing at least 59 host species (Kingery, 1998).

Population: National Breeding Bird Survey trend data indicate a significant an-nual population decrease of 1.0%. The North American population (north of Mexico) has been estimated at 50.96 million birds (Rich *et al.,* 2004).

Orchard Oriole (*Icterus spurius*)

Status: A local summer resident in eastern Montana and extreme eastern Wyo-ming, and a migrant farther south, with vagrants occasionally reaching the mountains.

Habitats and Ecology: Associated with lightly wooded riverbottoms, scattered trees in open country, shelterbelts, farmsteads, residential areas, and or-chards during the breeding season, and extending into sagebrush and ju-niper woodlands during the non-breeding season. In Colorado, these birds have been found breeding mostly in riparian woodlands, shelterbelts, and farmyard trees on the plains, these habitat types accounted for 96% of all reported breedings (Kingery, 1998). Nests are built in small to moderately large trees, from 5–70 feet above ground.

Suggested Viewing Locations: In Wyoming, breeding is now mostly confined to the region east of a line from Sheridan to Laramie, (Faulkner, 2010). Sites that are productive for seeing this species include Springer and Hawk Springs reservoirs and Rawhide Wildlife Unit in Goshen County, and the Laramie River bottomlands at Fort Laramie (Dorn and Dorn, 1990). The

Black Hills, Torrington, Goshen Hole, Pine Bluff and Cheyenne are also good locations (Scott, 1994). The densest regional breeding populations probably occur in eastern Montana, eastern Wyoming and eastern Colorado, where the birds are mostly distributed along riparian woodland corridors and plains reservoirs.

Population: National Breeding Bird Survey trend data indicate a nonsignificant annual population decrease of 0.3%. However trend estimates for the Rocky Mountains & Plains States region (U.S.F.W.S. Region 6) indicate a significant annual increase of 1.7%. The North American population (north of Mexico) has been estimated at 3.74 million birds (Rich *et al.*, 2004).

Bullock's Oriole (*Icterus bullockii*) and Baltimore Oriole (*Icterus galbula*)

Status: The Bullock's oriole is a local summer resident in wooded plains and foothills areas throughout most of the region, but becoming rare in the mountains, and absent from most of the northern montane parks. It breeds up to about 7,500 feet at the southern portions of the region. The Baltimore oriole (considered conspecific with Bullock's from 1983–1995, when they were collectively named the northern oriole) is limited to the eastern borders of Montana, Wyoming and Colorado. There it is in contact and sometimes hybridizes with the Bullock's oriole.

Habitats and Ecology: During the breeding season males of the Bullock's oriole especially favor riverbottom forests of willows and cottonwoods, but also occur in city parks, and on plains or foothill slopes and valleys with aspen, poplars, birches, and similar vegetation. In Colorado, mature native cottonwoods and exotic landscaping trees provide a major breeding habitat for both orioles, but all the Baltimore records came from only three habitat types, primarily cottonwoods plus some rural and urban habitats (Kingery, 1998). During summer and fall orioles are attracted to trees and bushes that provide berries.

Suggested Viewing Locations: The densest regional breeding populations of the Bullock's oriole occur in eastern Montana, eastern Wyoming and eastern Colorado. Some phenotypically pure Baltimore orioles may also breed in the easternmost parts of these states, although Faulkner (2010) reported that no evidence exists of breeding by phenotypically pure Baltimore orioles in Wyoming. In eastern Colorado, Baltimore-like individuals might be found at locations such as Bonny State Park and Tamarack State Wildlife Area (Kingery, 2007).

Population: National Breeding Bird Survey trend data indicate the Baltimore oriole had undergone a significant annual population decrease of 0.7%, and the Bullock's showed a significant annual population decrease of 0.9%. The North American population (north of Mexico) of Bullock's has been estimated at about 2.6 million, and of the Baltimore as six million birds (Rich *et al.*, 2004).

Scott's Oriole (*Icterus parisorum*)

Status: A local and rare to uncommon breeder along the western border of Colorado, and a very local breeder in southwestern Wyoming.

Habitats and Ecology: Associated with Utah juniper and pinyon-juniper woodlands in semidesert shrublands in plains and foothills. In Colorado these birds occur up to about 5,500 feet, in groves of sparse junipers and widespread yuccas and creosote bushes. They typically construct their nests with yucca leaves, and preferentially nest in these plants. Even when nesting in junipers or pinyons, the fibers obtained from the fibrous edges of yucca leaves are used for nest construction. In the Southwest the nest is placed under the live crown of the yucca, and below its drooping and bayonet-like leaf blades, providing the nest with visual and physical protection (Kingery, 1998).

Suggested Viewing Locations: In Wyoming, summer records have come from the vicinity of Anthill Reservoir on Powder Rim, in Sweetwater County (Faulkner, 2010). In Colorado, the records have been widely scattered, but birds were discovered in five breeding bird survey blocks in Montezuma County, west of Mesa Verde National Park. The other state records have been few and rather widely scattered (Kingery, 1998).

Population: National Breeding Bird Survey trend data indicate the Scott's oriole had undergone a nonsignificant annual population increase of 0.9%. The North American population (north of Mexico) has been estimated at about 800,000 birds (Rich *et al.*, 2004).

Finches

Gray-crowned, Black and Brown-capped Rosy-finches (*Leucosticte* spp.)

Status: All three rosy-finch species (sometimes regarded as subspecies) are limited to alpine areas of high mountains in the region, probably occurring on all such montane areas, including all the montane parks. In northern areas the birds (gray-crowned rosy-finch, *L. tephrocotis*) have gray crowns and reddish brown back and breast colors. In birds from central Idaho and west-central Montana south through central Wyoming the nape color is still gray, but the back and breast are dusky brown (black rosy-finch, *L. atrata*). Finally, in Colorado and adjacent southeastern Wyoming the crown and nape are dark brownish, and the breast and back are grayish brown, becoming reddish on the belly (brown-capped rosy-finch, *L. australis*). Hybrids occur in some areas (the Seven Devils area of western Idaho and the Bitterroot Mountains of eastern Idaho).

Habitats and Ecology: During the breeding season these birds inhabit cirques, talus slopes, alpine meadows with nearby cliffs, and adjacent snow and glacial surfaces (where foraging for frozen insects is common). Nesting is done in cliff crevices or among talus rocks. During fall and winter the birds move to lower elevations, and to habitats that include mountain meadows, grasslands, sagebrush, and agricultural lands.

Suggested Viewing Locations: In Wyoming, the black rosy-finch breeds in the alpine zone of the Greater Yellowstone region, and in the Bighorn Mountains. It has also been reported during summer in the Snowy Range, where the brown-capped rosy-finch is known to be the local breeder (Faulkner, 2010). Black rosy-finches nest in the alpine zone of the Teton and Wind River ranges and the Beartooth Plateau. From Jackson they can be easily seen in summer by taking the tramway at Teton Village to the top of Rendezvous Mountain, and can also be seen along the foothills of western Wyoming during winter. (Scott, 1993). McEneaney (1988) judged the Yellowstone population to be in the hundreds during the late 1980's, with the best observation area during summer months for the black rosy-finch being around Mount Washburn. In Montana, black rosy-finches can be seen in the alpine zone (Beartooth Pass) of the Red Lodge–Cooke City Highway (U.S. Hwy. 212). In Idaho the black rosy-finch my be found during summer at 11,000–12,000 feet on Borah and Leatherman peaks, or on Johnson Pass in the Sun Valley area (Svingen and Dumroese, 1997). Brown-capped rosy-finches can be seen in Wyoming during summer at Libby Flats, Medicine Bow peak, and along Brooklyn Ridge in the Medicine Bow Mountains of Albany County (Dorn and Dorn, 1990). In Colorado the brown-capped rosy-finch is widespread in alpine areas, and may be seen during summer along Trail Ridge Road in Rocky Mountain National Park, along the Mount Evans Highway (Summit Lake), at Loveland Pass (Pass Lake) and on Guanella Pass near Georgetown. Alpine areas in the vicinity of Breckenridge are also good viewing sites for the brown-capped form. The gray-crowned rosy-finch occurs at Logan Pass in Glacier National Park, and the nearby Hidden Lake Trail (McEneaney, 1993), as well as in the alpine zone of the Canadian montane parks durign summer. All three species of rosy-finches may be found wintering in Colorado's foothills, and both gray-crowned and black rosy-finches also disperse widely during winter over the lowlands of Idaho, Montana and Wyoming.

Population: National Breeding Bird Survey trend data are not available for these species. The black and brown-capped rosy-fiches are classified as Partners-in-Flight Watch List species of continental conservation importance. The North American population has been estimated as 20,000 birds for the black rosy-finch, of which 100% breed in the Intermountain West Avifaunal Biome. There is an estimated population of 45,000 for the brown-capped rosy-finch, of which 100% breed in the Intermountain West Avifaunal Biome. There is an estimated 200,000 gray-crowned rosy-finches in North America (Rich *et al.*, 2004).

Pine Grosbeak *(Pinicola enucleator)*

Status: A permanent resident in coniferous forests throughout the region, including all of the montane parks, and probably breeding in all of them.

Habitats and Ecology: Breeding occurs in the subalpine levels of the coniferous forest, primarily the alpine fir, Engelmann spruce zone. Nesting usually occurs in such conifers, especially in open or scattered woods near meadows or streams. In Colorado, spruce–fir forest are easily the most common breeding habitat (Kingery, 1998). In the interior Northwest, this species breeds in mixed coniferous forests in the mature and old-growth forest stages of succession (Sanderson, Bull and Edgerton, 1980). Outside the breeding season the birds descend to lower conifer zones, especially the pinyon–juniper zone, where the birds often feed on pinyon nuts. Some berries, grains, and other food sources are also used, but conifer seeds are primarily eaten.

Suggested Viewing Locations: Breeding in Wyoming occurs at high elevations in the Greater Yellowstone region, the Bighorn Mountains, and the Medicine Bow–Sierra Madre mountains. Breeding in the Bear Lodge Mountains and Laramie Range is less certain (Faulkner, 2010). During summer, Wyoming sites that might have pine grosbeaks include the west side of Blackhall Mountains in the Sierra Madre Mountains (Carbon County), Brooklyn Lake in the Medicine Bow Mountains (Albany County) and the Brooks Lake area (Fremont County)(Dorn and Dorn, 1990). They also can found at subalpine altitudes in the Tetons, Bighorns, and other ranges, and in Yellowstone National Park (Scott, 1993). In Idaho the birds are uncommon in the Upper Salmon drainage at altitudes of 6,000–9,400 feet (Roberts, 1992). The densest regional breeding populations probably occur in Colorado, where the San Juan Range has fairly large numbers of breeding birds, as do subalpine areas of the central and northern ranges (Kingery, 1998).

Population: National Breeding Bird Survey trend data indicate a nonsignificant annual population increase of 1.9%. The North American population (north of Mexico) has been estimated at 2.2 million birds (Rich *et al.*, 2004).

Purple Finch (*Carpodacus purpureus*)

Status: A local summer resident at the northern edge of the region, breeding south uncommonly to Jasper National Park and occasionally to Banff National Park. Farther south it is a migrant and wintering visitor in Montana, an occasional winter visitor In Wyoming, and an accidental winter visitor in Colorado.

Habitats and Ecology: Breeding in this species occurs in natural conifer forests, mixed forests, and conifer plantings, especially where moist and shaded habitats occur. In the interior Northwest, this species breeds in mixed coniferous forests from the pole-sapling to the old-growth forest stages of succession (Sanderson, Bull and Edgerton, 1980). Its preference seems to be for mixed forests, with the birds nesting in conifers, but feeding in deciduous trees. Buds and blossoms of a variety of broad-leaved trees are favored in spring, while in summer the birds consume a variety of berries, fruit, and insects. During the winter period they eat a variety of weed and grass seeds, and thus have a broad winter habitat distribution.

Suggested Viewing Locations: Purple finches are rare fall, winter and spring visitors to Wyoming, especially along the northwestern and western edges of the state. Breeding in this region occurs only in the Canadian montane parks.

Population: National Breeding Bird Survey trend data indicate a significant annual population increase of 1.2%. The North American population has been estimated at three million birds (Rich *et al.*, 2004).

Cassin's Finch (*Carpodacus cassinii*)

Status: A local summer or permanent resident in coniferous forests of the region north to extreme southern British Columbia and adjacent Alberta (breeding at Watertown Lakes National Park and reported rarely north to Jasper, but not known to breed there).

Habitats and Ecology: Breeding typically occurs in open, rather dry coniferous forests, including ponderosa pine forests, with the nests placed at considerable heights in large conifers. Generally it occurs at rather higher altitudes than do either the house finch or the purple finch, sometimes almost to timberline. In Colorado, nesting occur as altitudes of 6,000–11,000 feet, most frequently in subalpine spruce–fir forests, but with some nesting in mid-elevation confers, pinyon–juniper woodlands, aspen groves, and riparian trees (Kingery, 1998). In the interior Northwest, this species breeds in mixed coniferous forests from the young forest to the old-growth forest stages of succession (Sanderson, Bull and Edgerton, 1980). In central Idaho it breeds from 6,000 to more than 9,000 feet, usually in subalpine fir or lodgepole pine forests (Roberts, 1992). Throughout the year this species is primarily vegetarian, feeding on buds, berries, and seeds, especially those of conifers.

Suggested Viewing Locations: In Wyoming, this finch breeds commonly in all the major mountain ranges (Faulkner, 2010). Colorado birds may be found throughout the mountains and over much of the pinyon–juniper woodlands, but with the greatest abundance in the north central region, including the Flat Tops, Elks, Gore and Park ranges, and also the Grand Mesa (Kingery, 1998). The densest regional breeding populations probably occur in southern Idaho (*e.g.*, Grays Lake National Wildlife Refuge), the Greater Yellowstone region, and western Montana (Red Rock Lakes National Wildlife Refuge and Glacier National Park).

Population: National Breeding Bird Survey trend data indicate a significant annual population decrease of 2.2%. Trend estimates for the Rocky Mountains & Plains States region (U.S.F.W.S. Region 6) indicate a significant annual decline of 3.8%. The North American population (north of Mexico) has been estimated at 1.88 million birds of which 86% breed in the Intermountain West Avifaunal Biome (Rich *et al.*, 2004).

House Finch (*Carpodacus mexicanus*)

Status: A local summer resident from northern Idaho to southeastern Wyoming, mainly at lower altitudes, including plains and foothills up to 9,000 feet at the southern end of this region. Generally rare or absent from the montane parks, but an uncommon nester in Rocky Mountain National Park.

Habitats and Ecology: Now generally associated with human habitations over most of its range, nesting on buildings in such areas. Otherwise it nests in open woods, riverbottom woodlands, scrubby desert or semi-desert vegetation such as sagebrush, and tree plantings. In Colorado the birds breed from the plains up to at least 8,700 feet, and possibly to 10,000 feet (Kingery, 1998). In the interior Northwest, this species breeds in mixed coniferous forests from the shrub-seedling to the mature forest stages of succession (Sanderson, Bull and Edgerton, 1980). Deciduous underbrush, preferably close to water, is favored over dense coniferous woods, and sources of seeds, berries, or fruits are also needed throughout the year.

Suggested Viewing Locations: House finches are common town and city residents across Wyoming, at altitudes up to about 8,000 feet (Faulkner, 2010). Cities such as Cheyenne, Douglas, Cody, Rawlins and Sheridan have good populations (Dorn and Dorn, 1990). The densest regional breeding populations probably occur in Colorado (especially along the Front Range and the counties borderingUtah), southern Idaho (*e.g.,* Camas and Deer Flat national wildlife refuges) and Montana (eastern plains cities).

Population: National Breeding Bird Survey trend data indicate a nonsignificant annual population increase of 0.7%. Trend estimates for the Rocky Mountains & Plains States region (U.S.F.W.S. Region 6) indicate a significant annual increase of 6.9%. The North American population (north of Mexico) has been estimated at 16.59 million birds (Rich *et al.,* 2004).

Red Crossbill *(Loxia curvirostra)*

Status: A local resident in coniferous forest areas throughout the region, including all the montane parks, where it is a probable breeder in all.

Habitats and Ecology: Breeding is associated with coniferous forest habitats, especially those of pines, including ponderosa, lodgepole, and pinyon, but nesting in the region has also been observed in Engelmann spruces and subalpine firs, at elevations from 4,000–10,000 feet or more. Breeding in the Rocky Mountain region is associated with the higher levels of coniferous forests, but non-breeding birds often frequent the pinyon zone. In Colorado, crossbills with different bill shaoes and call types specialize on eating different conifer seeds (Douglas-fir, ponderosa pine, and lodgepole pine) during late winter and spring, but during nesting spruce–fir forests had the highest frequency of habitat use during breeding bird atlas surveys (Kingery, 1998). In Wyoming, five call-types (out of the nine known in North America) have been found; some of these different call-type populations may act as biologically distinct species (Smith and Beckman, 2007; Faulkner, 2010). In the interior Northwest, crossbills breed in mixed coniferous forests in mature and old-growth forest stages of succession (Sanderson, Bull and Edgerton, 1980). Although usually found in conifers, they also feed on ripe box elder seeds in late summer (Scott, 1993).

Suggested Viewing Locations: In Wyoming, breeding occurs in all the major mountain ranges (Faulkner, 2010), including the Black Hills. The densest regional breeding populations probably occur in northern Idaho, western Montana and the Greater Yellowstone region.

Population: National Breeding Bird Survey trend data indicate a significant annual population decrease of 1.3%. Trend estimates for the Rocky Mountains & Plains States region (U.S.F.W.S. Region 6) indicate a significant annual decline of 1.2%. The North American population has been estimated at 5.7 million birds (Rich *et al.*, 2004).

White-winged Crossbill *(Loxia leucoptera)*

Status: A local resident in coniferous forests of the northernmost part of the region (Banff and Jasper National Park, Cypress Hills). In Idaho, breeds in the Selkirk, Cabinet and Purcell mountains (Svingen and Dumroese, 1997). In Montana, it possibly breeds in Glacier National Park and the Whitefish Range west of the Park, and probably is resident through the mountains south almost to Yellowstone National Park. There are summer observations for many Wyoming locations, and reported breeding in the Jackson vicinity.

Habitats and Ecology: During the breeding season associated with coniferous forests and mixed forests containing spruces and tamarack, usually in the subalpine zone. Spruces and tamaracks seem to be this species' prime food sources, as their beaks are too weak to handle the larger cones of pines. Although nesting occurs most commonly during spring and summer or early fall, like the red crossbill it can apparently occur almost any time a rich seed source becomes available. In spring, catkins of aspens and poplars are sometimes eaten, and large weed seeds may be eaten during fall and winter.

Suggested Viewing Locations: In Wyoming, white-winged crossbills are rare. Three are only a few confirmed reports of breeding, which probably occurs from about 9,000 feet to timberline, in subalpine forests of Engelmann spruce (Faulkner, 2010). They are too rare and irregular through the entire region to predict reliable birding areas, but are apparently most likely to be seen in Idaho and Montana.

Population: National Breeding Bird Survey trend data are not available. The North American population has been estimated at 20.5 million birds (Rich *et al.*, 2004).

Common Redpoll *(Acanthis flammea)*

Status: A local wintering migrant almost throughout the region, both in montane areas and in plains or foothills, but probably commoner at lower altitudes in winter. The nearest breeding records are from central Alberta (Edmonton), but regular breeding occurs along the northern edges of the prairie provinces.

Habitats and Ecology: Breeding typically occurs in subarctic forests, typically nesting in dwarf spruces or in thickets of willows and alders. In the Rocky Mountain region the birds are associated with such open habitats as desert scrub, sagebrush, and grasslands. They also visit cities during winter to eat the seed cones of birches, visit bird feeders, and seek out weedy patches.

Suggested Viewing Locations: In Wyoming, and elsewhere in the region redpolls are migrants and winter visitors. They are mostly seen in the southeastern part of Wyoming, and regularly can be seen at bird feeders.

Population: National Breeding Bird Survey trend data are not available. The North American population has been estimated at 29.1 million birds (Rich *et al.*, 2004).

Hoary Redpoll (*Carduelis homemanni*)

Status: A rare wintering visitor in the region, with the nearest breeding areas in northeastern Manitoba, along the coast of Hudson Bay. In Alberta the species is generally but not invariably less common than the common redpoll in wintering flocks. In Montana it has been reported from about half as many latilongs as has the common redpoll. In Wyoming, and Idaho there is roughly a five-to-one ratio in reported latilong occurrences for the common redpoll relative to the hoary redpoll. It is apparently extremely rare in Colorado.

Suggested Viewing Locations: This redpoll is a very rare winter visitor to Wyoming, and mostly appears among flocks of common redpolls, especially in the northern parts of the state. The same is probably true of Montana and Idaho.

Population: National Breeding Bird Survey trend data are not available. The North American population has been estimated at 13 million birds (Rich *et al.*, 2004).

Pine Siskin (*Spinus pinus*)

Status: A local resident in coniferous forests virtually throughout the region, including all the montane parks, where common to abundant, and probably breeding in all.

Habitats and Ecology: Breeding occurs in coniferous or mixed forests, and rarely in deciduous woodlands. Nesting preferentially occurs in conifers of almost any type, but has also been observed in cottonwoods, lilacs, and willows in the Rocky Mountain region. In Colorado, breeding siskins make use of spruce–fir forests more often than pines during breeding; ponderosa, lodgepole and pinyon pines comprised only about 20% of observed breeding habitats, although conifers collectively accounted for 70% of the total (Kingery, 1998). In the interior Northwest, this species breeds in mixed coniferous forests from the pole-sapling to the old-growth forest stages of succession (Sanderson, Bull and Edgerton, 1980). Their foods are mainly conifer seeds, but also may include those of alders, birches, or various weeds, and they seasonally feed on flower buds and insects.

Suggested Viewing Locations: In Wyoming, this is the most common breeding finch, nesting in coniferous forests of all the major mountain ranges during summer (Faulkner, 2010), and at bird feeders in towns and cities dur-

ing winter (Dorn and Dorn, 1990). The densest regional breeding popula-
tions probably occur in the mountains of Idaho and western Montana, the
Greater Yellowstone region, and southwestern Colorado (San Juan and West
Elk Mountains).

Population: National Breeding Bird Survey trend data indicate a significant annual population decrease of 2.9%. Trend estimates for the Rocky Mountains & Plains States region (U.S.F.W.S. Region 6) indicate a significant annual decline of 2.8%. The North American population (north of Mexico)
has been estimated at 21.34 million birds (Rich *et al.*, 2004).

Lesser Goldfinch (*Spinus psaltria*)

Status: A very local summer resident in the southern part of the region, breed-
ing in southwestern Colorado (Colorado National Monument, Mesa Verde
National Park)and north to southeastern Wyoming along the foothills of the
Front Range,

Habitats and Ecology: Breeding occurs in sagebrush and riparian thicket areas,
as well as where scrub oaks merge with ponderosa pines. In Colorado it
has been found breeding over a broad altitudinal range but is common at
the lower elevations in oak-pinyon and pinyon–juniper woodlands. In Colorado breeding surveys, riparian cottonwood forests and pinyon–juniper
woodlands were the two most frequently reported breeding habitat types
(Kingery, 1998). It also nests commonly in cities and suburbs.

Suggested Viewing Locations: In Wyoming, this species breeds in the southwestern counties (Uinta, Sweetwater) and the southeastern corner of the
state (probably Albany, Laramie, Converse Platte and Goshen counties)
(Faulkner, 2010). The birds have most often been seen between Cheyenne
and Laramie (Scott, 1993). The densest regional breeding populations probably occur in southern and southwestern Colorado (*e.g.*, Carrizo Unit, Comanche National Grassland. Colorado River State Park).

Population: National Breeding Bird Survey trend data indicate a nonsignificant annual population decrease of 0.1%. The North American population (north of Mexico) has been estimated at 15.5 million birds (Rich *et al.*,
2004).

American Goldfinch (*Spinus tristis*)

Status: A seasonal or permanent resident almost throughout the entire region,
but absent from the northernmost montane areas, and apparently only a
rare breeder in the montane parks.

Habitats and Ecology: Breeding occurs in open grazing country, especially
where thistles are abundant, or where cattails are to be found. In Colorado, surveys produced a total of 26 different habitat codes used by breeding goldfinches, about two-thirds of the total were from riparian forests and
rural woodlots or shelterbelts (Kingery, 1998). The seeds of thistles and
other composites are used for feeding the young, and the "down" of thistles or cattails is used in nest construction. Riparian woodlands near weed-

infested fields provide an ideal nesting situation. During winter the birds range widely over weedy fields and farmlands, and often visit urban bird feeders.

Suggested Viewing Locations: In Wyoming, this is a common breeder throughout the state, up to 8,000 feet (Faulkner, 2010). It is most common eastwardly, and least common in the northwest. Some of the better locations to find them in summer include LAK Reservoir (Weston County), the North Platte River near Guernsey, the Green River, the Tongue River Canyon (Sheridan County) and the Yellowtail Reservoir (Big Horn County)(Dorn and Dorn, 1990). The densest regional breeding populations probably occur in eastern Montana, along riparian corridors.

Population: National Breeding Bird Survey trend data indicate a nonsignificant annual population decrease of 0.01%. The North American population has been estimated at 24 million birds (Rich *et al.*, 2004).

Evening Grosbeak (*Coccothraustes vespertina*)

Status: A local resident in coniferous forests almost throughout the region. Present in all the montane parks and probably breeding in all of them.

Habitats and Ecology: During the breeding season this species is primarily associated with mature coniferous forests, although nesting has been observed in riparian willow thickets and even in city parks and orchards. In Colorado, surveys indicated that coniferous forests, especially ponderosa pine forests, are the favorite breeding habitat there, where the long needles of this pine help camouflage nests (Kingery, 1998). In the interior Northwest, this species breeds in mixed coniferous forests in mature and old-growth forest stages of succession (Sanderson, Bull and Edgerton, 1980). Nesting in elms, maples, and box elders has also been reported. During the breeding season these grosbeak feed on insects, including spruce budworm larvae, which often reach high populations during outbreaks. During fall and winter the birds often occur in flocks that feed on such large and nutritious seeds as maples, ashes, and sunflowers. It might appear anywhere in the region during winter, when it regularly visits bird feeders.

Suggested Viewing Locations: Evening grosbeaks breed in Wyoming's northwestern mountain ranges, including Wyoming's national parks. Nesting records are few, but most summer reports are from coniferous and conifer-aspen forests at middle altitudes of Wyoming's major mountain ranges, where breeding is presumed to occur (Faulkner, 2010). The densest regional breeding populations probably occur in western Montana (*e.g.*, Glacier National Park; Swan River National Wildlife Refuge) and Idaho (Boise National Forest, McCrosky State Park).

Population: National Breeding Bird Survey trend data indicate a nonsignificant annual population increase of 1.5%. The North American population (north of Mexico) has been estimated at 5.7 million birds (Rich *et al.*, 2004).

Old World Sparrows

House Sparrow (*Passer domesticus*)

Status: A local permanent resident throughout the region in human, associated habitats. Generally locally common in the montane parks around developed areas, but rare or absent in more remote habitats.

Habitats and Ecology: Associated throughout the year with humans, and breeding occurs in cities, suburbs, farmsteads, ranches, developed campgrounds, *etc.* Nesting is usually done on artificial structures such as buildings that offer cavities or crevices, such as vine-covered buildings, billboard braces, birdhouses, or old nests of other species. Nesting also occurs in tree cavities and birdhouses.

Suggested Viewing Locations: This introduced species is abundant almost everywhere in the region.

Population: National Breeding Bird Survey trend data indicate a nonsignificant annual population decrease of 2.5%.

Appendix 1. Status and Abundance of Birds in Nine Rocky Mountain Parks

	United States Parks				Canadian Parks			
	Rocky Mtn.	Grand Teton	Yellowstone	Glacier	Waterton L.	Kootenay	Yoho	Banff/Jasper
Tundra Swan	M	M	M	M	M	V	M	M
Trumpeter Swan		BR(4)	BR(4)	M		V		
Snow Goose	V	M	M	M	M	V	M	M
Canada Goose	M	BR(4)	BR((4)	BR(4)	S(4)	M	S(3)	S(4)
Wood Duck	V	M	S(1)	S (3)	M		M	S(1)
Gadwall		BR(2)	BR(2)	S(3)	S(1)		M	M
American Wigeon	M	S(1)	S(2)	s (3)	s (3)	M	M	S(1)
Mallard	BR(4)	BR(4)	BR(4)	BR(5)	S(4)	S(3)	S(4)	R(4)
Blue-winged Teal	S(2)	S(2)	S(2)	s(3)	S(3)	S(3)	s(3)	S(2)
Cinnamon Teal	V	S(1)	S(2)	s(3)	S(1)		M	s(2)
Northern Shoveler	S	S (1)	S(1)	s(S)	M	M	M	M
Northern Pintail	M	BR(2)	R(2)	BR(4)	s(3)	s(3)	s(3)	S(2)
Green-winged Teal	s(1)	BR(1)	BR(2)	s(3)		S(3)	S(3)	S(2)
Canvasback	M	S(1)	M	M	M		M	M
Redhead	M	S(1)	S(1)	M	M		M	M
Ring-necked Duck	M	S(4)	S(2)	S(3)	M	S(3)	S(3)	S(O)
Greater Scaup		V	M	V	M			S(O)
Lesser Scaup	M	s(2)	S(4)	S(3)	M	M	M	S(O)
Harlequin Duck		S(2)	S(2)	S(3)	S(3)	S(3)	S(3)	S(2)
Long-tailed Duck	V		M		M			M
Surf Scoter		V		M				M
White-winged Scoter	V	V	V		M	M	M	M
Common Goldeneye	M	r	BR(4)	BR(4)	M	M	M	M
Barrow's Goldeneye	M	BR(4)	BR(4)	BR(4)	BR(3)	s(3)	S(3)	S(3)
Bufflehead	M	BR(2)	BR(2)	BR(4)	BR(3)	S(3)	S(3)	S(2)
Hooded Merganser	M	M	M	S	M		M	M
Common Merganser	BR(4)	BR(4)	BR(4)	BR(5)	S(4)	S(1)	S(3)	S(2)
Red-breasted Merganser	M	M	M	S(3)	M	M	M	M
Ruddy Duck	M	S(2)	S(2)	s(3)	M	M	M	M

Species									
Gray Partridge	BR(2)								
Chukar	V								V
Ring-necked Pheasant		R(1)		V					V
Ruffed Grouse	BR(4)	BR(3)		BR(1)	BR(5)	BR(4)	BR(4)	BR(3)	BR(3)
Greater Sage-Grouse	BR(4)	BR(2)			BR(2)				
Spruce Grouse		?			BR(5)	BR(3)	BR(3)	BR(3)	BR(2)
White-tailed Ptarmigan	BR(4)	V		V	BR(4)	BR(3)	BR(3)	BR(3)	R(3)
Dusky Grouse	BR(4)	BR(4)		BR(4)	BR(4)	BR(3)	BR(3)	BR(3)	BR(2)
Sharp-tailed Grouse		R(1)		R(1)	R(1)				
Common Loon	M	S(1)		S(1)	S(4)	s(3)	s(1)	S(3)	S(3)
Pied-billed Grebe	S(1)	S(3)		S(2)	s(1)	S(3)	s(1)	S(3)	S(1)
Horned Grebe		M		S(1)	s	M	M	M	s
Red-necked Grebe		M		V	s(3)	s(3)	M	M	S(2)
Eared Grebe		M		S(1)	s(4)	M	M	M	M
Western/Clark's Grebe	s(1)	S(1)		S(1)	S(4)	M	M	M	M
American White Pelican	s(1)	s(2)		S(4)		V			
Double-crested Cormorant		s(3)		S(4)	s(1)	V			
American Bittern	s(1)	S(2)		S(1)	S(1)	M	M	M	S(1)
Great Blue Heron	M	S(4)		S(4)	S(4)	s(3)	M	M	s(1)
Snowy Egret		M		s(1)	s(1)	s(3)	s(1)		
White-faced Ibis	M	M		V					
Turkey Vulture	s(3)	M(1)		M(1)	M(1)	M			V

Seasonal Status Symbols: BR: Breeding resident; R: resident, breeding assumed but unproven; S: breeding summer resident; s: summer resident, breeding unproven; M: migrant, including wintering visitors; V: vagrant, out of normal range.

Abundance Symbols: (5): Abundant; (4): Common; (3): Uncommon; (2): Occasional; (1): Rare. Information on Yoho, Kootenay & Watertown Lakes was based on overall status checklists; seasonal occurrence attributions for these parks are judgments of the author. Species known as no more than vagrants in any of these parks are excluded; for a complete list see Johnsgard (1986). Some changes on park status since 1986 have been incorporated.

Appendix 1. Status and Abundance of Birds in Nine Rocky Mountain Parks (continued)

	United States Parks				Canadian Parks			
	Rocky Mtn.	Grand Teton	Yellowstone	Glacier	Waterton L.	Kootenay	Yoho	Banff/Jasper
Osprey	S(3)	S(4)	S(4)	S(4)	S(1)	S(1)	s(4)	S(3)
Bald Eagle	M	BR(4)	BR(2)	BR(3)	S(3)	s(1)	M(3)	BR(O)
Northern Harrier	R(4)	BR(2)	BR(2)	s(4)	S(3)	M	M	s(2)
Sharp-shinned Hawk	R(3)	S(2)	BR(2)	BR(4)	S(1)	s(3)	s(3)	BR(1)
Cooper's Hawk	BR(3)	S(3)	S(2)	BR(3)	S(1)	s(1)	S(3)	S(1)
Northern Goshawk	BR(3)	BR(3)	BR(2)	BR(3)	BR	R(1)	R(1)	BR(1)
Swainson's Hawk	s(1)	S(4)	S(4)	s(3)	s(3)	V	V	V
Red-tailed Hawk	BR(4)	BR(4)	R(4)	S(4)	S(3)	S(4)	S(4)	S(3)
Ferruginous Hawk	M	S(1)	S(1)	S(1)	M		V	
Rough-legged Hawk	M	M	M	M	M		M	M
Golden Eagle	BR(3)	BR(2)	BR(2)	BR(4)	BR(3)	R(3)	R(3)	BR(3)
American Kestrel	R(4)	S(4)	S(4)	S(4)	S(3)	S(3)	S(3)	S(3)
Merlin	S(1)	M	M	s(1)	s(3)	S(1)	s(3)	s(1)
Peregrine Falcon	BR(1)	S(1)	S(1)	M	V	s(1)		M
Prairie Falcon	M(3)	R(2)	S(1)	R(1)	s(1)			M
Virginia Rail	S(1)	M	V				V	V
Sora	S(3)	S(2)	S(2)	S(3)	s(1)	s(3)	s(3)	S(2)
American Coot	BR(3)	S(2)	S(4)	S(4)	M(4)	S(4)	M(4)	s(2)
Sandhill Crane	M	S(3)	S(2)	M				V
Black-bellied Plover		M	M	M			V	
Semipalmated Plover		M	M	M	M		V	
Killdeer	R(4)	S(4)	S(4)	BR(4)	S(3)	s(3)	S(4)	M
Mountain Plover		M	M					S(2)
Black-necked Stilt		M						
American Avocet	M	M	S(1)	M	M		V	V

Species	1	2	3	4	5	6	7	8	9
Spotted Sandpiper	S(4)	S(4)	S(4)	S(4)	S(4)	S(4)	S(4)	S(4)	S(4)
Solitary Sandpiper	M	M	M	M	M	S(1)	s	s	S(3)
Greater Yellowlegs	M	M	M	M	M	M			S
Willet	s(1)	s(1)	s(1)	s(1)	s(1)				
Lesser Yellowlegs	M	M	M	M	M	M	M	M	V
Upland Sandpiper				M	M		V	V	V
Long-billed Curlew	s	S(3)	S(1)						V
Marbled Godwit	M	M	M	M	M	M			
Sanderling		M	M				V	V	V
Semipalmated Sandpiper	M	M			M		V	V	
Western Sandpiper	M	M			M		M		M
Least Sandpiper		M	M	M	M	M	M	M	M
Baird's Sandpiper	M	M	M		M	M	M	M	V
Pectoral Sandpiper		M	M	M			M	M	
Stilt Sandpiper		M					M	M	
Short-billed Dowitcher					M	M	M	M	M
Long-billed Dowitcher	M	M		M			M	M	M
Wilson's Snipe	BR(3)	BR(4)	S(1)	S(4)	s(3)	s(3)	S(4)	S(4)	S(2)
Wilson's Phalarope	s	S	S	M	S	M	M	M	s
Red-necked Phalarope	M	M	M	M	M	M	M	M	s
Parasitic Jaeger	V	V			M				V
Franklin's Gull	M	M	M	s(3)	M	V	V		
Bonaparte's Gull	M	M	M	M	M	V	V	V	V
Mew Gull							M	M	V
Ring-billed Gull	R(1)	M	S(2)	s(4)	s(4)	M	M	s(3)	s(3)
California Gull	M	M	S(4)	s(4)	s(4)	M	M	s(3)	s(3)
Herring Gull	M		M	M	M		M	M	M
Sabine's Gull			V					V	V

Seasonal Status Symbols: BR: Breeding resident; R: resident, breeding unproven; S: breeding summer resident; s: summer resident, breeding assumed but unproven; M: migrant, including wintering visitors; V: vagrant, out of normal range. Abundance Symbols: (5): Abundant; (4): Common; (3): Uncommon; (2): Occasional; (1): Rare. Information on Yoho, Kootenay & Watertown Lakes was based on overall status checklists; seasonal occurrence attributions for these parks are judgments of the author. Species known as no more than vagrants in any of these parks are excluded; for a complete list see Johnsgard (1986). Some changes on park status since 1986 have been incorporated.

Appendix 1. Status and Abundance of Birds in Nine Rocky Mountain Parks (continued)

	United States Parks				Canadian Parks			
	Rocky Mtn.	Grand Teton	Yellowstone	Glacier	Waterton L.	Kootenay	Yoho	Banff/Jasper
Caspian Tern		M	S(1)	V			V	
Common Tern	s	M	M	M				M
Forster's Tern		M	M	M	M			V
Black Tern	M	M	S(1)	S(3)	S(4)		M	M
Rock Pigeon		R(1)	BR(1)	M	R(1)	V	M	BR(2)
Band-tailed Pigeon	S(3)	M		V	V			
Mourning Dove	BR(3)	S(2)	S(2)	s(4)	s(3)	M	s(3)	S(2)
Eurasian Collared-Dove		R(2)	V					
Black-billed Cuckoo	V	V	V		s(1)			
Western Screech-Owl	R(3)	R(1)	BR(1)	BR(1)	M			V
Great Horned Owl	BR(3)	BR(2)	BR(4)	BR(4)	BR(3)	BR	R(1)	R(2)
Snowy Owl		V	V	M				V
Northern Hawk-Owl			V	V			M	R(1)
Northern Pygmy-Owl	BR(1)	R(1)	BR(1)	BR(4)	BR(1)	BR(1)	BR(3)	BR(1)
Burrowing Owl		s(1)	M					
Barred Owl		V		BR(3)		BR(1)	BR(3)	BR(1)
Great Gray Owl	BR(1)	BR(2)	BR(2)	BR(3)	R(1)	BR(1)	V	BR(1)
Long-eared Owl	BR(2)	BR(2)	BR(1)	s(1)		BR(1)	V	V
Short-eared Owl	V	BR(2)	s(1)	M	V	M	M	S(1)
Boreal Owl		R(1)	V	BR(1)	R(1)	R(1)	R(3)	BR(2)
Northern Saw-whet Owl	BR(3)	R(1)	BR(1)	BR(1)				BR(2)
Common Nighthawk	s(3)	S(4)	S(4)	S(4)	s(4)	S(4)	s(1)	S(2)
Common Poor-will	S(2)	V						
Black Swift	S(3)		V	S(1)	s(3)	s(3)	s(3)	S(3)
Vaux's Swift				S(4)	V	V	s(1)	

Species	Park 1	Park 2	Park 3	Park 4	Park 5	Park 6	Park 7	Park 8	Park 9
White-throated Swift	S(3)	s(1)	S(2)	M				S(1)	S(3)
Ruby-throated Hummingbird									
Black-chinned Hummingbird		M		V	s(1)				
Calliope Hummingbird	s(1)	S(4)	S(2)	S(4)	S(3)	S(1)	S(3)	S(1)	s(1)
Broad-tailed Hummingbird	S(4)	S(2)	S(1)	s(1)		s(3)		S(4)	s(2)
Rufous Hummingbird	s(2)	S(3)	S(1)	(4)	s(4)	S(3)	s(4)	S(3)	S(3)
Belted Kingfisher	R(3)	BR(4)	BR(4)	BR(4)	S(4)	S(3)	S(3)	S(3)	R(2)
Lewis' Woodpecker	s(3)	S(2)	S(1)	s(3)	s(1)		s(1)		S(1)
Red-headed Woodpecker	s(1)	M	s(1)	V	s(1)				
Red-naped Sapsucker	S(4)	S(4)	S(3)	S(4)	S(3)	S(3)	R(3)	BR(4)	S(2)
Williamson's Sapsucker	S(4)	S(2)	S(1)	S(3)	V				
Downy Woodpecker	BR(4)	BR(4)	BR(4)	BR(4)	BR(4)	R(1)	R(3)	R(3)	BR(2)
Hairy Woodpecker	BR(4)	BR(4)	BR(4)	BR(4)	BR(4)	BR(3)	BR(3)	R(3)	BR(2)
Am. three-toed Woodpecker	BR(3)	BR(2)	BR(2)	BR(4)	BR(3)	BR(4)	BR(3)	BR(3)	BR(3)
Black-backed Woodpecker		BR(2)	BR(1)	BR(3)	R(1)	BR	R(1)	R(1)	R(1)
Northern Flicker	S(4)	BR(4)	BR(4)	BR(4)	S(4)	S(4)	S(4)	BR(4)	S(3)
Pileated Woodpecker			V	BR(3)	R(3)	BR(3)	BR(3)	S(4)	BR(1)
Olive-sided Flycatcher	s(4)	S(4)	S(2)	S(4)	s(3)	s(4)	S(4)	S(4)	S(3)
Western Wood-Pewee	S(4)	S(4)	S(4)	S(3)	s(3)	s(3)	S(3)	S(3)	S(2)
Alder Flycatcher						s(3)	s(3)	s(3)	S(3)
Willow Flycatcher	S(4)	S(4)	S(4)	S(4)	s(3)	S(4)	S(4)	S(4)	S(4)
Least Flycatcher	M	M		s(1)	s(3)	s(3)	s(3)	s(4)	S(3)
Hammond's Flycatcher	S(3)	s(2)	S(2)	S(4)	s(3)	s(4)	s(4)	s(3)	S(4)
Dusky Flycatcher	S(4)	S(4)	S(2)	S(3)	s(3)	s(1)	s(3)	s(3)	S(3)
Cordilleran Flycatcher	S(4)	s(1)	S(1)	M	s(1)		s(3)	s(3)	s(1)
Say's Phoebe	s(1)	s(1)	S(1)	s(1)			V	V	s(1)

Seasonal Status Symbols: BR: Breeding resident; R: resident, breeding assumed but unproven; S: breeding summer resident; s: summer resident, breeding unproven; M: migrant, including wintering visitors; V: vagrant, out of normal range.

Abundance Symbols: (5): Abundant; (4): Common; (3): Uncommon; (2): Occasional; (1): Rare. Information on Yoho, Kooteney & Watertown Lakes was based on overall status checklists; seasonal occurrence attributions for these parks are judgments of the author. Species known as no more than vagrants in any of these parks are excluded; for a complete list see Johnsgard (1986). Some changes on park status since 1986 have been incorporated.

Appendix 1. Status and Abundance of Birds in Nine Rocky Mountain Parks (continued)

	UNITED STATES PARKS				CANADIAN PARKS			
	Rocky Mtn.	Grand Teton	Yellowstone	Glacier	Waterton L.	Kootenay	Yoho	Banff/Jasper
Western Kingbird	M	s(1)	S(2)	S(3)	s(1)	V		V
Eastern Kingbird	S(1)	s(2)	s(2)	S(4)	S(4)	S(3)	s(3)	S(3)
Northern Shrike	M	M	M	M	M	M	M	M
Loggerhead Shrike	s(1)	s(1)	M	s(3)	M			
Plumbeous Vireo	S(3)	M	M	S(4)				
Cassin's Vireo					s(1)	s(4)	s(3)	S(3)
Warbling Vireo	S(4)	S(5)	S(4)	S(4)	s(1)	S(4)	s(4)	S(4)
Red-eyed Vireo	s(1)	s(2)	s(1)	S(4)	s(1)	s(1)	s(1)	S(2)
Gray Jay	BR(4)	BR(4)	BR(4)	BR(4)	BR(4)	BR(4)	BR(4)	BR(3)
Steller's Jay	BR(4)	BR(4)	BR(2)	BR(4)	R(4)	BR(3)	BR(3)	V
Blue Jay	V			R(1)	R(1)			V
Western Scrub-Jay	M							
Pinyon Jay	M	s(1)	s(1)					
Clark's Nutcracker	BR(4)	BR(4)	BR(4)	BR(4)	BR(3)	BR(4)	BR(4)	BR(4)
Black-billed Magpie	BR(4)	BR(4)	BR(3)	BR(4)	BR(4)	R(3)	R(3)	BR(3)
American Crow	R(3)	BR(2)	BR(2)	BR(4)	R(4)	R(1)	BR(4)	BR(4)
Common Raven	BR(4)	BR(4)	BR(4)	BR(4)	BR(4)	BR(4)	BR(4)	BR(4)
Horned Lark	R(4)	R(2)	BR(2)	BR(4)	R(3)		R(1)	R(4)
Tree Swallow	S(4)	S(5)	S(4)	S(5)	S(4)	S(3)	S(3)	S(4)
Violet-green Swallow	S(5)	S(3)	S(4)	S(4)	s(4)	S(3)	S(4)	S(3)
N. Rough-winged Swallow	s(1)	S(1)	S(2)	S(3)	s(4)	S(4)	S(3)	S(3)
Bank Swallow		S(4)	S(5)	S(4)	s(3)	S(3)	s(1)	S(2)
Cliff Swallow	S(3)	S(3)	S(5)	S(4)	S(4)	S(3)	S(4)	S(4)
Barn Swallow	S(4)	S(4)	S(2)	S(4)	S(4)	S(4)	S(4)	S(3)

Species								
Black-capped Chickadee	BR(4)	BR(4)	BR(3)	BR(5)	R(4)	BR(4)	BR(3)	BR(4)
Mountain Chickadee	BR(4)	BR(4)	BR(4)	BR(5)	R(4)	BR(4)	BR(3)	BR(4)
Boreal Chickadee				BR(3)	R(3)	BR(4)	BR(4)	BR(4)
Chestnut-backed Chickadee				BR(1)	V			
Red-breasted Nuthatch	BR(1)	BR(2)	BR(4)	BR(4)	BR(4)	BR(4)	BR(4)	BR(3)
White-breasted Nuthatch	BR(4)	BR(4)	BR(2)	BR(1)	R(3)	R(1)	V	R(1)
Pygmy Nuthatch	BR(4)	V	R(1)					V
Brown Creeper	BR(3)	BR(2)	BR(2)	BR(4)	R(3)	R(1)	BR(3)	BR(2)
Rock Wren	S(4)	S(2)	S(3)	S(3)	s(1)			s(1)
Canyon Wren	S(4)							
House Wren	S(4)	S(4)	S(2)	S(3)	S(3)			s
Winter Wren	M	M		S(4)	S(3)	S(4)	S(3)	S(3)
Marsh Wren		S(2)	S(1)	V	s(1)			
American Dipper	BR(4)	BR(4)	BR(4)	BR(4)	BR(4)	BR(4)	BR(3)	BR(2)
Golden-crowned Kinglet	BR(3)	R(2)	BR(1)	BR(5)	R(3)	R(4)	BR(4)	BR(4)
Ruby-crowned Kinglet	S(4)	S(4)	S(4)	S(4)	s(4)	s(4)	s(4)	S(4)
Western Bluebird	S(3)	M	V	s(1)				
Mountain Bluebird	S(4)	S(4)	S(4)	S(4)	S(3)	S(3)	S(3)	s(1)
Townsend's Solitaire	s(1)	s(2)	S(2)	S(3)	s(3)	V	V	s
Swainson's Thrush	S(4)	S(4)	S(2)	S(4)	s(3)	S(4)	S(3)	S(3)
Hermit Thrush	S(4)	S(4)	S(4)	S(4)	S(3)	S(4)	S(3)	S(3)
American Robin	R(5)	S(5)	S(5)	S(5)	S(4)	S(4)	S(4)	S(3)
Varied Thrush	V	V	V	S(4)	S(3)	S(4)	S(4)	S(4)

Seasonal Status Symbols: BR: Breeding resident; R: resident, breeding assumed but unproven; S: breeding summer resident; s: summer resident, breeding unproven; M: migrant, including wintering visitors; V: vagrant, out of normal range.

Abundance Symbols: (5): Abundant; (4): Common; (3): Uncommon; (2): Occasional; (1): Rare. Information on Yoho, Kootenay & Watertown Lakes was based on overall status checklists; seasonal occurrence attributions for these parks are judgments of the author. Species known as no more than vagrants in any of these parks are excluded; for a complete list see Johnsgard (1986). Some changes on park status since 1986 have been incorporated.

Appendix 1. Status and Abundance of Birds in Nine Rocky Mountain Parks (continued)

	UNITED STATES PARKS				CANADIAN PARKS			
	Rocky Mtn.	Grand Teton	Yellowstone	Glacier	Waterton L.	Kootenay	Yoho	Banff/Jasper
Gray Catbird	s(3)	s(1)	S(1)	S(1)	S(3)		V	V
Sage Thrasher	s(3)	S(2)	S(1)					
Brown Thrasher	s(1)							
American Pipit	S(4)	S(4)	S(2)	S(4)	s(3)	S(4)	S(4)	S(4)
Bohemian Waxwing	M	M	M	M?	M	S(3)	R(3)	R(3)
Cedar Waxwing	V	R(2)	S(3)	S(4)	S(4)	s(1)	S(3)	s(3)
European Starling	BR(4)	BR(4)	BR(3)	BR(3)	BR(4)	BR(4)	BR(4)	BR(3)
Tennessee Warbler	M	M		M	s(1)	s(3)	s(1)	S(2)
Orange-crowned Warbler	S(3)	s(2)	S(1)	S(3)	s(1)	s(3)	S(4)	S(4)
Nashville Warbler	M	V	V	M				M
Virginia's Warbler	S(4)							
Yellow Warbler	S(4)	S(5)	S(4)	S(4)	s(4)	s(3)	s(3)	S(3)
Magnolia Warbler	V					s(1)	V	M
Yellow-rumped Warbler	S(4)	S(5)	S(4)	S(4)	S(4)	s(4)	S(5)	S(4)
Townsend's Warbler	M	s(1)	s(1)	S(4)	s(3)	s(4)	S(5)	S(4)
Black-throated Green Warbler	V	V			M			V
Palm Warbler	V		V		M		V	
Blackpoll Warbler			s(1)		s(1)	s(1)	S(3)	S(4)
Black-and-white Warbler	M			V				s(3)
American Redstart	M	s(2)	S(1)	S(4)	s(3)	s(3)	s(3)	S(4)
Northern Waterthrush	M	s(1)	s(1)	S(3)	s(3)	s(3)	s(3)	S(2)
MacGillivray's Warbler	S(4)	S(4)	S(2)	S(4)	S(3)	s(4)	S(3)	S(3)
Common Yellowthroat	s(1)	S(4)	S(4)	S(4)	S(3)	S(4)	S94)	S(4)
Wilson's Warbler	S(4)	S(4)	S(2)	S(4)	s(3)	s(4)	S(4)	S(4)
Western Tanager	S(4)	S(4)	S(4)	S(4)	s(4)	S(3)	S(3)	S(2)
Rose-breasted Grosbeak	M	M						

Species									
Black-headed Grosbeak	S(4)	S(4)	S(4)	S(1)	s(1)	s(3)			V
Lazuli Bunting	s(1)	S(2)	S(2)	S(4)	S(2)	s(3)	M		s(1)
Indigo Bunting	V	V	V	V	V				M
Green-tailed Towhee	S(4)	S(4)	S(4)	S(1)	S(3)	V	M		V
Spotted & Eastern Towhees	s(1)	s(1)	S(1)	s(3)	M	s(3)	S(5)		M
American Tree Sparrow	M	M	M	M	M		M		
Chipping Sparrow	S(4)	S(4)	S(4)	S(4)	S(4)	S(4)	S(5)	S(4)	S(4)
Clay-colored Sparrow	M	S?(1)	S(4)	S(4)	S(4)	s(1)	s(1)	s(1)	S(3)
Brewer's Sparrow	S(1)	S(4)	S(3)	s(1)	s(1)		s(3)	s(1)	S(3)
Vesper Sparrow	S(4)	S(4)	S(4)	S(4)	S(4)	s(3)	V		S(2)
Lark Sparrow	S(1)	M	S(2)	M	M				
Sage Sparrow		V	V	s(1)					
Lark Bunting	M	M	M	S(1)	s(1)	V	S(4)		V
Savannah Sparrow	S(4)	S(4)	S(4)	S(4)	S(4)	s(3)	S(4)	S(4)	S(4)
LeConte's Sparrow				s(1)	V				
Fox Sparrow	S(1)	S(2)	S(1)	S(4)	S(3)	s(3)	s(3)	S(3)	S(3)
Song Sparrow	BR(4)	R(4)	BR(4)	BR(4)	S(3)	R(1)	R(3)	R(3)	R(3)
Lincoln's Sparrow	S(4)	S(4)	S(3)	S(3)	s(3)	S(3)	s(3)	S(3)	S(3)
White-throated Sparrow	M	M	M	M	M				S(2)
Golden-crowned Sparrow				s					S(3)
White-crowned Sparrow	S(4)	S(5)	S(4)	S(4)	s(4)	S(3)	S(4)	S(4)	S(4)
Harris' Sparrow	M	M	M	V			M		M
Dark-eyed Junco	BR(4)	BR(5)	BR(5)	BR(4)	BR(4)	BR(4)	BR(4)	BR(4)	BR(4)
McCown's Longspur		V	V	V	V		M		M
Lapland Longspur		V	M	M	M	M		M	V
Chestnut-collared Longspur			M	M	s(1)				M
Snow Bunting	M	M	M	M	M	M	M	M	M

Seasonal Status Symbols: BR: Breeding resident; R: resident, breeding assumed but unproven; S: breeding summer resident; s: summer resident, breeding unproven; M: migrant, including wintering visitors; V: vagrant, out of normal range.

Abundance Symbols: (5): Abundant; (4): Common; (3): Uncommon; (2): Occasional; (1): Rare. Information on Yoho, Kootenay & Watertown Lakes was based on overall status checklists; seasonal occurrence attributions for these parks are judgments of the author. Species known as no more than vagrants in any of these parks are excluded; for a complete list see Johnsgard (1986). Some changes on park status since 1986 have been incorporated.

Appendix 1. Status and Abundance of Birds in Nine Rocky Mountain Parks (continued)

	United States Parks				Canadian Parks			
	Rocky Mtn.	Grand Teton	Yellowstone	Glacier	Waterton L.	Kootenay	Yoho	Banff/Jasper
Bobolink	M	M	S(1)	s(1)	s(1)		M	V
Red-winged Blackbird	BR(4)	BR(4)	BR(4)	S(4)	S(4)	S(3)	S(4)	S(4)
Western Meadowlark	BR(3)	BR(2)	BR(4)	S(4)	S(4)	s(3)	M	M
Yellow-headed Blackbird	s(1)	S(4)	S(3)	M	S(3)	M	M	M
Rusty Blackbird	V	V		s(1)			s(3)	s(1)
Brewer's Blackbird	S(4)	S(4)	S(4)	s(3)	S(3)	S(3)	S(3)	S(3)
Common Grackle	S(1)	s(2)	V	s(1)	s(1)		V	V
Brown-headed Cowbird	S(1)	S(4)	S(2)	S(3)	s(3)	s(1)	S(4)	S(4)
Baltimore/Bullock's Oriole (1)	S(2)	S(1)	S(1)	s(1)			V	
Rosy Finch spp,	BR(4)	BR(4)	BR(2)	BR(4)	R(3)	R(3)	BR(3)	BR(4)
Pine Grosbeak	BR(3)	R(2)	BR(4)	BR(4)	R(3)	BR(3)	BR(3)	BR(3)
Purple Finch					s(1)	s(1)	s(1)	S(2)
Cassin's Finch	BR(4)	BR(4)	BR(4)	S(4)	(3)		s(1)	V
House Finch	BR(3)	V	V			s(1)	s(1)	
Red Crossbill	BR(3)	R(2)	BR(2)	BR(4)	R(1)	R(3)	R(3)	BR(3)
White-winged Crossbill	M	V	M	R(1)	R(3)	R(1)	R(3)	BR(3)
Common Redpoll	M	M	M	M	M	M	M	M
Hoary Redpoll		V			M		M	M
Pine Siskin	BR(4)	BR(4)	BR(5)	BR(4)	s(4)	s(4)	S(4)	S(4)
Lesser Goldfinch	s(1)							
American Goldfinch	s(3)	S(2)	S(1)	s(3)	s(3)	s(3)	s(3)	s(2)
Evening Grosbeak	S(1)	R(2)	BR(2)	BR(4)	R(3)	R(3)	R(3)	BR(2)
House Sparrow		BR(4)	BR(1)	R(1)	S(4)	V	BR(4)	BR(4)

Seasonal Status Symbols: BR: Breeding resident; R: resident, breeding assumed but unproven; S: breeding summer resident, breeding unproven; M: migrant, including wintering visitors; V: vagrant, out of normal range.

Abundance Symbols: (5): Abundant; (4): Common; (3): Uncommon; (2): Occasional; (1): Rare. Information on Yoho, Kooteney & Watertown Lakes was based on overall status checklists; seasonal occurrence attributions for these parks are judgments of the author. Species known as no more than vagrants in any of these parks are excluded; for a complete list see Johnsgard (1986). Some changes on park status since 1986 have been incorporated.

References

Colorado

Andrews, R., and R. Righter. 1992. *Colorado Birds: A Reference to Their Distribution and Habitat.* Denver, CO: Denver Museum of Natural History.

Bailey, A. M., and R. J. Niedrach, R. 1965. *Birds of Colorado.* 2 vols. Denver, CO: Denver Museum of Natural History.

Chase, C. A., III, S. J. Bissell, H. E. Kingery, and W. D. Graul. 1982. *Colorado Bird Latilong Study.* Revised ed. Denver, CO: Denver Museum of Natural History.

Colorado Birding Society. This society has a useful website with local checklists, recent sightings of rare birds, and up-to-date information of birding in the state. URL: http://:home.att.net/~birdertoo.

Gildart, B., and J. Gildart. 2005. *A Falconguide to Dinosaur National Monument.* 2nd ed. Guilford, CT: Globe Pequot Press.

Holt, H. R. 1997, *A Birder's Guide to Colorado.* Denver, CO: American Birding Association.

Holt, H. R., and J. A. Lane. 1988, *A Birder's Guide to Colorado.* Denver, CO: American Birding Association.

Kingery, H. 2007. *Birding Colorado.* Helena, MT: Morris Book Pub. Co. (Describes 93 birding sites in detail, and lists "specialty sites" for locating particular species.)

Kingery, H. (ed.). 1998. *Colorado Breeding Bird Atlas.* Denver, CO: Colorado Division of Wildlife.

Lane, J. A., and H. R. Holt. 1979. *A Birder's Guide to Eastern Colorado.* Denver, CO: L & P. Press.

Righter, R. *et al.* 2004. *Birds of Western Colorado Plateau and Mesa Country.* Grand Junction, CO: Grand Valley Audubon Society.

Scott, V. E., and G. L. Crouch. 1988. Breeding birds in uncut aspen and 6- to 10-year-old clearcuts in southwestern Colorado. Denver, CO: USDA Forest Service Res. Note RM-485.

Weber, W. A. 1967. *Rocky Mountain Flora.* Boulder: Univ. of Colo. Press. (Specific to Front Range of Colorado.

Young, M. T. 2000. *Colorado Wildlife Viewing Guide.* 2nd ed. Helena, MT: Falcon Press. 216 pp. (An outstanding guide in the Watchable Wildlife series, with 201 locations described.)

Rocky Mountain National Park

Collister, A. E. 1965. *A List of Birds of Rocky Mountain National Park.* 16 pp. Estes Park, CO: Rocky Mountain Nature Association.

_____. 1970. *Annotated Checklist of Birds of Rocky Mountain National Park and Shadow Mountain Recreation Area in Colorado.* 64 pp. Denver, CO: Denver Natural History Museum Pictorial No. 18.

Packard, F. M. 1950. *The Birds of Rocky Mountain National Park.* 81 pp. Estes Park, CO: Rocky Mountain Nature Association.

Idaho

Arvey, M. D. 1947. A check-list of the birds of Idaho. Lawrence, KS: University of Kansas, *Museum of Natural History Publications* 1:193– 216.

Bureau of Land Management. 1979. *Snake River Birds of Prey Special Research Report to the Secretary of Interior.* Boise, ID: Bureau of Land Management, Boise District.

Burleigh, T. D. 1972. *Birds of Idaho.* Caldwell, Idaho: Caxton Printers.

Carpenter, L. B. 1990. *Idaho Wildlife Viewing Guide.* Helena, MT: Falcon Press. 104 pp. (Describes 94 wildlife and birding sites.)

Idaho Dept. of Fish & Game. 2006 *Idaho Birding Trail.* Boise, ID: Idaho Dept. of Fish & Game. 135 pp.

Larrison, E. J., J. L. Tucker, and M. T. Jollie. 1967. *Guide to Idaho Birds.* Ricks College, Rexburg, ID: Idaho Academy of Science.

McClung, J. M. 1992. Boreal owls in Idaho. *Birding* 24:78-834.

Pope, A. L. 2003. *Idaho Wildlife Viewing Guide.* Boise, ID: Idaho Dept. of Fish & Game. 104 pp. (Describes 100 wildlife and birding sites)

Roberts, H. B. 1992. *Birds of East Central Idaho.* Boise, ID: Idaho Dept. of Fish & Game. 119 pp. (Annotated bird list, and nine birding sites or routes.)

Stephens, D. A., and S. Sturts. 1998. *Idaho Bird Distribution: Mapping by Latilong.* Pocatello, ID: Museum of Natural History, Special Publication No. 11. See website: http://www.idahobirds.net/distribution/dist_index.html

Sturts, S. H., and S. Schultz. 1993. *Birds and Birding Routes in the Idaho Panhandle.* Boise, ID: Idaho Dept. of Fish & Game. 64 pp. (Describes ten northern Idaho birding routes.)

Svingen, D., and K. Dumroese. (eds.). 1997. *A Birder' Guide to Idaho.* Colorado Springs, CO: American Birding Association. 339 pp. (Describes 110 birding sites.)

U. S. Forest Service. 1994. *Birds of the Sawtooth National Recreation Area.* Ogden, UT: U. S. Forest Service, Intermountain Region. 18 pp. (Out of print.)

Montana

Bonham, D., and D. Cooper. 1986. *Birds of West-Central Montana.* Missoula, MT: Five Valleys Audubon Society.

Davis, C. V., 1961. A distributional study of the birds of Montana. Ph.D. dissertation, Oregon State University, Corvallis, OR.

Fischer, C., and H. Fischer. 1995. *Montana Wildlife Viewing Guide.* Helena, MT: Falcon Press. (Describes 97 wildlife and birding sites.)

Last Chance Audubon Society. 1986. *Birding in the Helena Valley.* Helena, MT: Last Chance Audubon Society.

Lenard, S., J. Carlson, J. Ellis, C. Jones, C. Tilly and P. D. Skaar. 2003. *P. D. Skaar's Montana Bird Distribution.* 6th ed. Helena, MT: Montana Natural Heritage Program, Special Publication No. 3.

Marks, Jeff, *et al. Birds of Montana.* In preparation.

McEneaney, T. 1993. *The Birder's Guide to Montana.* Helena, MT: Falcon Press. 314 pp. (Describes 45 birding sites or local areas, and provides seasonal range maps and suggested birding locations for 137 species.)

Phister, R. D. B. L. Kovalchik, S. F. Arno & R. C. Presby. 1977. *Forest Habitat Types of Montana.* Ogden, UT: USDA Forest Service Gen. Tech. Rpt. INT–34.

Saunders, A. A. 1921. A distributional list of the birds of Montana. Cooper Ornithological Society, *Pacific Coast Avifauna* No. 14.

Tobalske, B. W., R. C. Shearer and R. L. Hutto. 1991. *Bird populations in Logged and Unlogged Western Larch/Douglas-fir forests in Northwestern Montana.* Ogden, UT: USDA Forest Service Res. Pap. INT-442.

Glacier National Park

Beaumont, G. 1978. *Many Splendored Mountains: the Life of Glacier National Park.* 138 pp. Washington, DC: U.S. National Park Service, Department of Interior.

Parrat, L. P. 1970. *Birds of Glacier National Park.* 86 pp. Washington, DC: U.S. National Park Service, Department of Interior.

Wyoming

Baker, M. C., and J. T. Boylan. 1999. Singing behavior, mating associations and reproductive success in a population of hybridizing lazuli and indigo buntings. *Condor* 101:493-504.

Bureau of Land Management. 1994. *Birds of Southeast Wyoming,* Rawlins District. Rawlins, WY: Bureau of Land Management. 8 pp.

Cary, M. 1917. Life zone investigations in Wyoming. U.S. Dept. of Agriculture, Bureau of Biological Survey, *North American Fauna No. 40.*

Cerovski, A. O., M. Grenier, B. Oakleaf. J. Van Fleet ad S. Patla. 2004. *Atlas of Birds, Mammals, Reptiles and Amphibians in Wyoming.* Lander, WY: Wyoming Game & Fish Department. (See also 2009 update on Wyoming Game & Fish Dept. website: http://state.wyo.us/wildlife/nongame/index.asp

Dorn, J. L. 1978. *Wyoming Ornithology – A History & Bibliography with Species and Wyoming Area Indexes.* U.S. Bureau of Land Management and Wyoming Game & Fish Dept. 369 pp.

Dorn, J. L., and R. D. Dorn. 1990. *Wyoming Birds.* Cheyenne, WY: Mountain West Publishing Co.

Downing, H. (ed.) 1990. *Birds of North-central Wyoming and the Bighorn National Forest, 1966–1990.* Sheridan, WY, Privately published. 98 pp. (revised edition in preparation, J. Canterbury and P. A. Johnsgard)

Faulkner, D. 2010. *Birds of Wyoming*. Greenwood Village, CO: Ben Roberts & Co. 404 pp.

Grave, B. H., and E. P. Walker. 1916. Wyoming birds. *Univ. of Wyo. Bull.* (Laramie) 12(6):1–137.

Knight, D. 1994. *Mountains and Plains: The Ecology of Wyoming Landscapes*. New Haven, CT: Yale Univ. Press.

Long, C. A. 1965. The Mammals of Wyoming. Lawrence, KS: *Univ. of Kansas, Museum of Nat. Hist. Publication* 14(18): 493–754.

McCreary, O. 1939. *Wyoming Bird Life*. Revised ed. Minneapolis, MN: Burgess Publishing Co.

Pettingill, O. S., and N. R. Whitney, Jr. 1965. *Birds of the Black Hills*. Ithaca, NY: Cornell Laboratory of Ornithology, Spec. Pub. No. 1.

Porter, C. L. 1962. Vegetation zones in Wyoming. *Univ. Wyo. Publ.* 27:6–12.

Salt, G. 1957. An analysis of avifauna in the Teton Mountains and Jackson Hole, Wyoming. *Condor* 59:373–393.

Scott, O. K. 1993. *A Birder's Guide to Wyoming*. Colorado Springs, CO: American Birding Association. (Describes the birding sites in 19 locations or regions, and includes an annotated checklist with suggested locations for finding most species.)

Stephenson, S. 2011. Under the bark: Pine beetles are changing the face of Wyoming's mountains. *Wyoming Wildlife* 73(1):17–21.

Wyoming Game and Fish Department. 1996. *Wyoming Wildlife Viewing Guide*. Lander, WY: Wyoming Game and Fish Department. (Describes more than 51 viewing sites, with an emphasis on game species.)

Wyoming Game and Fish Department. 2008. *Wyoming Bird Checklist*. Lander, WY: Wyoming Game and Fish Department. 15 pp.

Grand Teton National Park

Houston, D. B. 1969. The bird fauna of Grand Teton National Park. 4 pp. mimeo. *Special Study No. 1, Grand Teton National Park Tour Guide*. Lander, WY: Wyoming Game and Fish Department.

Johnsgard, P. A. 1982. *Teton Wildlife: Observations by a Naturalist*. Boulder: Colorado Associated University Press. This title is available online through the University of Nebraska's Digital Commons website: http://digitalcommons.unl.edu/biosciornithology/52/.

Nye, D. L., M. Back, and H. Hinchman. undated. *Birds of the Upper Wind River Valley*. Cody, WY: USDA Forest Service. 34 pp.

Raynes, B. 1984. *Birds of Grand Teton National Park and the Surrounding Area*. Moose, WY: Grand Teton Natural History Association. 90 pp.

Raynes, B. and D. Wile. 1994. *Birds of the Jackson Hole area, including Grand Teton National Park*. Jackson, WY: Darwin Wile

Taylor, D. L., and W. J. Barmore, Jr. 1980. Post-fire succession of avifauna in coniferous forests of Yellowstone and Grand Teton National Parks, Wyoming. Pp. 130–144, in DeGraaf, R. M., and N. G. Tilghman (eds.), *Management of Western Forests and Grasslands for Nongame Birds*. Ogden, UT: USDA Forest Service, U.S. Forest Service Gen. Tech. Report INT– 86.

Wuerther, G. (ed.) 2006. *Wild Fire: A Century of Failed Forest Policy*. Sausalito, CA: Foundation for Deep Ecology and Island Press.

Yellowstone National Park

Brodrick, H. J. 1952. *Birds of Yellowstone National Park*. U.S. National Park Service, Yellowstone Interpretive Series No. 2. 58 pp.

Follet, D. 1986. *Birds of Yellowstone and Grand Teton National Parks*. U.S. National Park Service. Yellowstone Library and Museum Association, and Roberts Rinehart. 71 pp.

McEneaney, T. 1988. *Birds of Yellowstone*. Boulder, CO: Roberts Rinehart, Inc. 171 pp.

Skinner, M. P. 1925. The birds of Yellowstone National Park. *Roosevelt Wildlife Bulletin*, Syracuse University School of Forestry 3:11–189.

Canada

Anonymous, 1982. Checklist of Alberta birds. 9 pp. 4th ed. Edmonton, AB: Provincial Museum of Alberta.

Bird Studies Canada. In preparation. *British Columbia Breeding Bird Atlas*. URL: http://birdatlas.bc.ca/english/index.jsp

Clarke, C. H. D., and Cowan, I. McT., 1945. Birds of Banff National Park, Alberta. *Canadian Field-Naturalist* 59:83–103.

Cowan, I. McT., 1955. Birds of Jasper National Park, Alberta, Canada. Canadian Wildlife Service, *Wildlife Management Bulletin, Series 2*.

Federation of Alberta Naturalists. 2007. *The Atlas of Breeding Birds of Alberta: A Second Look*. Edmonton, AB: Federation of Alberta Naturalists.

Gadd, B. 1995. *Handbook of the Canadian Rockies*. 2nd. ed. Jasper: Corax Press.

Godfrey, W. E., 1950. Birds of the Cypress Hills and Flotten Lake region, Saskatchewan. Ottawa: *National Museums of Canada Bulletin* 120:1– 96.

Hardy, W. G. (ed.). 1967. *Alberta, a Natural History*. Edmonton. AB: Hurtig Publishers.

Kondla, N. G. 1978. The birds of Dinosaur Provincial Park, Alberta. *Blue Jay* 36:103–14.

Rand, A. L. 1948. The birds of southern Alberta. Ottawa, Ontario: *National Museum of Canada Bulletin* 111: 1–105.

Sadler, T. S., and M. T. Myres. 1976. Alberta birds, 1961– 1970, with particular reference to migration. Edmonton, AB: Provincial Museum of Alberta, Natural History Section. *Occasional Paper No. 1: 1– 314*.

Salt, W. R., and J. R. Salt. 1976. *The Birds of Alberta*. Edmonton, AB: Hurtig Publishers.

Soper, J. D. 1947. Observations on mammals and birds in the Rocky Mountains of Alberta. *Canadian Field-Naturalist* 61:143–73.

Van Tighem, K., and G. Holroyd. 1981. A birder's guide to Jasper National Park, Alberta. *Alberta Naturalist* 11:134– 40.

Other Regional References

Anderson, S. H. 1980. Habitat selection, succession and bird community organization. Pp. 13–225, in *Workshop Proceedings: Management of Western Forests and Grasslands for Nongame Birds*(R. M. DeGraaf and N. G. Tilghman, eds.). Ogden, UT: USDA Forest Service Gen. Tech. Rep. INT-86.

Austin, J. E., and A. L. Richert. 2001. *A Comprehensive Review of Observational and site Evaluation data of Migrant Whooping Cranes in the United States, 1943–99*. Jamestown, ND: Technical Report, Northern Prairie Research Center. 157 pp.

Balda, R. P., and N. Masters. 1980. Avian communities in the pinyon-juniper woodlands: A descriptive analysis. Pp. 146–167, in *Workshop Proceedings: Management of Western Forests and Grasslands for Nongame Birds* (R. M. DeGraaf and N. G. Tilghman, eds.). Ogden, UT: USDA Forest Service Gen. Tech. Rep. INT-86.

Boyle, W. J., Jr., and H. Wauer. 1994. *Birdfinding in Forty National Forests and Grasslands*. Colorado Springs, CO: American Birding Association. (Includes ten Rocky Mountain forests and two national grasslands)

Braun, C. E. 1980. Alpine bird communities of western North America: Implications for management and research. Pp. 280–291, in *Management of Western Forests and Grasslands for Nongame Birds* (R. M. DeGraaf, and N. G. Tilghman, eds.). Ogden, UT: USDA Forest Service, U.S. Forest Service Gen. Tech. Report INT– 86.

Cable, T. T., S. Seltman and K. J. Cook, 1996. *Birds of Cimarron National Grassland*. USDA Forest Service Gen. Tech. Rpt. RM-GTR-281.

DeGraaf, R. M. (tech. coordinator). 1978. *Proceedings of the Workshop on Nongame Bird Habitat Management in the Coniferous Forests of the Western United States*. Portland, OR: USDA Forest Service Gen. Tech. Rep. PNW-65.

DeGraaf, R. M., and R. N. Conner. 1979. *Management of North-central and Northeastern Forests for Nongame Birds*. Minneapolis, MN: USDA Forest Service Gen. Tech. Rep. NC-51.

DeGraaf, R. M., and N. G. Tilghman (eds.). 1980. *Workshop Proceedings: Management of Western Forests and Grasslands for Nongame Birds*. Ogden, UT: USDA Forest Service Gen. Tech. Rep. INT-86.

Diem, K., and S. L. Zeveloff. 1980. *Ponderosa pine bird communities*. Pp. 170–197, in *Management of Western Forests and Grasslands for Nongame Birds* (R. M. DeGraaf and N. G. Tilghman, eds.). Ogden, UT: USDA Forest Service, U.S. Forest Service Gen. Tech. Report INT– 86.

Dobkin, D. S. 1994. *Conservation and Management of Neotropical Migrant Landbirds in the Northern Rockies and Great Plains*. Moscow, ID: Univ. of Idaho Press.

Dugger, B. D., K. M. Dugger and L. H. Fredrickson. 1994. Hooded merganser. In *The Birds of North America*, No. 98 (A. Poole and F. Gill, eds.). Philadelphia, PA: The Birds of North America, Inc. 24 pp.

Engeline, S. 1980. Wildlife relationships and forest planning. Pp. 379–389, in *Workshop Proceedings: Management of Western Forests and Grasslands for Nongame Birds* (R. M. DeGraaf and N. G. Tilghman, eds.). Ogden, UT: USDA Forest Service Gen. Tech. Rep. INT-86.

Evans, K. E., and R. N. Conner. 1979. Snag management. Pp. 215–224, in *Management of North-central and Northeastern Forests for Nongame Birds* (R. M. DeGraaf, and R. N. Conner, eds.). Minneapolis, MN: USDA Forest Service Gen. Tech. Rep. NC-51.

Finch, D. M. 1992. *Threatened, Endangered, and Vulnerable Species of Terrestrial Vertebrates in the Rocky Mountain Region.* Denver, CO: USDA Forest Service, Rocky Mt. Forest & Range Exp. Station. Gen. Tech. Rept. RM– 215.

Flack, J. A. D. 1976. Bird populations of aspen forests in western North America. American Ornithologists' Union: *Ornithological Monographs* No. 19.

Hein, D. 1980. Management of lodgepole pine for birds. Pp. 238–246, in *Workshop Proceedings: Management of Western Forests and Grasslands for Nongame Birds.* (R. M. DeGraaf and N. G. Tilghman, eds.). Ogden, UT: USDA Forest Service Gen. Tech. Rep. INT-86.

Hutto, R. L., and J. S. Young. 1999. *Habitat Relationships of Landbirds in the North Region, USDA Forest Service.* Ogden, UT: USDA Forest Service RMRS–GTR-32. 72 pp.

Johnson, R. R., L. T. Haight, M. F. Riffey, and J. M. Simpson. 1980. Brushland/steppe bird populations. Pp. 98–112, in *Workshop Proceedings: Management of Western Forests and Grasslands for Nongame Birds* (R. M. DeGraaf and N. G. Tilghman, eds.). Ogden, UT: USDA Forest Service Gen. Tech. Rep. INT-86.

Johnsgard, P. A. 1979. *Birds of the Great Plains: Breeding Species and their Distribution.* Lincoln: University of Nebraska Press.

Johnsgard, P. A. 1986. *Birds of the Rocky Mountains.* Boulder: Colorado Associated Univ. Press.

Johnsgard, P. A. 2001. *Prairie Birds: Fragile Splendor in the Great Plains.* Lawrence, KS: Univ, Press of Kansas.

Johnsgard, P. A. 2005. *Prairie Dog Empire: A Saga of the Great Plains.* Lincoln, NE: Univ. of Nebraska Press.

Johnsgard, P. A. 2009. *Birds of the Rocky Mountains.*
Revised edition: http://digitalcommons.unl.edu/bioscibirdsrockymtns/1,
and Supplement: http://digitalcommons.unl.edu/bioscibirdsrockymtns/3

Koplin, J. R. 1972. Measuring predation impact of woodpeckers on spruce beetles. *J. Wildl. Manage.* 36:308–320.

Kotliar, N. B., S. J. Heil, R. L. Hutto, V. A. Saab, C. P. Melcher, and M. E. McFadzen. 2002. Effects of fire and post-fire salvage logging on avian communities in conifer-dominated forests of the western United States. Pp 49–64, in *Effects of Habitat Fragmentation on Birds in Western Landscapes: Contrasts with Paradigms for the Eastern United States* (T. George and D. S. Dobkin, eds.). Cooper Ornith. Soc., Studies in Avian Biology No. 25.

Koplin, J. R., and P. H. Baldwin. 1970. Woodpecker predation on an endemic population of Engelmann spruce beetles. *Amer. Midl. Nat.* 83:510–515.

Larrison, E. J. 1981. *Birds of the Pacific Northwest: Washington, Oregon, Idaho & British Columbia*. Moscow, ID: Univ. Press of Idaho.

Mannan, R. W. 1980. Assemblages of bird species in western coniferous old-growth forests. Pp. 357–368, in *Workshop Proceedings: Management of Western Forests and Grasslands for Nongame Birds* (R. M. DeGraaf and N. G. Tilghman, eds.). Ogden, UT: USDA Forest Service Gen. Tech. Rep. INT-86.

Mutel, C. F., and J. C. Emerick. 1992. *From Grassland to Glacier: The Natural History of Colorado and Surrounding Regions*. Boulder, CO: Johnson Books.

Paige, L. C. 1990. Population trends of songbirds in western North America. M.S. thesis, Univ. of Montana, Missoula, MT.

Pettingill, O. S. 1981. *A Guide to Bird Finding West of the Mississippi*. 2nd ed. New York, NY: Oxford Univ. Press.

Raynes, B. 2003. *Winter Wings: Birds of the Northern Rockies*. Omaha, NE: Images of Nature.

Reel, S., L Schassberger, and W. Ruederger. 1989. *Caring for our Natural Community: Region 1 – Threatened, Endangered and Sensitive Species Program*. Missoula, MT: USDA Forest Service, Northern Region.

Sanderson, H. R., E. L. Bull, and P. J. Edgerton. 1980. Bird communities in mixed conifer forests of the interior Northwest. Pp. 224–237, in *Workshop Proceedings: Management of Western Forests and Grasslands for Nongame Birds* (R. M. DeGraaf and N. G. Tilghman, eds.). Ogden, UT: USDA Forest Service Gen. Tech. Rep. INT-86.

Sauer, J. R., and S. Droege. 1992. Geographic patterns of population trends of Neotropical migrants in North America. Pp, 26–42, in *Ecology and Conservation of Neotropical Migrant Landbirds* (J. M. Hagen & D. W. Johnstone, eds.). Washington, D.C.: Smithsonian Inst. Press.

Scott, V. E, J. A. Whelan and P. L. Svoboda. 1980. Cavity-nesting birds and forest management. Pp. 310–324, in *Workshop Proceedings: Management of Western Forests and Grasslands for Nongame Birds* (R. M. DeGraaf and N. G. Tilghman, eds.). Ogden, UT: USDA Forest Service Gen. Tech. Rep. INT-86.

Smith, K. G. 1980. Nongame birds of the Rocky Mountain spruce-fir forests and their management. Pp. 258–279, in *Workshop Proceedings: Management of Western Forests and Grasslands for Nongame Birds* (R. M. DeGraaf and N. G. Tilghman, eds.). Ogden, UT: USDA Forest Service, U.S. Forest Service Gen. Tech. Report INT– 86.

Verner, J, 1980. Bird communities in mixed-conifer forests of the Sierra Nevada. Pp. 198–237, in *Workshop Proceedings: Management of Western Forests and Grasslands for Nongame Birds* (R. M. DeGraaf and N. G. Tilghman, eds.). Ogden, UT: USDA Forest Service Gen. Tech. Rep. INT-86.

Zwinger, A. H., and B. E. Willard. 1972. *Land Above the Trees: A Guide to American Alpine Tundra*. New York, NY: Harper & Row.

National References

Alderfer, J. (ed.). 2006. *National Geographic Complete Birds of North America*. Washington, D.C.: National Geographic Society.

Baughman, M. (ed.). 2006. *Reference Atlas to the Birds of North America*. Washington, D.C.: National Geographic Society.

Bildstein, K., J. Smith, E. R. Inzunza, and R. E. Veit (eds.). 2008. *The State of North American Birds of Prey*. Nuttall Ornithological Club and American Ornithologists' Union.

Burger, J., and M. Gochfeld. 2002. Bonaparte's Gull. In *The Birds of North America*, No. 634 (A. Poole and F. Gill, eds.). Philadelphia, PA: The Birds of North America, Inc. 24 pp.

Dixon, R. D., & V. A. Saab. 2000. Black-backed Woodpecker. In *The Birds of North America*, No. 509 (A. Poole and F. Gill, eds.). Philadelphia, PA: The Birds of North America, Inc. 20 pp.

Crump, D. (ed.). 1984. *A Guide to Our Federal Lands*. Washington, DC: National Geographic Society.

Dalsten, D. L. 1982. Relationships between bark beetles and their natural enemies. Pp. 140–182, in *Bark Beetles in North American Conifers*, J. B. Milton and K. B. Sturgeon (eds.). Austin, TX: Univ. of Texas Press.

De Graaf, R. M., V. E. Scott, R. H. Hamre, I. Earnst, and S. H. Anderson. 1991. *Forest and Rangeland Birds of the United States: Natural History and Habitat Use*. Washington, D.C.: USDA, Forest Service Agricultural Handbook 688.

Fellows, S. D., and S. L. Jones. 2009, *Status Assessment and Conservation Action Plan for the Long-billed Curlew in 2009*. Washington, D.C.: U. S. Fish & Wildlife Service.

Johnsgard, P. A. 2011. *The Sandhill and Whooping Cranes: Ancient Voices over America's Wetlands*. Lincoln: University of Nebraska Press.

Jones, J. O. 1990. *Where the Birds Are: A Guide to all 50 States and Canada*. New York, NY: Wm. Morrow. (Includes checklists for nearly all of the national wildlife refuges, as well as for some state and miscellaneous preserves.)

Kear, J. 2005. *Ducks, Geese and Swans*. 2 vol. Oxford, UK: Oxford Univ. Press.

Kessel, B., D. A. Rocque and J. S. Barclay. 2002. Greater Scaup. In *The Birds of North America*, No. 650 (A. Poole and F. Gill, eds.). Philadelphia, PA: The Birds of North America, Inc. 23 pp.

Leonard, D. L., Jr. 2001. Three-toed Woodpecker. In *The Birds of North America*, No. 558 (A. Poole and F. Gill, eds.). Philadelphia, PA: The Birds of North America, Inc. 24 pp.

Morrison, R. I., R. E. Gill, Jr., B. A. Harrington, S. Skagen, G. W. Page, G. L. Gatto-Trevor, and S. M. Haig. 2001. Estimates of shorebird populations in North America. Ottawa: Canadian Wildlife Service, Environment Canada, *Occasional Paper 104*:1–64.

Olsen, K. M., and K. Larson. 2004. *Gulls of North America, Europe and Asia*. Princeton, NJ: Princeton University Press.

Omernik, J. M. 1987. Ecoregions of the coterminous United States. *Annals of the Association of American Geographers*, 77:118– 125.

Poole, A. (ed.). Varied dates. *The Birds of North America Online*, http://bna. birds.cornell.edu/ (Includes the entire 715-part series of *The Birds of North America* monographs.)

Price, J., S. Droege, and A. Price. 1995. *The Summer Atlas of North American Birds*. San Diego, CA: Academic Press.

Rich, T. C. *et al.* (eds.). 2004. *North American Landbird Conservation Plan*. Ithaca, NY: Partners in Flight and Cornell University Laboratory of Ornithology.

Rose, P. M., & D. A. Scott. 1997. *Waterfowl Population Estimates* (2nd. ed.). Wagenigen, Netherlands: Wetlands International, Publ. No. 44.

Sauer, J. R., J. E. Hines and J. Fallon. 2008. The North American Breeding Bird Survey, Results and Analysis, 1966 – 2007. Version 5.15.2008. Laurel, MD: Patuxent Wildlife Research Center. See associated website: http://www.mbr-pwrc.usgs.gov/bbs/bbs.html

Scott, V. E., K. E. Evans, D. R. Patton and C. P. Stone. 1977. *Cavity-nesting Birds of North American Forests*. USDA Forest Service, Agricultural Handbook No. 511. 112 pp.

Smith, J. W., and C. W. Beckman. 2007. A coevolutionary arms race causes speciation in crossbills. *American Naturalist* 169:455–465.

Stark, R. W. 1982. Generalized ecology and life cycles of bark beetles. Pp. 21–45, in *Bark Beetles in North American Conifers,* J. B. Milton and K. B. Sturgeon (eds.). Austin, TX: Univ. of Texas Press.

U. S. Fish & Wildlife Service. 2009. *2009 Waterfowl Population Status*. Washington, D.C.: Administrative Report, U.S. Dept. of Interior. URL: *http.//www. fws.gov/migratorybirds/html*.

Wetlands International. 2002. *Waterfowl Population Estimates – Third Edition*. Wageningen, The Netherlands: Wetlands International Global Series No. 12.

White, M. 1999. *Guide to Birdwatching Sites: Western U.S.* Washington, D.C.: National Geographic. (Describes birding opportunities in 13 Colorado, 10 Montana, 9 Idaho and 6 Wyoming locations.)

Zimmer, K. J. 2000. *Birding in the American West*. Ithaca, NY: Cornell Univ. Press.

www.ingramcontent.com/pod-product-compliance
Lightning Source LLC
Chambersburg PA
CBHW031502270326
41930CB00006B/212